AF471817

QUANTUM SCATTERING THEORY FOR SEVERAL PARTICLE SYSTEMS

Dear MyCopy Customer,

This Springer book is a monochrome print version of the eBook to which your library gives you access via SpringerLink. It is available to you at a subsidized price since your library subscribes to at least one Springer eBook subject collection.

Please note that MyCopy books are only offered to library patrons with access to at least one Springer eBook subject collection. MyCopy books are strictly for individual use only.

You may cite this book by referencing the bibliographic data and/or the DOI (Digital Object Identifier) found in the front matter. This book is an exact but monochrome copy of the print version of the eBook on SpringerLink.

Mathematical Physics and Applied Mathematics

Editors:

M. Flato, *Université de Bourgogne, Dijon, France*

The titles published in this series are listed at the end of this volume.

Volume 11

Quantum Scattering Theory for Several Particle Systems

by

L.D. FADDEEV

and

S.P. MERKURIEV †

Institute of Theoretical Physics,
University of St. Petersburg,
St. Petersburg, Russia

† Professor S.P. Merkuriev died on May 18, 1993

SPRINGER-SCIENCE+BUSINESS MEDIA, B.V.

Library of Congress Cataloging-in-Publication Data

Merkur'ev, S. P. (Stanislav Petrovich)
[Kvantovaia teoriia rasseianiia dlia sistem neskol'kikh chastits. English]
Quantum scattering theory for several particle systems / by L.D. Faddeev and S.P. Merkuriev.
p. cm. -- (Mathematical physics and applied mathematics ; v. 11)
Translation of: Kvantovaia teoriia rasseianiia dlia sistem neskol'kikh chastits.
Author's names in reverse order in original Russian ed.
Includes index.

DOI 10.1007/978-94-017-2832-4

1. Scattering (Physics) 2. Quantum theory. 3. Few-body problem. I. Faddeev, L. D. II. Title. III. Series.
QC794.6.S3M4713 1993
539.7'58--dc20 93-11377

This is the translation of the original Russian work,
Kvantovaja teoria rasseivania dlja sistem neskolkih pastid,
Published by Nauka Publishers, Moscow, © 1985.

Printed on acid-free paper

All Rights Reserved
© 1993 Springer Science+Business Media Dordrecht
Originally published by Kluwer Academic Publishers in 1993
MyCopy version of the original edition 1993

No part of the material protected by this copyright notice may be reproduced or utilized in any form or by any means, electronic or mechanical, including photocopying, recording or by any information storage and retrieval system, without written permission from the copyright owner.
www.springer.com/mycopy

Contents

Introduction

The last decade witnessed an increasing interest of mathematicians in problems originated in mathematical physics. As a result of this effort, the scope of traditional mathematical physics changed considerably. New problems especially those connected with quantum physics make use of new ideas and methods. Together with classical and functional analysis, methods from differential geometry and Lie algebras, the theory of group representation, and even topology and algebraic geometry became efficient tools of mathematical physics. On the other hand, the problems tackled in mathematical physics helped to formulate new, purely mathematical, theorems.

This important development must obviously influence the contemporary mathematical literature, especially the review articles and monographs. A considerable number of books and articles appeared, reflecting to some extend this trend. In our view, however, an adequate language and appropriate methodology has not been developed yet. Nowadays, the current literature includes either mathematical monographs occasionally using physical terms, or books on theoretical physics focused on the mathematical apparatus.

We hold the opinion that the traditional mathematical language of lemmas and theorems is not appropriate for the contemporary writing on mathematical physics. In such literature, in contrast to the standard approaches of theoretical physics, the mathematical ideology must be utmost emphasized and the reference to physical ideas must be supported by appropriate mathematical statements. Of special importance are the results and methods that have been developed in this way for the first time.

This monograph is intended to present an example of an up-to-date and difficult problem of theoretical physics, namely the quantum mechanical problem of scattering of N particles. We have made this choice for several reasons: First, the N-body problem is a traditional and difficult problem of mathematical physics. Its quantum analog has a number of interesting applications in atomic and nuclear physics. Second, the problem is very interesting from the mathematical point of view. The mathematical treatment of this problem has led to considerable enrichment of methods of classical and functional analysis. Finally, for many years both authors worked and continue to work in it, so that this book can reflect their views, approaches, and results.

As a mathematical task, the problem considered is too cumbersome and, moreover, has not been completely solved yet. A strict presentation according to the standards of theoretical mathematics requires a large number of technical details which may often be very tedious. hence, in accordance with the aforementioned methodology, we will give a basic explanation of the formal level of the theory, without giving the full proofs. However, in contrast to theoretical physics papers, we admit the necessity of proofs and will explicitly show that the results can be justified. In this way the more mathematically oriented readers will be freed from tedious estimates and physicists will get ideas on the mathematical methods in situations they are familiar with. Therefore we hope that this book will be useful both for mathematicians and physicists. This book is mainly focused on the new and rapidly growing group of specialists working in contemporary mathematical physics.

The book can be divided into three parts: the general formulation of the problem of N particle scattering, its justification on the basis of compact integral equations and description of basic objects of scattering theory: wave functions, their asymptotics and scattering amplitudes. Chapters 1 and 2 are devoted to the first part, the second part is contained mainly in Chapters 3 and 6, and partially in Chapter 4 and 5. Chapter 7 and the main part of Chapter 4 and 5 are devoted to the last circle of problems. In Chapter 1, the basic dynamical concepts are introduced, the wave and scattering operators are defined and their general properties are described. In Chapter 2, we turn to the stationary formalism of the scattering theory. here we describe general properties of the resolvent of the energy operator and we obtain expressions for the kernels of wave operators and scaterring operators in terms of singularities of the resolvent kernel. Chapter 3 is devoted to the method of integral equations. The Fredholm-type equations are obtained for few-body systems and properties of their resolvent kernels are investigated also in the momentum representation. The next two chapters are devoted to the study of wave functions in the configuration space: Chapter 4 treats neutral particles and Chapter 5 treats systems of charged particles. The Fredholm-type integral equations for systems of charged particles are obtained in Chapter 5 too. Some problems of the mathematical foundation of scattering theory are considered in Chapter 6. Here, the general features of the wave operators discussed in Chapter 1 are proved. A number of applications of the stationary scattering theory is discussed in Chapter 7.

The aim of each chapter is more precisely described in short introduction at the beginning. Similar introductory comments are presented at the beginning of each section.

We do not refer to any specific bibliographic sources in the main text. References of special importance are given at the and of the book.

Finally, we introduce some basic notation which will be used throughout the book. The term variable and the letters x, y, X, k, p, P with or without indices denote vectors in N-dimensional space. The symbol(k,p) denotes the scalar product of two vectors k and p, $k^2 = (k,k)$, $|k| = (k^2)^{1/2}$, $\hat{k}$ is the unit vector in the direction of k: $\hat{k} = \frac{k}{|k|}$, and dk and $d\hat{k}$ are the volume and surface elements on the unit sphere. The symbol $\int$ with no indication of the integration range means the integral over the whole range of integration variables.

The letters x, y, X denote vectors in the configuration space, k, p, P, q are vectors in the momentum space. Transformation from the coordinate representation to the momentum representation is carried out by Fourier transformation, to wit

$$\hat{f}(k) = (2\pi)^{-n/2} \int \exp\{-i(k,x)\} f(x) dx.$$

As it is common in the physical literature, we will often use the same symbols for functions and their Fourier transforms $f(x) \leftrightarrow f(k)$ with the argument indicating the actual case.

CHAPTER 1

General Aspects of the Scattering Problem

1.1 Formulation of the Problem

This book deals with the quantum mechanical N-body problem. In the main part of the text the particles are supposed to be non-relativistic, structureless and interacting only pairwisely. All characteristic features of the described theory can be shown on this somewhat simplified case and possible generalizations do not meet with principal difficulties.

Each particle is described by its coordinate $r_i \in R^3$ and its mass m_i, $i = 1, 2, \ldots, N$. The N-particle energy operator acts in the Hilbert space $L_2(R^{3N})$ of functions $\Psi(r_1, r_2, \ldots, r_N)$ depending on coordinates $r_1, r_2, \ldots, r_N$ and is given by a differential operator

$$H_N = -\sum_{i=1}^{N} \frac{1}{2m_i}\Delta_i + \sum_{i<j} v_{ij}(r_i - r_j).$$

Here Δ_i is the three-dimensional Laplace operator in coordinates r_i and the functions $v_{ij}(r)$ are the potentials of the pair interaction which depend on the relative positions $r_i - r_j$ of i - th and j - th particles.

We will deal with two types of pair interactions which we call short-range interactions and long-range interactions respectively.

In the first case, all functions $v_{ij}(r)$ tend to zero sufficiently fast for $|r| \to \infty$. The meaning of this has to be specified in each particular case. The condition

$$v_{ij}(r) = o(|r|^{-1-\epsilon}), \quad \epsilon > 0$$

is a typical requirement which is sufficient in many considerations; however, $\epsilon > 1$, $\epsilon > 2$ or even a stronger bound must be assumed to get more detailed results.

The long-range potential differs from the short-range potential by the Coulomb component

$$v_{ij} = \frac{z_{ij}}{|r|} + \tilde{v}_{ij}(r)$$

where $\tilde{v}_{ij}(r)$ is a short-range potential and z_{ij} is a real constant, which differs from $q_i q_j$, the product of charges of the particles i and j, by the universal factor $z_{ij} = \gamma q_i q_j$ only.

From the mathematical point of view more general cases of long range interactions can be studied. However, the Coulomb potential represents the most interesting case and therefore we confine ourselves to it. Consequently, the terms "neutral particles" and "charged particles" will mean the particles with short and long ranges of interaction.

The short-range potential may be singular in a finite region, e.g., it may have a singularity of the form

$$v(r) = o(|r - r_0|^{-\beta}), \quad \beta < 2$$

at the point r_0. Under the conditions given above the operator H_N is a self-adjoint operator in $L_2(R^{3N})$ - the N-particle Hamiltonian.

The dynamics of the quantum mechanical system is given by the Schrödinger equation:

$$i\frac{\partial \Psi}{\partial t} = H_N \Psi \tag{1.1}$$

The time development is described by the unitary evolution operator

$$U(t) = \exp\{-iH_N t\}.$$

The scattering problem is fully specified if appropriate initial conditions for the solution of the Schrödinger equation are imposed.

An idealized physical formulation of the scattering problem is as follows: a long time before the scattering process ($t \to -\infty$) the system is in a state representing a set of bound clusters of particles far apart and moving towards one another. During a finite time the clusters interact and then for $t \to \infty$ bound clusters are formed once again. The outcoming clusters differ, in general, from the original ones. Hence, the scattering process consists in restructuring the clusters and in changing their internal state.

In this formulation the states of the system before $(t \to -\infty)$ and after $(t \to +\infty)$ the scattering are treated symmetrically.

In realistic experiments the situation is slightly different. Nevertheless, it can be formally described in the terms introduced above. The calibration of a free particle beam corresponds to the first stage $(t \to -\infty)$ and the analysis of particles scattered after impinging a target corresponds to the second stage $(t \to +\infty)$. Therefore, in this book we give a formal description of the above mentioned formulation of the scattering problem.

A first approximation to the above mentioned description goes as follows. Let us consider l bound clusters. This means that the following data are given:

1. The set of coordinates $y_1, y_2, \ldots, y_l$ of the centres of masses of clusters.
2. Their masses $M_1, M_2, \ldots, M_l$, each being the sum of the masses of all particles constituting the cluster.
3. The relative coordinates x_k, $k = 1, 2, \ldots, l$ of particles constituting the cluster (details of the description of the clusters are given in the next section).
4. The wave functions $\psi_k(x_k)$, $k = 1, 2, \ldots, l$ describing the internal states of the individual clusters.

Different sets of these data will be distinguished by special subscripts A, B, and so on.

The wave function

$$\Psi_A(r_1, r_2, \ldots, r_N; t) = \chi_A(y_1, y_2, \ldots, y_l; t) \prod_{k=1}^{l} \psi_k(x_k) \tag{1.2}$$

describes the free motion of the clusters if the function $\chi_A(y_1, y_2, \ldots, y_l; t)$ solves the Schrödinger equation

$$i\frac{\partial \chi_A}{\partial t} = H_A^{(N)} \chi_A \tag{1.3}$$

where $H_A^{(N)}$ is the differential operator

$$H_A^{(N)} = -\sum_k \frac{1}{2M_k}\Delta_k - E_A$$

acting in $L_2(R^{3N})$. The constant E_A is the sum of the binding energies of the clusters. It is a well-known property of the solutions of the Schrödinger equation for a free motion that for $t \to \infty$, the wave function spreads into the neighbourhood of infinity in R^{3N}, so that the clusters are really scattered over large distances.

The general asymptotic motion is given by the superposition of these states:

$$\Psi_{as}(r_1, r_2, \ldots, r_N; t) = \sum_A c_A \Psi_{A,as}(r_1, r_2, \ldots, r_N; t).$$

The scattered state $\Psi(t)$ which corresponds to the given initial state $\Psi_{as}^{(-)}(t)$ is determined by the solution of the Schrödinger equation (1.1) which tends to $\Psi_{as}^{(-)}$ as $t \to -\infty$:

$$\|\Psi(t) - \Psi_{as}^{(-)}(t)\| \to 0 \quad \text{as } t \to -\infty$$

If we alter the scattering process, this state must pass into an asymptotic state in the sense that there exists a function $\Psi_{as}^{(+)}(t)$ such that the relation

$$\|\Psi(t) - \Psi_{as}^{(+)}(t)\| \to 0 \quad \text{as } t \to +\infty$$

holds.

The scattering consists in the transformation of the state $\Psi_{as}^{(-)}(t)$ into $\Psi_{as}^{(+)}(t)$. The relation between $\Psi_{as}^{(-)}(t)$ and $\Psi_{as}^{(+)}(t)$ is obviously linear. Consequently, there exists a linear operator S satisfying the relation

$$\Psi_{as}^{(+)}(t) = S\Psi_{as}^{(-)}(t). \tag{1.4}$$

The operator S does not depend on t. This not obvious property expresses the energy conservation law; we will discuss it in Chapter 2. The operator S acts, in fact, on the set of functions $\chi_A(t)$, so that it is given by a matrix S_{AB}. This operator contains all information about all possible results of scattering and therefore it is called the scattering operator.

This formulation of the scattering problem is based on the following assumptions:

1. We are able to prove that the solution $\Psi(t)$ satisfying the initial condition (1.2) for $t \to -\infty$ does exist;

2. We can show that the asymptotic behaviour of $\Psi(t)$ for $t \to \infty$ is given by the formula (1.4).

In spite of the fact that at first sight the formulations of both problems seem nearly symmetric, the first one is easier. The asymmetry follows from the fact that in the first case we deal with $\Psi_{as}^{(-)}$ explicitly given, while in the second case we have to deal with the much less controllable $\Psi_{as}^{(+)}$.

More formally, suppose we have solved the first problem. Then there are two subspaces $\mathfrak{H}^{(-)}$ and $\mathfrak{H}^{(+)}$; the vector $\Psi(t)$ obtained as a solution, for $t \to -\infty$ and $t \to +\infty$ belongs to the subspaces $\mathfrak{H}_{as}^{(-)}$ and $\mathfrak{H}_{as}^{(+)}$ in $L_2(R^{3N})$, respectively. These subspaces are invariant under action of the evolution operator $U(t)$. Let the condition

$$\mathfrak{H}_{as}^{(+)} = \mathfrak{H}_{as}^{(-)} \tag{1.5}$$

be satisfied. Then each vector $\Psi(t)$ which we obtain by solving the initial problem for $t \to -\infty$ can also be obtained if we start from $t \to \infty$, and it has the required asymptotics. This requirement means that of all possible asymptotic motions none is missing. Therefore, this condition is called the condition of asymptotic completeness. The mathematical proof of this statement is one of the most difficult tasks of the scattering theory and it has not been yet completed in full generality. Now, we consider charged particles. In this case, the description of the asymptotic dynamics of the wave vectors corresponding to the centres of mass of the clusters must be slightly modified. The physical reason is that charged particles move asymptotically along hyperbola which do not approach straight lines sufficiently fast.

The corrections to the asymptotic dynamics can be found by the following considerations. First, we consider the classical motion of two charged particles. The Hamiltonian is given by

$$H = \frac{1}{2m_1}p_1^2 + \frac{1}{2m_2}p_2^2 + \frac{z_{12}}{|r_1 - r_2|}$$

where p_1 and p_2 are the momenta of the particles 1 and 2. The classical motion along the straight lines is given by the formulae

$$r_1 = \frac{p_1}{m_1}t + q_1, \quad r_2 = \frac{p_2}{m_2}t + q_2$$

and the Coulomb potential on these trajectories is

$$\frac{m_1 m_2 z_{12}}{|m_2 p_1 - m_1 p_2|} \cdot \frac{1}{|t|} + O(1/t^2).$$

The first component does not go to zero for $|t| \to \infty$ fast enough. This spoils the actual trajectories, as mentioned above. However, if the last expression is added to the Hamiltonian of the free motion so that it assumes the form

$$H_{as}(t) = \frac{1}{2m_1} p_1^2 + \frac{1}{2m_2} p_2^2 + \frac{m_1 m_2 z_{12}}{|m_2 p_1 - m_1 p_2|} \cdot \frac{1}{|t|} + O(1/t^2)$$

the trajectories given by this Hamiltonian approximate, for $|t| \to \infty$, the actual trajectories sufficiently well. Similar considerations also work in the case of several charged particles. Each pair of charged particles or bound clusters contributes to the Hamiltonian of the asymptotic motion.

Returning to quantum mechanics, we see that Eq. (1.3) must be replaced by the equation

$$i\frac{\partial \chi}{\partial t} = H_A^{as}(t)\chi_A$$

where

$$H_A^{as} = H_A^{(N)} + \frac{1}{|t|} \sum_{i \le k}^{l} B_{ik}$$

and the operators B_{ik} are

$$B_{ik} = \frac{\zeta_{ik} M_i M_k}{\sqrt{(M_i \nabla_k - M_k \nabla_i)^2}}.$$

The constants z_{ik} are proportional to the product $\sum_l q_l \sum_j q_j$ where the summation is over the particles in the clusters i and k, respectively.

1.2 Kinematics

In this section we introduce kinematic variables which are used for the description of states of a N-body system in the configuration and momentum spaces.

1.2.1 Subsystems and partitions

Let us consider a system of N bodies, labelled from 1 to N. We call a set of l particles, $1 \leq l \leq N$ with indices n_1, n_2, ..., n_l a subsystem of it. We denote subsystems by the multiindices ω_l, $\omega_l(n_1, n_2, \ldots, n_l)$ whose components run over the indices of particles which constitute the subsystem. By definition, we take as the subsystem $\omega_1(n_1)$ the set consisting of one particle with index n_1, while the subset ω_N is identical with the original system of N-bodies. Sometimes we omit the index l denoting the number of particles.

We call a division of an N-body system into k subsystems with $k \leq N$ the partition. We distinguish different partitions by the Latin letters $a_k^{(N)}$, $b_k^{(N)}$. The label N denoting the number of particles in the system will usually be omitted.

We describe the partition in detail by giving explicitly the contained subsystems. For example, the symbol

$$a_3 = (132)(4)(65)$$

means the partition of six particles into three subsystems $\alpha_3(132)$, $\beta_1(4)$, $\gamma_2(56)$. It is clear that the order of the subsystems in the decomposition and the order of the particles in the subsystems is unimportant.

Let us note that the partition is uniquely determined by the specification of the pair of particles (ij) joined in the subsystem α_2. Therefore, we will often use the symbol of the partition α_{N-1} for the corresponding pair of particles (ij) of the subsystem α_2.

If the partition a_k is obtained from the partition a_i $(i < k)$ by division of its subsystems into parts, we say that a_k follows a_i or that a_i precedes a_k

and we write $a_k \subset a_i$ or $a_i \supset a_k$. We use the same letter only for partitions which are related in this way, i.e., $a_2 \supset a_3 \supset \cdots \supset a_{N-1}$.

A succession of such partitions which starts with some a_k, $2 \leq k \leq N-1$ is called the chain of partitions A_k:

$$A_k = (a_k, a_{k+1}, \ldots, a_{N-1}).$$

The last partition in a chain is always a_{N-1}. The chain of partitions represents a method of step-by-step division of a_{k-1} into individual particles. In particular, the full chain $A_2 = (a_2, a_3, \ldots, a_{N-1})$ determines a partition of the N-body system itself. In the notation of chains, it is sometimes useful to distinguish the preceding partition, i.e., to write

$$A_k = (a_k, A_{k-1}) = (a_k, a_{k-1}, A_{k-2}) \quad \text{ect.}$$

The chain can be pictured as a "tree". Every branch corresponds to the switching on an interaction between the particles. A succession of the branches from below upwards determines a succession of divisions of subsystems into two parts.

For the system of three particles there is only one type of a tree (Figure 1), which corresponds to switching on an interaction between the particles 1 and 2 and then between both of them and the particle 3. All other possible trees are obtained by a simple renumbering of branches.

In the case of four particles, the picture is more complicated. In this case we have two possibilities of switching on an interaction (Figures 2 and 3). There exist twelve trees of the first type and six trees of the second type.

The number of different types of trees grows fast with the number N.

For each subsystem $\omega_l(n_1, n_2, \ldots, n_l)$ we have a corresponding energy operator

$$H_{\omega_l} = -\sum_{i=1}^{l} \frac{1}{2m_{n_i}} \Delta_{r_{n_i}} + \sum_{i<j} v_{n_i n_j}(r_{n_i} - r_{n_j}) \tag{1.6}$$

acting in the Hilbert space $L_2(R^{3l})$ of functions $\varphi(r_{n_1}, r_{n_2}, \ldots, r_{n_l})$ which depend on the coordinates of the particles in the subsystem. According to our convention, the operator H_{ω_N} is identical with the energy operator H_N of the whole system and H_{ω_n} is the operator of the of the n-th particle $-\sum_{i=1}^{l} \frac{1}{2m_n} \Delta_{r_n}$.

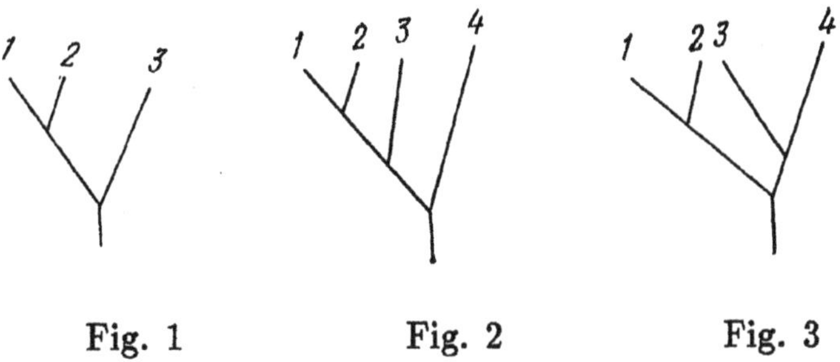

Fig. 1 Fig. 2 Fig. 3

To the partition a_k, there corresponds a decomposition of the space of states $L_2(R^{3N})$ into a tensor product of spaces $L_2(R^{3l_j})$, which describe the states of the subsystems of a_k, $j = 1, 2, \ldots, k$:

$$L_2(R^{3N}) = \prod_{j=1}^{k} \otimes L_2(R^{3l_j}) \tag{1.7}$$

In accordance with this expression, we define the energy operator of the partition as the sum of the energy operators of the subsystems:

$$H_{a_k} = \sum_{\omega} H_{\omega}.$$

In this expression every term in the sum acts non-trivially in the corresponding space of states of the subsystem and it is supposed to act as the unit operator on the other components of (1.7).

Below, we will often encounter the similar situation considering a tensor product and operators acting actively on one of its terms only. Without especially saying so, we will always use the same notation for the operator acting on the corresponding term and for the operator which we obtain from the first one by its trivial continuation.

1.2.2 Reduced coordinates

Let us have a set of numbers k_1, k_2, ..., k_N. We form the succession of subsystems ω_2, ω_3, ..., ω_{N-1}, ω_N in such a way that their members are obtained by successive joining of the particles k_2, ..., k_N to the particle k_1 i.e., $\omega_2 = (k_1, k_2)$, $\omega_3 = (k_1, k_2, k_3)$ etc. Let y_{ω_j} be the coordinate of the centre of mass of the subsystem with mass m_{ω_j}

$$y_{\omega_j} = \frac{1}{m_{\omega_j}} \sum_{i-1}^{j} m_{k_i} r_{k_i}, \quad m_{\omega_j} = \sum_{i-1}^{j} m_{k_i} \tag{1.8}$$

and let $\mu_{\omega_j k_{j+1}}$ be the reduced mass of the particle k_{j+1} with respect to the subsystem ω_j

$$\mu_{\omega_j k_j+1} = \frac{m_{\omega_j} m_{k_{j+1}}}{m_{\omega_{j+1}}}.$$

The reduced coordinate $x_{\omega_j k_{j+1}}$ of the particle k_{j+1} with respect to the centre of mass of the subsystem ω_j is given by

$$x_{\omega_j k_j} = (2\mu_{\omega_j k_j})^{1/2} (y_{\omega_j - r_{k_{j+1}}}). \tag{1.8'}$$

The set of variables $x_{\omega_j k_{j+1}}$ and the coordinate of the centre of mass of the system $y_{\omega_N} = \rho$ determine the coordinates of the point in the configuration space in terms of a new orthogonal basis called the Jacobi basis. There exist $N!$ different Jacobi bases corresponding to all possible permutations of N particles.

We introduce a notation which expresses the ordering of particles with respect to which the basis is constructed.

For example, the three body system characterized by the symbol (312) can be described by the Jacobi coordinates

$$x_{31} = \sqrt{2\mu_{31}}(r_3 - r_1), \quad x_{31,2} = \sqrt{2\mu_{31,2}} \left(\frac{m_3 r_3 + m_1 r_1}{m_3 + m_1} - r_2 \right) \tag{1.9}$$

where

$$\mu_{31} = \frac{m_3 m_1}{m_3 + m_1}, \quad \mu_{31,2} = \frac{(m_3 + m_1) m_2}{m_1 + m_2 + m_3}.$$

It is helpful to have in mind an illustrative scheme which corresponds to these coordinates (Figure 4).

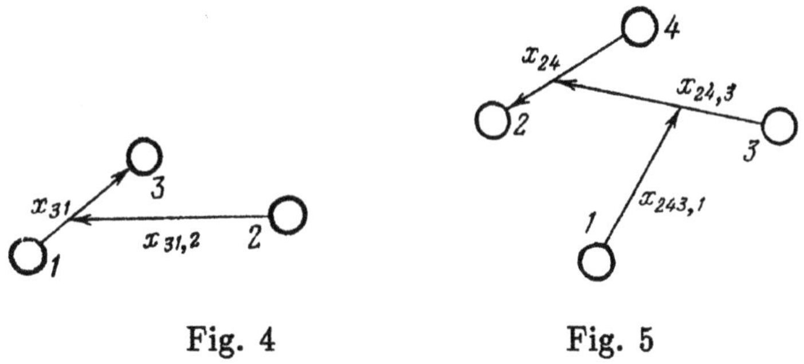

Fig. 4 Fig. 5

The system of coordinates which corresponds to the symbol (2341) of four bodies is represented in the Figure 5. Here

$$x_{24} = \sqrt{2\mu_{24}}(r_2 - r_4)$$

$$x_{24,3} = \sqrt{2\mu_{24,3}}\left(\frac{m_2 r_2 + m_4 r_4}{m_2 + m_4} - r_3\right)$$

$$x_{243,1} = \sqrt{2\mu_{243,1}}\left(\frac{m_2 r_2 + m_4 r_4 + m_3 r_3}{m_2 + m_4 + m_3} - r_1\right)$$

$$\mu_{243,1} = \frac{(m_2 + m_4 + m_3)m_2}{m_1 + m_2 + m_3 + m_4}.$$

We will often regard the set of $N-1$ relative coordinates $x_{\omega_j k_{j+1}}$ ($j = 1, 2, \dots, N-1$) as a vector X_{ω_N} in the $3(N-1)$-dimensional configuration space of relative motion of the particles:

$$X_{\omega_N} = \left\{x_{\omega_1\varkappa_2}, x_{\omega_2\varkappa_3}, \dots, x_{\omega_{N-1}\varkappa_N}\right\}. \tag{1.10}$$

At the same time the configuration space R^{3N} can be written as an orthogonal sum,

$$R^{3N} = R_c^{3(N-1)} \oplus R^3$$

where the 3-dimensional space R^3 corresponds to the coordinates of the centre of mass

$$\rho \equiv y_{\omega_N} = \frac{1}{M}\sum_{i=1}^{N} m_i r_i$$

where $\sum_{i=1}^{N} m_i$ is the total mass of the system. The coordinates X_{ω_N} are called relative coordinates. For the vector X_{ω_N} we will often omit the subscript ω_N.

Together with the Jacobi basis, we will often use the relative coordinates associated with the partition a_k. This new basis is constructed in the following way. In every subsystem ω_l of the partition a_k, we introduce a system of reduced Jacobi coordinates $\{X_{\omega_l}, y_{\omega_l}\}$, where the vector $X_{\omega_l} \in R^{3(l-1)}$ is determined by (1.10) and y_{ω_l} is the coordinate of the centre of mass of the subsystem (1.8). Further, we consider the set of k vectors y_{ω_l} as the coordinates of point particles of masses m_{ω_l} and construct the Jacobi basis

$$\left\{y_{\omega_{l_1}\omega_{l_2}}, y_{\omega_{l_{12}}\omega_{l_3}}, \ldots, y_{\omega_{l_1 l_2 \ldots l_{k-1}}\omega_{l_k}}\right\}$$

where

$$y_{\omega_{l_1 l_2 \ldots l_m}\omega_{l_{m+1}}} = \sqrt{2\mu_{\omega_{l_1 l_2 \ldots l_m}\omega_{l_{m+1}}}}\left(y_{\omega_{l_1 l_2 \ldots l_m}} - y_{\omega_{l_{m+1}}}\right) \tag{1.11}$$

and $\omega_{l_1 l_2 \ldots l_m}$ is the subsystem obtained as the union of the subsystems ω_{l_1}, $\omega_{l_2}, \ldots, \omega_{l_m}$

$$\omega_{l_1 l_2 \ldots l_m} = (\omega_{l_1}, \omega_{l_2}, \ldots, \omega_{l_m}).$$

Finally, we obtain a set of vectors $X_{\omega_{l_j}}$ $(j = 1, 2, \ldots, k)$ and $y_{\omega_{l_1 l_2 \ldots l_m}\omega_{l_{m+1}}}$ $(m = 1, 2, \ldots, k-1)$ which determine a point in the configuration space $R_c^{3(N-1)}$ in terms of the basis induced by the partition a_k.

We combine the set of relative coordinates of the centres of mass of the subsystems into a $3(k-1)$-dimensional vector y_{a_k}:

$$y_{a_k} = \left\{y_{\omega_{l_1}\omega_{l_2}}, y_{\omega_{l_1 l_2}\omega_{l_3}}, \ldots, y_{\omega_{l_1 l_2 \ldots l_{k-1}}\omega_{l_k}}\right\}$$

and the set of vectors $X_{\omega_{l_j}}$ $(j = 1, 2, \ldots, k)$ into the $3(N-k)$-dimensional vector x_{a_k},

$$x_{a_k} = \left\{x_{\omega_{l_1}}, x_{\omega_{l_2}}, \ldots, x_{\omega_{l_k}}\right\}.$$

At the same time, the configuration space of the relative motion $R_c^{3(N-1)}$ can be written as an orthogonal sum

$$R_c^{3(N-1)} = R_{a_k} \oplus \tilde{R}_{a_k}$$

where

$$x_{a_k} \in R_{a_k} \quad y_{a_k} \in \tilde{R}_{a_k}.$$

We call the Jacobi coordinates of the subsystems X_{ω_l} and x_{a_k} the intrinsic coordinates. Similarly, the extrinsic coordinates of the partition a_k are the reduced coordinates of the centres of mass of the subsystems y_{a_k} and $y_{\omega_{l_1 l_2 \ldots l_j l_{j+1}}}$.

In the subspaces $\tilde{R}_{a_k}$, it is convenient to use the bases associated with the chain of partitions $A_k = (a_1, a_2, \ldots, a_k)$ $(k = 3, 4, \ldots, N-1)$. We introduce the corresponding relative coordinates in the following way. As the first coordinate we choose the vector $y_{a_2} \in R^3$ for the first partition $a_2 = (\alpha, \beta)$ in the chain. If the partition $a_3 \subset a_2$ is obtained from a_2 by division of the subsystem α into two parts $\alpha = (\alpha_1, \alpha_2)$, we denote by $y_{a_2 a_3}$ the reduced relative coordinate of the subsystems a_1 and α_2 determined by relation (1.11).

In this way we obtain two variables y_{a_2} and $y_{a_2 a_3}$ which determine the coordinates of the vector y_{a_3} in terms of the basis corresponding to the chain $A_3 = (a_2, a_3)$. This procedure can be repeated. In every step $a_j \to a_{j+1}$ $(j = 2, 3, \ldots, k-1)$, we introduce a reduced relative coordinate $y_{a_2 a_3 \ldots a_{j+1}}$ between subsystems into which some subsystem splits as a result of the partition a_{j+1}. Finally, we obtain $k-1$ three-dimensional vectors

$$y_{a_2}, y_{A_2}, \ldots, y_{A_k}$$

which determine the basis

$$y_{a_k} = \{y_{A_2}, y_{A_3}, \ldots, y_{A_k}\} \tag{1.12}$$

associated with the chain A_k.

We give now a number of examples of relative coordinates of the types x_{a_k} and y_{a_k}. In the case of three bodies, the latter coordinates coincide with the former, i.e., with the Jacobi coordinates:

$$x_{12,3} = y_{12,3}, \quad x_{23,1} = y_{23,1}, \quad x_{31,2} = y_{31,2} \tag{1.12'}$$

In a 4-body system two types of relative coordinates exist which correspond to two ways of two-cluster partition. The first one is generated by the partition of the form $a_2 = (243)(1)$, the second one by the partition $b = (12)(34)$. The coordinates of the types x and y coincide with the coordinates schematically shown in Figure 5 if we put

$$x_{243,1} = y_{243,1}.$$

The same coordinates correspond to the chain $A_3 = (a_2, a_3)$, where $a_3 = (12)(3)(4)$ and $y_{a_2a_3} = x_{24,1}$. The partition b_2 and the chain $B_3 = (b_2, b_3)$, where $b_3 = (12)(3)(4)$ induce the coordinates x_{12}, x_{34}, $y_{12,34}$, and $y_{b_2b_3} = x_{34}$ shown in Figure 6, where

$$x_{12} = \sqrt{2\mu_{12}}(r_1 - r_2), \quad x_{34} = \sqrt{2\mu_{34}}(r_3 - r_4),$$

$$y_{12,34} = \sqrt{2\frac{(m_1 + m_2)(m_3 + m_4)}{m_1 + m_2 + m_3 + m_3}} \left(\frac{m_1r_1 + m_2r_2}{m_1 + m_2} - \frac{m_3r_3 + m_4r_4}{m_3 + m_3} \right).$$

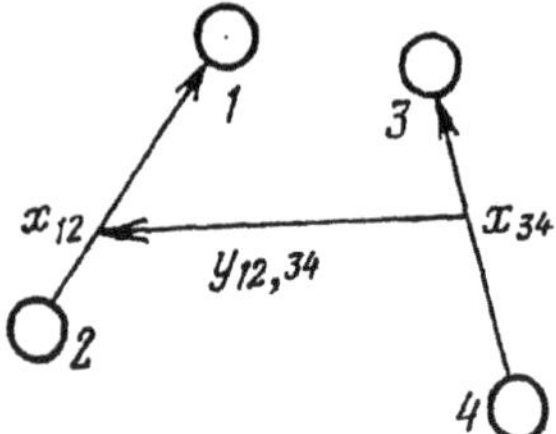

Fig. 6

A similar procedure can be applied to systems of five or more particles; however, already in these cases it is difficult to enumerate all different combinations of the relative coordinates. We will not complicate our exposition by details of the combinatorics, but mention only two important facts. If the partition a_{N-1} is generated by the pair $\alpha = (ij)$, the vector $x_{a_{N-1}}$ coincides with the reduced relative coordinate of this pair

$$x_{a_{N-1}} = x_\alpha.$$

In a more general case when the partition consists of a subsystem ω_k and of $N - k + 1$ free particles, the coordinate x_{a_k} coincides with the vector X_{ω_k}:

$$x_{a_k} = X_{\omega_k}, \quad a_k = (\omega_k, j_1, j_2, \ldots, j_{N-k}).$$

The reduced relative coordinates described above differ from the Jacobi coordinates usually employed in classical mechanics by factors depending on the masses of the particles. We use such coordinates for the following reason.

First of all, it can be seen from (1.8) - (1.12) that transition from one basis to another is equivalent to an orthogonal transformation of the coordinate system in the space $R_c^{3(N-1)}$. Secondly, and this is the main merit of these coordinates, in these bases the kinetic energy operators of the subsystems reduce to multidimensional Laplace operators in the corresponding subspaces. Thus, the form of these operators does not change with a change of the basis and this fact substantially facilitates the work with the differential and integral equations of the theory of scattering. Later, we will see many other examples of advantages of the reduced coordinates introduced in this section.

1.2.3 The momentum space

Passage to the momentum representation is carried out by means of the Fourier transformation in the space R^{3N}:

$$\hat{f}(p_1, p_2, \ldots, p_N) = (2\pi)^{-3N/2} \int dr_1 dr_2 \cdots dr_N f(r_1, r_2, \ldots, r_N) \tag{1.13}$$

where $p = (p_1, p_2, \ldots, p_N)$ is a point in the momentum space R_m^{3N} and $(r, p) = \sum_{i=1}^{N} (r_i, p_i)$.

Next, we describe the reduced momentum variables induced by the bases in the coordinate space introduced above. We denote by $k_{\omega_j l_{j+1}}$ the reduced relative momentum coordinates conjugate to the intrinsic coordinates . They are given in terms of the momenta p_i of the particles by

$$k_{\omega_j l_{j+1}} = (\mu_{\omega_j l_{j+1}})^{-1/2} \left(\frac{m_{l_{j+1}} p_{\omega_1} - m_{\omega_1} p_{l_{j+1}}}{m_{l_{j+1}} m_{\omega_j}} \right) \tag{1.14}$$

where p_{ω_j} is the total momentum of the subsystem ω_j:

$$p_{\omega_j} = \sum_{i=1}^{j} p_{k_i}, \quad \omega_j = (k_1, k_2, \ldots, k_j).$$

The reduced relative momenta conjugate to the extrinsic coordinates $y_{\omega_{k_1 k_2 \ldots k_+ l} \omega_{k_{l+1}}}$ are given in terms of the momenta of the particles by

$$p_{\omega_{k_1 k_2 \ldots k_+ l} \omega_{k_{l+1}}} =$$

$$= \left(\mu_{\omega_{k_1 k_2 \ldots k_+ l}\omega_{k_{l+1}}}\right)^{-1/2} \left(\frac{m_{\omega_{k_1 k_2 \ldots k_+ l}} p_{\omega_{k_{l+1}}} - m_{\omega_{k_{l+1}}} p_{\omega_{k_1 k_2 \ldots k_+ l}}}{m_{\omega_{k_1 k_2 \ldots k_+ l}} + m_{\omega_{k_{l+1}}}}\right). \quad (1.14')$$

Let P_{ω_j} be a vector in the momentum space conjugate to the coordinate X_{ω_j}

$$P_{\omega_j} = \{k_{\omega_1 l_2}, k_{\omega_2 l_3}, \ldots, k_{\omega_{j-1} l_j}\}, \quad \omega_i = (l_1, l_2, \ldots, l_i).$$

We denote by k_{a_l} the vector conjugate to the intrinsic coordinate x_{a_l}

$$k_{a_l} = \{p_{\omega_{i_1}}, p_{\omega_{i_2}}, \ldots, p_{\omega_{i_l}}\}$$

and by p_{ω_l} the vector conjugate to the extrinsic coordinate

$$p_{a_l} = \{p_{\omega_{i_1}\omega_{i_2}}, p_{\omega_{i_{12}}\omega_{i_3}}, \ldots, p_{\omega_{i_1 i_2 \ldots i_{l-1}}\omega_l}\}.$$

Similarly to the case of the configuration space, the momentum space can be represented by the orthogonal sum

$$R^{3N} = R_{a_l} \oplus \tilde{R}_{a_l} \oplus R^3$$

where $k_{a_l} \in R_{a_l}$, $p_{a_l} \in \tilde{R}_{a_l}$, and R^3 is the space of the total momentum $q = p_1 + p_2 + \cdots + p_N$. The scalar product (r, p) is equal to the sum of the scalar products in the subspaces R_{a_l}, $\tilde{R}_{a_l}$, and R^3:

$$(r, p) = (x_{a_l}, k_{a_l}) + (y_{a_l}, p_{a_l}) + (\rho, q)$$

for an arbitrary a_l.

To the chain of partitions A_l corresponds the momentum variables p_{A_2}, p_{A_3}, ..., p_{A_l} conjugate to the coordinates y_{A_2}, y_{A_3}, ..., y_{A_l}

$$p_{a_l} = \{p_{A_2}, p_{A_3}, \ldots, p_{A_l}\}.$$

Note that the replacement of one set of momenta by another complete set is realized by the same rotation which connects the corresponding variables in the configuration space.

1.3 Fundamental Concepts of Dynamics

In this section we introduce the fundamental concepts which are needed for the description of the dynamics of the N-body system and its subsystems.

1.3.1 The energy operators

Let (ρ, X) be the coordinates of the system of N particles. Here ρ is the coordinate of the centre of mass and X is the set of relative coordinates introduced in the preceding section. By the change of variables

$$(r_1, r_2, \ldots, r_N) \to (\rho, X)$$

the kinetic energy operator

$$H_0 = -\sum_{i=1}^{N} \frac{1}{2m_i} \Delta_i$$

is transformed into

$$H_0 = -\frac{1}{2M}\Delta_\rho - \Delta_X$$

where M is the total mass of the system and Δ_X the Laplace operator in the $3(N-1)$ variables which constitute the vector X. Indeed, no mixed derivatives are introduced by the considered linear transformation and the mass-independent factors in relations similar to (1.8′) were chosen in such a way that the coefficients at all the terms with quadratic derivatives are equal to one.

In accordance with the partition

$$L_2(R^{3N}) = L_2(R^3) \oplus \mathfrak{H}$$

where $\mathfrak{H} = L_2(R^{3(N-1)})$ is the space of states for the relative motion, we introduce the kinetic energy operator of the relative motion

$$H_0 = -\Delta_X.$$

The form of this operator does not depend on the particular choice of relative coordinates.

In the basis corresponding to the partition a_l, the operator H_0 is given by

$$H_0 = -\Delta_{x_{a_l}} - \Delta_{y_{a_l}}$$

where $\Delta_{x_{a_l}}$ and $\Delta_{y_{a_l}}$ are the Laplace operators in the variables x_{a_l} and y_{a_l}, respectively. This decomposition is in accordance with the decomposition of the space $\mathfrak{H}$ as the tensor product

$$\mathfrak{H} = \mathfrak{H}_{int} \otimes \mathfrak{H}_{ext} \tag{1.15}$$

of the spaces of states of intrinsic motion in the subsystems and of relative motion of the centres of mass of the subsystems. Hence, it is natural to denote

$$-\Delta_{x_{a_l}} = h^{(a_l)}_{0\,int}, \qquad -\Delta_{y_{a_l}} = h^{(a_l)}_{0\,ext}$$

and to consider these operators on the spaces

$$\mathfrak{H}_{int} = L_2(R^{3(N-l)}), \qquad \mathfrak{H}_{ext} = L_2(R^{3(l-1)})$$

respectively.

The action of the kinetic energy operators in the momentum representation reduces to multiplication by the sum of the squares of the corresponding momenta:

$$h^{(a_l)}_{0\,int}\varphi(k_{a_l}) = k^2_{a_l}\varphi(k_{a_l}),$$
$$h^{(a_l)}_{0\,ext}\varphi(p_{a_l}) = p^2_{a_l}\varphi(p_{a_l}).$$

The interaction potential

$$V(r_1, r_2, \ldots, r_N) = \sum_{i<j} v_{ij}(r_i - r_j)$$

depends on the relative coordinates only. It determines an operator in $\mathfrak{H}$ which reduces to multiplication by the function $V(X)$.

In accordance with our convention that the pair $(i\,j)$ is identified with the partition a_{N-1}, we have $V = \sum_{a_{N-1}} V_{a_{N-1}}$; here, every term is the operator of multiplication by a function

$$V_{a_{N-1}} f(X) = v_{a_{N-1}}(x_{a_{N-1}}) f(X).$$

In the momentum representation, this operator is expressed as an integral with kernel

$$v_{a_{N-1}}(P,P') = v_{a_{N-1}}(k_{a_{N-1}} - k'_{a_{N-1}})\delta(p_{a_{N-1}} - p'_{a_{N-1}})$$

Here $v(k)$ is the Fourier transform of the function $v(x)$

$$v(k) = (2\pi)^{-3/2} \int e^{-i(k,x)} v(x)\, dx.$$

The total energy operator with the separated centre of mass term reads

$$H = H_0 + V$$

Let us consider now similar objects for partitions. We will denote by H_{ω_l} the energy operator for the subsystem ω_l with the centre of mass separated

$$H_{\omega_l} = -\Delta_{x_{\omega_l}} + V_{\omega_l}$$

acting in $L_2(R^{3(l-1)})$. Here, V_{ω_l} is the interaction operator of particles making up the subsystem ω_l.

For the partition a_l, we denote by H_{a_l} the energy operator which we obtain after separating of the centre of mass of N particles. In accordance with decomposition (1.15), this can be written as

$$H_{a_l} = h_{0\,ext}^{(a_l)} + h_{int}^{(a_l)} \tag{1.16}$$

where

$$h_{int}^{(a_l)} = \sum_{j=1}^{l} H_{\omega_{k_j}} = h_{int}^{(a_l)} + \sum_{a_{N-1} \subset a_l} V_{a_{N-1}}. \tag{1.16$'$}$$

Here, the summation runs over all partitions a_{N-1} which follow a_l. Recall that the operator H_{a_l} describes the relative motion of the system of N particles in which the interactions between the particles belonging to different subsystems of the partition a_l were switched off.

By definition, we will assume that

$$V_{a_1} = V, \quad V_{a_N} = 0$$

so that the operator H_{a_1} coincides with the original Hamiltonian and the operator H_{a_N} coincides with the kinetic energy operator:

$$H_{a_1} = H, \quad H_{a_N} = H_0.$$

1.3.2 Clusters

Let us consider the energy operators of the subsystems consisting of two or more particles. We will assume that their spectra satisfy the following important condition which seems plausible from the physical point of view: The principal part of the spectrum is continuous and covers the semi-axis $[\lambda_{\omega_k}, \infty)$, where λ_{ω_k} is a non-positive number; left to λ_{ω_k} the operator H_{ω_k} has a point spectrum only. Let $\epsilon^j_{\omega_k}$ and $\psi^j_{\omega_k}(X_{\omega_k})$ be the corresponding eigenvalues and eigenfunctions.

If a discrete spectrum exists, we say that the subsystem ω_l possesses bound states. The corresponding eigenvalues are called binding energies.

If every many particles subsystem ω_{l_j} of the partition a_k has a bound state, we call a_k a cluster partition. Because of separation of variables in (1.16) the energy operator $h^{(a_l)}_{int}$ of the intrinsic motion of the cluster partition a_k has the discrete spectrum with eigenvalues

$$-\varkappa_I^2 = \sum_{j=1}^{k} \epsilon^{i_j}_{\omega_{k_j}}$$

where $I = (i_1, i_2, \ldots, i_l)$ is the set of numbers of the bound states. The corresponding eigenfunctions are given by the products

$$\psi_I = \prod_{j=1}^{k} \psi^{i_j}_{\omega_{k_j}}(X_{\omega_{k_j}})$$

A cluster partition with a fixed set I is called detailed partition and we denote it by I_{a_l}. Each number $-\varkappa^2_{I_{a_l}}$ represents the origin of the branch of the continuous spectrum of the total energy operator H_{a_l} of the partition a_l, which is generated by the operator of free motion $h^{(a_l)}_{ext}$. Indeed, the functions

$$\chi_{I_{a_l}}(X) = \psi_{I_{a_l}}(x_{a_l})e^{i(p_{a_l}, y_{a_l})} \tag{1.17}$$

are eigenfunctions of H_{a_l} with the eigenvalues

$$E_{I_{a_l}}(p_{a_l}) = p^2_{a_l} - \varkappa^2_{I_{a_l}}.$$

The numbers just introduced must satisfy the following consistency condition: for every subsystem ω_k, the origin of the continuous spectrum λ_{ω_k} coincides with the lowest eigenvalue of the continuous spectrum of any partition of ω_k.

In particular, for the total energy operator H, the origin of the continuous spectrum coincides with the smallest of the numbers $-\varkappa_{I_{a_l}}$.

We stress once more that this character of the spectrum, though very reasonable from the physical point of view, is considered as a fundamental postulate. In this book this assumption will be justified only in part.

Further, we introduce the projector $P_{I_{a_l}}$ projecting on the subspace spanned by the eigenfunctions $\chi_{I_{a_l}}$. It is convenient to express $P_{I_{a_l}}$ by an integral operator with kernel

$$P_{I_{a_l}}(X, X') = \psi_{I_{a_l}}(x_{a_l})\psi^*_{I_{a_l}}(x_{a_l})\delta(y_{a_l} - y'_{a_l}).$$

The index I_{a_l} introduced above seems to be a little bit clumsy. Below, we replace it by a single capital letter from the beginning of the Latin alphabet: A, B, etc. In accordance with this convention we also use the notation x_A, y_A, k_A, p_A, ψ_A, χ_A, $-\varkappa_A$, and P_A; the meaning of these symbols will be clear from the context. For N free particles we use the subscript $A = 0$ and, by definition, we let P_0 be the identity transformation. The function ψ_A is put equal to one for $A = 0$.

1.3.3 Channels of reactions

From the physical point of view, every detailed cluster partition can be considered as some initial or final state of the scattering process. For the formal description of this assertion, we introduce the concept of channel connected with such a partition.

Channels can be defined in many ways. For instance, we can take channel to mean the subspace $\hat{\mathfrak{H}}_A$ onto which the operator P_A projects,

$$\hat{\mathfrak{H}}_A = P_A\mathfrak{H}.$$

However, the subspaces $\hat{\mathfrak{H}}_A$ are not mutually orthogonal which is formally disadvantageous. For this reason we consider spaces $\mathfrak{H}_A$ isomorphic to $\hat{\mathfrak{H}}_A$, and we assume that $\mathfrak{H}_A$ consists of the functions $f(p_A)$. Next, we introduce the operatorL_A: $L_A\mathfrak{H}_A = \hat{\mathfrak{H}}_A$.

We call the space $\mathfrak{H}_A$ the space of channel A or simply the channel and we call the direct orthogonal sum

$$\mathfrak{H}_{as} = \sum_A \oplus \mathfrak{H}_A \tag{1.18}$$

the space of channels or the space of asymptotic states. Let us stress that the space $\mathfrak{H}_{as}$ is larger than the space $\mathfrak{H}$, the latter being isomorphic to the space $\mathfrak{H}_0$ of the channel A which corresponds to the partition into separated particles.

By means of L_A we can carry over some operators from $\mathfrak{H}$ to $\mathfrak{H}_{as}$, and vice versa. In particular, the operator $L_A^{-1} H_A L_A$ acts in $\mathfrak{H}_A$ as the operator of multiplication by the function $E_A(p_A)$. For $A = 0$, this operator acts as the operator of multiplication by the function $E_0(P) = P^2$.

1.4 Wave Operators

The concepts introduced in the two preceding sections allow us to state more precisely and formally the scattering problem and to formulate the problem justifying it.

1.4.1 Definition of wave operators

In Section 1, we considered the Hilbert space $L_2(R^{3N})$ of states of all particles. Now we start with motion of the centre of mass separated, so that we work with the space $\mathfrak{H}$. What we previously called the collection of bound clusters, we now call the detailed partition A.

The subspace $\hat{\mathfrak{H}}_A$ corresponds to the wave function $\psi_A(r_1, r_2, \ldots, r_N, t)$, and the Schrödinger equation (1.3) determines the evolution operator

$$\hat{U}_A(t) = e^{-iH_A t} P_A. \tag{1.19}$$

It is clear that the function χ_A is determined uniquely by the element f_A, so that the evolution can be given also by the operator

$$U_A(t) = L_A^{-1} \hat{U}_A(t) L_A$$

which is explicitly given by the formula

$$U_A(t) f(p_A) = e^{-ip_A^2 t + i\varkappa_A^2 t} f(p_A)$$

In other words, the asymptotic state corresponding to the channel A is given by

$$\psi_{as}^A(t) = L_A U_A(t) f,$$

where $f \in \mathfrak{H}_A$. The limit

$$\lim_{t \to \infty} \|\psi(t) - \psi_{as}^A(t)\| = 0 \tag{1.20}$$

exists if the limit $t \to \infty$ exists in the strong sense for the operator

$$V_A(t) = e^{iHt} L_A U_A(t) \tag{1.21}$$

which acts from $\mathfrak{H}_A$ to $\mathfrak{H}$. In fact, the left hand side of (1.20) can be written as

$$\|e^{-iHt}\Psi_0 - L_A U_A(t) f\| = \|\Psi_0 - V_A(t) f\|$$

where $\Psi_0 = \Psi(t)|_{t=0}$. In the last step we used unitarity of the operator e^{-iHt}. It follows that (1.20) is fulfilled if the operators

$$U_A^{(\pm)} = \lim_{t \to \mp\infty} V_A(t) \tag{1.22}$$

exist. The operators $U_A^{(\pm)}$ are called wave operators. The notation $(\pm)$ which corresponds to the limits $t \to \mp\infty$ is in agreement with the convention generally accepted in the physical literature.

Given the definitions of the operators $V_A(t)$ and $U_A(t)$, we get the representation $U_A^{(\pm)} = \tilde{U}_A^{(\pm)} L_A$, where the operator in $\mathfrak{H}$,

$$\tilde{U}_A^{(\pm)} \lim_{t \to \mp\infty} e^{iHt} e^{-iH_A t} P_A \tag{1.23}$$

is also called the wave operator.

Thus, as the first step in the justification of the formulation of the scattering problem discussed in Section 1, the existence of wave operators must be proved.

The spaces $\mathfrak{H}_{as}^{(\pm)}$ which are to be compared in the second justification step are the union of ranges of the operators $U_A^{(\pm)}$. Therefore, it is necessary to show that this union is correctly defined and to study its construction. This question will be the first problem considered in the next section.

In the case of long-range interaction the definitions introduced must be slightly modified. The definition of evolution operator changes for every channel containing charged bound states. We denote by H_A^{as} the operator of multiplication by the function

$$H_A^{as}(p_A, t) = p_A^2 - \varkappa_A^2 + \frac{\eta_A(p_A)}{t},$$

where

$$\eta_A(p_A) = \sum_{\substack{i,j \\ i>j}} \frac{q_{\omega_{l_i}} q_{\omega_{l_j}}}{2|p_{\omega_{l_i}\omega_{l_j}}|} \sqrt{2\mu_{\omega_{l_i}\omega_{l_j}}} \tag{1.24}$$

Here q_ω is the total charge of the cluster ω; the sum runs over all clusters of the detailed partition A.

The asymptotic evolution operator is given by

$$U_A^c(t) = \exp\{-ip_A^2 t + i\varkappa_A^2 t - i\,\mathrm{sign}\,t\eta_A \log|t|\}.$$

We obtain the wave operators from the definitions given above by replacing $U_A(t)$ in (1.21) by $U_A^c(t)$:

$$V_A(t) = e^{iHt} L_A U_A^c(t). \tag{1.25}$$

These wave operators will be also called generalized (or Coulomb) wave operators.

1.4.2 Existence of wave operators

Now we consider the first stage of justification of the formulation of the scattering problem. First, we prove that the wave operators (1.22) exist if both the potentials and the Fourier transforms of their squares $v^2(k)$ are smooth bounded functions.

The proof makes use of the following criterion:

CRITERION A. *Let $\{f_A\}$ be a set of smooth bounded functions which is dense in the space $\mathfrak{H}_A$. Then the strong limits (1.22) exist for $t \to \infty$, provided the functions*

$$\varphi_A(t) = \left\| \frac{d}{dt} V_A(t) f_A \right\| \tag{1.26}$$

are absolutely integrable on the interval $[1, \infty)$ ($(-\infty, -1)]$).

To prove this assertion, we use the formula

$$(V_A(t_1) - V_A(t_2)) f_A = \int_{t_1}^{t_2} \frac{\mathrm{d}}{\mathrm{d}t} V_A(t) f_a \, dt$$

and we estimate the norm of the integral on the right hand side. By assumption, the integral $\int_{t_1}^{t_2} \| \frac{\mathrm{d}}{\mathrm{d}t} V_A(t) f_a \| dt$ tends to zero for $t_1, t_2 \to \pm\infty$. Thus

$$\|(V_A(t_1) - V_A(t_2)) f_A\| = o(1)$$

for $t_1, t_2 \to \pm\infty$ and the strong limits $\lim_{t\to\pm\infty} V_A(t)$ exist.

Let us consider the function $\varphi_A(t)$. Evaluating the derivative in (1.26), we obtain

$$\varphi_A(t) = \|\bar{V}_A L_A e^{-iE_A t} f_A\| \tag{1.27}$$

where $\bar{V}_A$ denotes the operator describing the interaction between the clusters of the detailed fragmention A,

$$\bar{V}_A = V - V_A.$$

We estimate of expression for large values of $|t|$.

First we consider a two-body system. In this case the quadratic form φ_A^2 can be written in the momentum representation as the integral

$$\varphi_A^2 = \int dp dp' \, e^{i(p^2 - p'^2)t} \hat{v}^2(p - p') \overline{f_A(p)} f_A(p') \tag{1.28}$$

where we denote by $\hat{v}^2(p - p')$ the Fourier transform of the square of the potential $v^2(x)$. For $|t| \to \infty$ the integrand is a rapidly oscillating function. Such integrals are called rapidly oscillating integrals. Their behaviour can be studied by the method of stationary phase. In our case the function $\hat{v}^2(p-p')$ is smooth, so the principal contribution is given by the critical points $p, p' = 0$. Evaluating this contribution, we obtain the estimate $|\varphi'_A(t)| \leq C(t + |t|)^{-3/2}$. Thus, the function $\varphi_A(t)$ is absolutely integrable and consequently the limits (1.22) exist.

Next we consider the general case. Let ω, ω' be subsystems of the detailed partition A and let Ω be the subsystem obtained as the union of these subsystems $\Omega = (\omega, \omega')$. Let $V_{\omega\omega'}$ be the operator describing the interaction between the subsystems ω and ω',

$$V_{\omega\omega'} = V_\Omega - (V_\omega + V_{\omega'}).$$

It is clear that the operator describing the interaction between the clusters can be written as the sum of the interaction operators between the subsystems,

$$\bar{V}_A = \sum_{\omega, \omega'} V_{\omega\omega'}$$

Thus, it suffices to apply the criterion to each component $V_{\omega\omega'}$ separately. We show that the quadratic forms corresponding to these components can be expressed as integrals of the type (1.28).

Let $\{x_A, y_A\}$ be the reduced Jacobi coordinates corresponding to the detailed partition A and let $\{k_A, p_A\}$ be the corresponding momenta. We express the vectors y_A and p_A in the basis induced by the chain of partitions A_l,

$$y_A = \{y_{a_l a_{l-1}}, y_{A_{l-1}}\} \quad p_A = \{p_{a_l a_{l-1}}, p_{A_{l-1}}\},$$

where the partition a_{l-1} is obtained from a_l as the union of the subsystems ω and ω'. In this basis the function $E_A(p_A)$ is given by the sum

$$E_A(p_A) = p^2_{a_l a_{l-1}} + p^2_{A_{l-1}} - \chi^2_A.$$

Note that the potential $V_{\omega\omega'}$ depends only on the intrinsic coordinates of the clusters X_ω, $X_{\omega'}$ and on the extrinsic coordinate $y_{a_l a_{l-1}}$ only. Thus, the interaction operator in the momentum space acts with respect to the coordinates $p_{A_{l-1}}$ as the unit operator. Thus, the nontrivial part of the exponential function depends only on the variables $p_{a_l a_{l-1}}$. The quadratic form corresponding to the component $V_{\omega\omega'}$ in this basis is given by

$$\varphi^2_{\omega\omega'} = \int dp_{A_{l-1}} dp'_{a_l a_{l-1}} dp_{a_l a_{l-1}} \exp\{i(p^2_{a_l a_{l-1}} - p'^2_{a_l a_{l-1}})t\} \times$$

$$\times \tilde{v}^A_{\omega\omega'}(p_{a_l a_{l-1}} - p'_{a_l a_{l-1}}) f(p_{a_l a_{l-1}}, p_{A_{l-1}}) f(p'_{a_l a_{l-1}}, p_{A_{l-1}}) \qquad (1.29)$$

where $\tilde{v}^A_{\omega\omega'}$ is the Fourier transform of the square of the "mean average" of the potential, which is defined by

$$\tilde{v}^A_{\omega\omega'}(y_{a_l a_{l-1}}) = \int |\psi_\omega(X_\omega)|^2 |\psi_{\omega'}(X_{\omega'})|^2 v_{\omega\omega'}(X_\omega X_{\omega'}, y_{a_l a_{l-1}})\, dX_\omega dX_{\omega'}.$$

It is obvious that both the functions $\tilde{v}^A_{\omega\omega'}$ and $\hat{v}_{a_{N-1}}$ are smooth. Thus, we can estimate the value of the integral (1.29) similarly as in the case of the quadratic form considered above. Finally, we find that the functions $\varphi_{\omega\omega'}(t)$ are absolutely integrable on the intervals $(\pm 1, \infty)$.

Thus, we have established that the wave operators exist, provided that the Fourier transforms of the squares of the pair potentials are smooth bounded functions. The corresponding conditions can be formulated also in the x-representation. However, we will not consider this question here. Let us note only that these conditions are satisfied for potentials decreasing as a power $|x|^{-3-\epsilon}$ for $|x| \to \infty$.

The consideration above can be generalized to the broad class of short-range potentials characterized in Section 1. In general, their Fourier transforms will have singularities; however, these singularities are not very strong. In contrast to these potentials, Fourier transforms of Coulomb-like potentials possess singularities which cannot be considered within the given scheme.

As we have seen above, the definition of wave operators must be modified in these cases.

Thus we have to prove the existence of generalized wave operators for Coulomb potentials. To this end, we consider the expression

$$\Phi_A^c(t) = \frac{\mathrm{d}}{\mathrm{d}t} e^{iHt} L_A U_A^c(t) f_A = e^{iHt}(\bar{V}_A L_A - L_A \bar{V}_A(p_A, t)) e^{-iE(p_A)} f_A(p_A)$$

and we estimate its norm for $|t| \to \infty$. Let the Coulomb potential be given in the x-representation. In the variables being used, the problem of investigating of the function $||\Phi_A^c(t)||^2$ reduces to the study of the behaviour of the integral

$$I_A^c(t) =$$

$$= \int_{|x|>1} dx \left| \int dp e^{-itp^2 + i(x,p)} \left(\frac{1}{|x|} - \frac{1}{2pt} \right) e^{-i\mathrm{sign} t \eta_{\omega\omega'} \log|t|} f(p) \right|^2,$$

$$\eta_{\omega\omega'} = \gamma \frac{q_\omega q_{\omega'}}{2p} \sqrt{2\mu_{\omega\omega'}}.$$

We make the substitution $x = ty$ and then we perform the first integration with respect to the variable p. The rapidly oscillating exponential takes the form $\exp\{itg(p)\}$, where the function $g(p) = p^2 - (p, y)$ has a single critical point $p_0 = y/2$, $\nabla g(p_0) = 0$ at which $g(p_0) = -y^2/4$.

Note that at this critical point the function $(|y|^{-1} - (2pt)^{-1})$ in the integrand vanishes. Thus, for $y > 0$ this integral decreases as $t^{-5/2}$, i.e., faster than the power $|t|^{-3/2}$ which appears when the integrand is non-zero. This is the point w difference between the modified definition of wave operators (1.25) and the definition (1.22) for the short-range potentials shows up.

On the other hand, at the points $y = 0$ and $p = 0$ the integrand grows to infinity. It can be proved that the integral decreases as $t^{-3/2-\epsilon}$ for $y = O(t^{-1})$, $0 < \epsilon < 1/2$.

Thus, for the remaining integral with respect to y the estimate

$$I_A^c(t) = |t|^{-2-2\epsilon} \int dy |f(y)|^2 |y|^{-2-2\epsilon}, \quad 0 < \epsilon < 1/2$$

is valid. Consequently, for $t > 1$ the integral $I_A^c(t)$ is bounded and is of the order $O(|t|^{-2-2\epsilon})$, i.e., $\sqrt{I_A^c(t)}$ is an absolutely integrable function of t on the interval $(\pm 1, \pm\infty)$.

Thus, in the case of Coulomb potentials the generalized wave operators exist. The whole procedure can be repeated, with a slight modification, in the case of long-range potentials with a short-range component. It shows that generalized wave operators exist in this case, too.

1.5 Properties of the Wave Operators

Next we describe the general relations holding for the wave operators. In stating them, it is not important whether we deal with rapidly decreasing or Coulomb potentials. The technique in the proofs is the same in both cases. We give proofs for the case of rapidly decreasing potentials.

To make the formulae less obscure, we will omit the symbols $(\pm)$ occurring in the wave operators and we will give the proof only for the case $(-)$. However, unless stated otherwise, the final formulae are valid both for the operators with the label $(-)$ and $(+)$.

PROPERTY A. *The relations*

$$HU_A = U_A E_A \tag{1.30}$$

are satisfied (the so called intertwining property).

To prove it we use the equality

$$e^{iHt}U_A = U_A e^{iE_A t}$$

which follows from the relations

$$\lim_{\tau\to\infty} e^{iHt}e^{iH\tau}e^{-iH_A\tau}L_A = \lim_{\tau'\to\infty} e^{iH\tau'}e^{-iH_A\tau'}L_A e^{iE_A t}.$$

Differentiating this equality with respect to t, we obtain the relation

$$e^{iHt}HU_A = U_A E_A e^{iE_A t}$$

which gives (1.30) for $t = 0$.

PROPERTY B. *The operators U_A are partially isometric*

$$U_A^* U_A = I_A \tag{1.31}$$

and their ranges are mutually orthogonal

$$U_A^* U_B = 0 \quad \text{for } A \neq B. \tag{1.32}$$

In order to prove this property, it suffices to prove that

$$(U_A f_A, U_B f_B') = \delta_{AB}(f_A, f_B'),$$

where δ_{AB} is the Kronecker symbol:

$$\delta_{AB} = 1 \quad \text{for } A = B \quad \text{and } \delta_{AB} = 0 \quad \text{for } A \neq B.$$

We write the bilinear form $I_{AB} = (U_A f_A, U_B f_B')$ as a limit

$$I_{AB} = \lim t \to \infty (e^{-iH_A t} L_A f_A, e^{-iH_B t} L_B f_B) \tag{1.33}$$

and consider three cases successively.

Let $A = B$. Then the equality $I_{AA} = (f_A, f_A')$ follows immediately from the unitarity of the operator $e^{-H_A t}$.

Let $A \neq B$ and let the corresponding partitions be different $a_l \neq b_j$. In this case the scalar product (1.33) can be written in the form of a rapidly oscillating integral

$$I_{AB} = \int dP f_A(p_A) \overline{f_B(p_B)} \psi_A(k_A) \overline{\psi_B(k_B)} e^{-it(E_A(p_A) - E_B(p_B))}.$$

The limit of such integrals is equal to zero for $t \to \infty$.

If $a_l \neq b_j$, rapidly oscillating functions do not appear, because $E_A(p_A) = E_B(p_B)$ in this case. Both functions $\psi_A(k_A)$ and $\psi_B(k_B)$ are eigenfunctions of the operator H_A, so that the integral I_{AB} vanishes as a result of their orthogonality. Thus, property B is proved.

The orthogonality of the ranges of the wave operators has a simple physical interpretation. Note that the vectors $L_A U_A(t) f_A$, $L_B U_B(t) f_B'$ appearing in the scalar product in (1.33) describe asymptotic states in channels A and B. Therefore, vanishing of the limit (1.33) expresses orthogonality of asymptotic wave packets. The larger t, the better the condition of their orthogonality is satisfied. It is interesting that this property takes place despite the fact that the channels $\hat{\mathfrak{H}}_A$ are $\hat{\mathfrak{H}}_B$ not orthogonal.

We formulate the final property of the wave operators. We determine the union of the ranges of these operators and thereby answer the question which appears at the second stage of justification of the statement of the scattering problem.

First of all, we note that the operators $\pi_A^{(\pm)} = U_A^{(\pm)} U_A^{(\pm)*}$ are projectors onto orthogonal spaces. In fact, in accordance with equalities (1.31) and (1.32), the relations

$$\pi_A^2 = \pi_A, \quad \pi_A \pi_B = 0 \quad A \neq B$$

are satisfied.

Thus, the sum of these operators

$$P_c^{(\pm)} = \sum_A U_A^{(\pm)} U_A^{(\pm)*}$$

is a projector too. Therefore, our problem reduces to determining the ranges of the operators $P_c^{(\pm)}$.

We denote by P_c the projector onto the subspace $\mathfrak{H}_c$ corresponding to the continuous spectrum of the energy operator H. We assume that the ranges of operators $P_c^{(\pm)}$ satisfy the following important hypothesis:

PROPERTY C. *The ranges of the operators $P_c^{(\pm)}$ coincide with the subspace $\mathfrak{H}_c$:*

$$P_c^{(+)} = P_c^{(-)} = P_c \tag{1.34}$$

Note that Property C guarantees the fulfilment of the condition (1.5) of asymptotic completeness. In this case we say the wave operators have the property of the asymptotic completeness.

A problem of proving of this property is one of the principal problems of the theory of scattering. Despite the great efforts by many people, this problem has not been solved for the general case yet. Existing methods described in the next section have been proved successful only in the case of two and three bodies. Nevertheless, all the constructions which are used in the theory of scattering assume that the asymptotic completeness condition is satisfied, because it is a necessary condition for a correct physical picture of scattering.

Properties of the wave operators are sometimes formulated by means of the continuous spectrum eigenfunctions of the Hamiltonian. In this case, the wave operators are regarded as transforms of the channels $\mathfrak{H}_A$ in the configuration space. Their kernels in this so-called mixed representation are functions of coordinate variables X and momentum variables p_A. In this approach it is natural to take the momentum variable as a parameter. These functions are called wave functions.

The intertwining property is expressed in terms of wave functions by the stationary Schrödinger equation:

$$\left(-\Delta + \sum_{\alpha_{N-1}} v_{A_{N-1}} - E_A(p_A)\right) U(X, p_A) = 0 \tag{1.35}$$

Thus the wave functions may be regarded as the eigenfunctions of the energy operator corresponding to its continuous spectrum. Equality (1.31) which takes a form

$$\int \overline{U_A(X, p_A)} U_A(X, p'_B)\, dX = \delta_{AB}\delta(p_A - p'_B)$$

can be considered as the orthogonality condition for these functions. Finally, the asymptotic completeness condition says that any function from the subspace $\mathfrak{H}_c$ can be expressed as a generalized Fourier integral of wave functions

$$f_c = \sum_a \int U_A(X, p_A) g_A(p_A)\, dp_A$$

where the Fourier coefficients are given by

$$g_A(p_A) = \int \overline{U_A(X, p_A)} f_c(X)\, dX.$$

Finally, we express the properties of wave operators in a compact form. To this end, we introduce the matrix wave operator $\hat{U}^{(\pm)}$ mapping the channel space $\mathfrak{H}_{as}$ to the space $\mathfrak{H}$. We define

$$\hat{U}^{(\pm)}\hat{f} = \sum_A U_A^{(\pm)} f_A$$

and denote by $\hat{H}_0$ the operator

$$\hat{H}_0 = \sum_A \oplus L_A^{-1} H_A L_A$$

which is reducible in $\mathfrak{H}_{as}$.

The operator $\hat{H}$ is called the asymptotic Hamiltonian. It acts in each channel $\mathfrak{H}_A$ as a multiplication by the function $E_A(p'_A)$,

$$(L_A^{-1} H_A L_A) f_A(p_A) = (p_A^2 - \varkappa_A^2) f_A(p_A) \equiv E_A(p_A) f_A(p_A).$$

The identity transformation in $\mathfrak{H}_{as}$ is denoted by $\hat{I}$. Then the properties of the wave operator can be summarized by the following assertion:

PROPOSITION U. *The energy operator H of an N-body system in the subspace $\mathfrak{H}_c$ is unitary equivalent to the asymptotic Hamiltonian $\hat{H}_U$. There exists an isometric operator $\hat{U}$ from the channel space $\mathfrak{H}_{as}$ to the subspace $\mathfrak{H}_c$ such that the relation*

$$\hat{U}^*\hat{U} = \hat{I}, \quad \hat{U}\hat{U}^* = P_c \quad H\hat{U} = \hat{U}\hat{H}_0 \tag{1.36}$$

are satisfied. The matrix wave operators $\hat{U}^{(\pm)}$ can be taken as the operator $\hat{U}$.

The first relation is the orthogonality condition for the ranges of the wave operators. The second expresses the asymptotic completeness condition in a closed form and it must be regarded as a conjecture. The last one is equivalent to equalities (1.35) which are written in the form of the Schrödinger equation for the wave functions.

Let us stress that Proposition U is valid for both short-range and long-range potentials. For wave operators, we must take either the limits (1.22), or the generalized wave operators (1.25).

1.6 The Scattering Operator

In this section we give a rigorous definition of the scattering operator. We describe its general properties and discuss the physical meaning of its matrix kernels.

Let us consider the operator

$$\hat{S} = \hat{U}^{(-)*}\hat{U}^{(+)} \tag{1.37}$$

in $\mathfrak{H}_{as}$; it is so-called scattering operator. This operator acts on elements of the channel space according to the formulae

$$\hat{f}' = \hat{S}\hat{f}, \quad \hat{f} = \{f_B\}, \quad \hat{f}' = \{f'_A\}$$

$$f'_A = \sum_B S_{AB} f_B,$$

where the matrix elements S_{AB} determine the scattering operator from the channel $\mathfrak{H}_B$ to the channel $\mathfrak{H}_A$:

$$S_{AB} = U_A^{(-)*} U_B^{(+)} \tag{1.38}$$

The definition of wave operators leads to the following non-stationary representation of the operators S_{AB}:

$$S_{AB} = \lim_{\substack{t_1 \to \infty \\ t_2 \to -\infty}} e^{iE_A t_1} L_A^{-1} e^{iH(t_2 - t_1)} L_B e^{-iE_B t_2}. \tag{1.39}$$

In the case of charged particles, the matrix elements S_{AB} are given by

$$S_{AB} = \lim_{\substack{t_1 \to \infty \\ t_2 \to -\infty}} e^{iE_A t_1 + i\eta_A \log|t_1|} L_A^{-1} e^{iH(t_2 - t_1)} L_B e^{-iE_B t_2 + i\eta_B \log|t_2|}. \tag{1.40}$$

General properties of the scattering operator are summarized in the following proposition.

PROPOSITION $\hat{S}$. *Let the condition of asymptotic completeness be satisfied. Then the operator $\hat{S}$ is unitary and commutes with the energy operator in the channel space $\hat{H}_0$:*

$$\hat{S}\hat{S}^* = \hat{S}^*\hat{S} = \hat{I}, \quad \hat{S}\hat{H}_0 = \hat{H}_0\hat{S}$$

First of all we prove the equality $\hat{S}\hat{S}^* = 1$. Note it follows from the definition of scattering operator and the condition of asymptotic completeness that

$$\hat{S}\hat{S}^* = \hat{U}^{(-)}P_c\hat{U}^{(-)}.$$

The range of the operator $\hat{U}^{(-)}$ is identical with $\mathfrak{H}_c$ and the relation $\hat{S}\hat{S}^* = 1$ follows immediately from the preceding equality. Similarly we prove that $\hat{S}^*\hat{S} = 1$.

In order to prove that the operators $\hat{S}$ and $\hat{H}_0$ commute, we use the intertwining property (1.36). First we interchange the order of the operators $\hat{H}_0$ and $\hat{U}^{(-)*}$, $\hat{H}_0\hat{U}^{(-)*}\hat{U}^{(+)} = \hat{U}^{(-)*}\hat{H}\hat{U}^{(+)}$; then the interchange of the operators $\hat{H}$ and $\hat{U}^{(+)}$ gives the desired result. Now, we will describe the general properties of kernels of the scattering operator $S_{AB}(p_A, p'_B)$. We have shown that the operator $\hat{S}$ commutes with the Hamiltonian of channels $\hat{H}_0$ which acts on them as multiplication by the function $E_A(p_A)$. Thus, the kernels S_{AB} must be proportional to δ-functions $\delta(E_A(p_A) - E_B(p'_B))$:

$$S_{AB}(p_A, p'_B) = \delta(E_A(p_A) - E_B(p'_B))\bar{S}_{AB}(\hat{p}_A, \hat{p}'_B, E_A).$$

The kernels $\bar{S}_{AB}$ define an integral operator which is called the scattering matrix. This operator acts from the Hilbert space of functions on the unit sphere $|p_B| = 1$ to the Hilbert space of functions on the unit sphere $|p_A| = 1$.

Let us clarify the physical meaning of elements of S-matrix. We consider the simplest case of the two-body scattering. We will see in Chapter 4 that the kernel of scattering operator has the form

$$s(p,p') = \delta(p - p') - \delta(p^2 - p'^2)\bar{t}(\hat{p}, \hat{p}', p^2)$$

where the kernel $\bar{t}$ is already free of singularities. Thus $|\bar{t}|^2$ is well-defined and this is the expression which has the physical meaning.

Let us consider the initial state with approximately exact values of energy and with the direction of motion

$$\Omega_{as}^{(-)}(x,t) =$$

$$= \left(\frac{1}{2\pi}\right)^{3/2} \int \epsilon_E \bar{\delta}(E - E_0)\epsilon_\alpha \bar{\delta}(\alpha - \alpha_0)\frac{\sqrt[4]{E}}{\sqrt{2}} e^{i\sqrt{E}(x,\alpha)} e^{-iEt}\, dE d\alpha$$

where α represents a point on the unit sphere. The functions $\bar{\delta}(E - E_0)$ and $\bar{\delta}(\alpha - \alpha_0)$ which determine this state satisfy the conditions

$$\int \bar{\delta}(E - E_0)\, dE = 1, \quad \int \bar{\delta}(\alpha - \alpha_0)\, d\alpha = 1$$

and

$$\epsilon^{-2}2_E = \int (\bar{\delta}(E - E_0))^2\, dE, \quad \epsilon^{-2}2_\alpha = \int (\bar{\delta}(\alpha - \alpha_0))^2\, d\alpha.$$

Let the axis x_3 be in the direction α_0. The expression

$$dW = \int_{-\infty}^{\infty} |\Omega_{as}^{(-)}(x, t)|^2 dx_3 d\Theta \equiv$$

$$\cong \left(\frac{1}{2\pi}\right)^2 \epsilon_\alpha^2 \int_{-\infty}^{\infty} dx_3 \left| \left(\frac{1}{2\pi}\right) \int_{-\infty}^{\infty} \epsilon_E \bar{\delta}(E - E_0) \frac{\sqrt[4]{E}}{\sqrt{2}} e^{i\sqrt{E}x_3 - iEt}\, dE \right|^2 dQ$$

gives the probability of finding a free particle described by the wave packet in the cylinder with the cross section dQ perpendicular to the x_1, x_2-plane for arbitrary t, provided the values of x_1 and x_2 are not too large. For large t and x, the integral with respect to E can be computed by the method of stationary phases. Putting $x_3 = 2\lambda t$, we get

$$dW = \left(\frac{1}{2\pi}\right)^2 \epsilon_\alpha^2 \int_{-\infty}^{\infty} \epsilon_\lambda^2 (\bar{\delta}(E - E_0))^2 \lambda\, d\lambda dQ = \left(\frac{1}{2\pi}\right)^2 \epsilon_\alpha^2 E_0 dQ.$$

Thus the probability does not depend on x_1, x_2, and t. If a source produces N particles per unit time and if these particles form a beam carrying a flux of j particles per unit time, we get $j dQ = N dW$, i.e.,

$$j = \left(\frac{1}{2\pi}\right)^2 \epsilon_\alpha^2 E_0 N \tag{1.41}$$

Further, we will see that the probability density of the scattering in the direction $\alpha \neq \alpha_0$ is determined by the amplitude

$$h_+(E, \alpha) \cong (2\pi)^{-2} \epsilon_E \bar{\delta}(E - E_0) \epsilon_\alpha \tilde{t}(\alpha, \alpha_0, E_0), \quad E_0 = p^2 \tag{1.42}$$

and thus the number of particles scattered per unit time per solid angle $d\Omega$ is given by

$$dN = N \epsilon_\alpha^2 |\tilde{t}(\alpha, \alpha_0, E_0)|^2 d\Omega. \tag{1.43}$$

Comparing (1.41) and (1.43) we see that the ratio

$$\frac{1}{j}\frac{\mathrm{d}N}{\mathrm{d}\Omega} = \frac{\mathrm{d}\sigma}{\mathrm{d}\Omega} = \left(\frac{1}{2\pi}\right)^2 \frac{|\tilde{t}(\alpha, \alpha_0, E_0)|^2}{E_0}$$

does not depend on accuracy with which the direction of momentum is given. This quantity has the dimension of area and is called the effective cross section. Such a relation exists for any scattering process, provided there are only two clusters in the initial state. In particular, the kernel of the scattering operator in this case is

$$S_{AB} = \delta_{AB}\delta(p_A - p_B) + \delta(E_A - E_B)\tilde{T}_{AB}(\hat{p}_A, \hat{p}_B, E_A)$$

where $|\tilde{T}_{AB}|^2$ is well-defined. This object represents the effective cross section, i.e., it characterizes the transition probability from channel B to channel A. If the initial state consists of three or more clusters, the definition of the effective cross section more involved. In these cases the kernel $\tilde{T}_{AB}$ may have δ-like singularities. We will see this effect in Chapter 4.

Finally, we introduce some terms useful for the physical interpretation of elements of the S-matrix. The function $E_A(p_A)$ will be called the scattering energy in the channel A. The value of this function for zero momentum determines the channel threshold:

$$E_A(0) = -\varkappa_A^2.$$

The channel is called open if the energy of the system exceeds the threshold. The channel of decay into N free particles opens for zero energy $E_0(0) = 0$. Note that if one of the channels A or B is closed, the corresponding S-matrix element is supposed to be equal to zero.

This concludes our description of general features of the scattering theory.

CHAPTER 2

Stationary Approach to Scattering Theory

In this chapter we express the wave operator in terms of singularities of kernel of the resolvent. These results form a basis for an alternative approach to the scattering problem.

This approach allows to seek the solution of the original dynamical problem by means of the perturbation theory of linear operators. In addition, the computational methods of the scattering theory can be most efficiently developed in this framework.

2.1 Resolvent and Wave Operators

In order to give stationary definitions of the wave operators, we make use of the following obvious proposition.

Let $U(t)$ be an operator function whose range is in the set of bounded operators in the Hilbert space and possesses limits for $t \to \mp\infty$,

$$U_{\pm} = \lim_{t \to \mp\infty} U(t).$$

Then these limits are equal to

$$U_{\pm} = \lim_{\epsilon \downarrow 0} \pm\epsilon \int_{\mp\infty}^{0} e^{\pm\epsilon t} U(t)\, dt \tag{2.1}$$

where the symbol $\lim\limits_{\epsilon \downarrow 0}$ means that the parameter ϵ is non-negative.

Now we perform the transformation of the wave operators. Because of (2.1), they can be written as the integral

$$U_A^{(\pm)} = \lim_{\epsilon \downarrow 0} \pm\epsilon \int_{\mp\infty}^{0} e^{it(H \mp i\epsilon)} L_A e^{-iE_A t}\, dt. \tag{2.2}$$

Computing this integral, we get

$$U_A^{(\pm)} = \lim_{\epsilon\downarrow 0} \mp i\epsilon R(E_A \pm i\epsilon)L_A \tag{2.3}$$

where $R(z)$ is the resolvent of the energy operator,

$$R(z) = (H - z)^{-1}.$$

It is useful to express (2.3) in terms of wave functions.

Note that for $A = 0$, the kernel of L_A reduces to $L_0(P, P') = \delta(P - P')$ and for $A \neq 0$

$$L_A(P', p_A) = \psi_A(k_A)\delta(p'_A - p_A) \tag{2.4}$$

where $\psi_A(k_A)$ is the Fourier transform of the eigenfunction of the operator h_A^{int}. This kernel is an eigenfunction of the continuous spectrum of the operator H_A. Thus, in agreement with our interpretation, it describes the asymptotic state of the system in channel A. In arbitrary representation, L_A can be defined to be a function of the first variable and the variable p_A may be considered as a parameter. In the X-representation we get e.g.,

$$L_0(X, P) = (2\pi)^{(3N-3)/2} \exp\{i(X, P)\},$$

$$L_A(X, p_A) = (2\pi)^{-(3l_A-3)/2}\psi_A(x_A) \exp\{i(p_A, y_A)\}, \quad A \neq 0.$$

Using this notation, the wave function $U_A(X, p_A)$ can be written as

$$U_A^{(\pm)}(X, p_A) =$$

$$= \lim_{\epsilon\downarrow 0}(\mp i\epsilon)(2\pi)^{-(3l_A-3)/2} \int R(X, X', E_A \pm i\epsilon)\psi_A(x'_A)e^{i(p_A, y_A)}dX', \; A \neq 0,$$

$$U_0^{(\pm)}(X, P) =$$

$$= \lim_{\epsilon\downarrow 0}(\mp i\epsilon)(2\pi)^{-(3l_A-3)/2} \int R(X, X', P^2 \pm i\epsilon)e^{i(X', P)}dX',$$

where $R(X, X', z)$ is the kernel of the resolvent in X-representation. This kernel is called the Green function. The subscript l_A is the number of clusters belonging to the detailed partition A.

In momentum representation, the kernel is given by

$$U_A^{(\pm)}(P', p_A) =$$

$$= \lim_{\epsilon\downarrow 0}(\mp i\epsilon)\int dk''_A R(P', P(k''_A, p_A), E_A(p_A) \pm i\epsilon)\psi_A(k''_A), \quad A \neq 0 \qquad (2.5)$$

Here, the vector $P = (k''_A, p_A)$ is given in the basis k''_A, p_A induced by the detailed partition A. Then the kernel U_0 is obtained for $\psi_0 \equiv 1$, i.e.,

$$U_A^{(\pm)}(P', P) = \lim_{\epsilon\downarrow 0}(\mp i\epsilon)R(P', P, P^2 \pm i\epsilon).$$

Note that kernels of the resolvent of wave operators are, in general, distributions. Thus, relations (2.5) and similar expressions we will encounter later have to be taken in a space of distributions. However, we will not mention this, unless it is of principal importance.

Now, we consider Coulomb wave operators. In the first step, we use (2.1). Instead of (2.2), we get

$$U_A^{(\pm)} = \lim_{\epsilon\downarrow 0} \pm\epsilon \int_{\mp\infty}^{0} e^{it(H\mp i\epsilon)} L_A e^{-iE_A t - i\,\mathrm{sign}\, t\eta_A \log|t|}\, dt.$$

To evaluate this integral we use the following proposition.

Let $\varphi(z)$ be holomorphic in a neighbourhood of the real axis. Then for every self-adjoint operator A

$$\varphi(A) = \frac{1}{2\pi i}\oint_\gamma \varphi(z)R_A(z)\, dz \qquad (2.6)$$

where $R_A(z)$ is the resolvent of the operator A, $R_A(z) = (A - z)^{-1}$. The integral is continued in the negative direction over a contour γ around the spectrum of A.

Using this relation to compute the t integral, we obtain

$$U_A^{(\pm)} = \frac{1}{2\pi i}\lim_{\epsilon\downarrow 0} \mp\epsilon \oint_\gamma R(z)L_A \frac{\Gamma(1 \pm i\eta_A)e^{-\pi\eta_A/2}}{(z - E_A \mp i\epsilon)^{1\pm i\eta_A}}\, dz \qquad (2.6')$$

where η_A is given by (1.24).

Let us consider the kernels present in this expression in momentum representation. Integrating with respect to z we obtain

$$U_A^{(\pm)}(P', p_A) = e^{-\pi\eta_A/2}\Gamma(1 \pm i\eta_A)\times$$

$$\times \lim_{\epsilon\downarrow 0} \mp\epsilon \int R^{1\pm i\eta_A}(P, k''_A, p_A, E_A(p_A) \pm i\epsilon)\psi_A(k''_A)\, dk''_A. \qquad (2.7)$$

Here, by $R^{1\pm i\eta_A}(P, P', z)$ we denote the kernel of the operator $R^{1\pm i\eta_A}$; the index A runs over all values including $A = 0$.

Note that the representation (2.7) is much less useful for computational purposes than (2.3) because of the complex exponent of the resolvent $R^{1\pm i\eta_A}$. In Section 4, we will study principal singularities of the resolvent. This will enable us to transfer the complex exponent to the parameter ϵ.

Formulae (2.5) and (2.7) are fundamental for the considerations in the rest of this chapter. From now on, we will use the time-independent form of the wave operators. We will be interested in evaluation of limits (2.3) and (2.7) without making use of the original dynamical definitions of wave operators.

It is clear that the result of taking the limit in (2.5) and (2.7) is determined by singularities of the kernel of resolvent $R(P, P', z)$ for real z. In other words, if this kernel is expressed as the sum of bounded terms which are singular for $\operatorname{Im} z = 0$, then the contributions of all non-singular components vanish in the limit $\epsilon \downarrow 0$ because of the ϵ factor. Only the terms which tend to infinity like ϵ^{-1} give nontrivial contributions. Thus the problem consists in determining these singularities.

2.2 Singularities of the Resolvent. Neutral Particles

In this section, we describe the principal singularities of the kernel of the resolvent $R(z)$ in the case of short-range potentials. We will work in momentum space where these singularities are poles and δ-functions. The properties of the kernel of resolvent in configuration space will be described in Chapters 4 and 5.

A consistent and mathematically rigorous treatment of singularities is a hard problem. Its full solution can be given by means of a Fredholm-type integral equation. However, one can guess the structure of principal singularities with the help of heuristic arguments based on a conjecture describing the character of the point spectrum of energy operator of subsystems.

The corresponding scheme is interesting in itself. It makes it possible to determine explicit expressions for physically interesting quantities without analysing complicated integral equations. However, the integral equation methods are necessary to obtain rigorous results.

2.2.1 Equations of perturbation theory

First, we find some identities which the resolvent R(z) satisfies. We insert $A = H - z$ and $B = H_0 - z$ into the operator identity

$$A^{-1} - B^{-1} = A^{-1}(B - A)B^{-1}. \tag{2.8}$$

Denoting by $R_0(z)$ the resolvent of the kinetic energy operator $R_0(z) = (H_0 - z)^{-1}$, we get

$$R(z) = R_0(z) - R_0(z)VR(z), \quad R(z) = R_0(z) - R(z)VR_0(z). \tag{2.9}$$

On the other hand, we can treat these identities as equations for the resolvent $R(z)$. They are called the equations of the perturbation theory. It can be proved that they determine this resolvent uniquely for z outside the spectrum of the operator H.

We can also take the operators $H - z$ and $H_{a_k} - z$ for A and B in (2.8) with any a_k. Then we obtain the new set of equations

$$R(z) = R_{a_k}(z) - R_{a_k}(z)\bar{V}_{a_k}R(z), \tag{2.10}$$

$$R(z) = R_{a_k}(z) - R(z)\bar{V}_{a_k}R_{a_k}(z). \tag{2.10'}$$

where $R_{a_k}(z) = (H_{a_k} - z)^{-1}$. The operator of perturbation $\bar{V}_{a_k}$ describes the interaction between the subsystems belonging to the partition

$$\bar{V}_{a_k} = V - V_{a_k}.$$

In analogy with the preceding case, relations (2.10) regarded as equations determine the resolvent uniquely for z not belonging to the spectrum of H. Note that equations (2.9) are special cases of (2.10) for $k = N$.

Finally, we can take the operators $H_{a_l} - z$ and $H_{a_k} - z$ corresponding to the partitions a_l and a_k related by $a_k \supset a_l$. Inserting these operators for A and B in (2.8), we obtain the set of equations

$$R_{a_k}(z) = R_{a_l}(z) - R_{a_k}(z)\bar{V}_{a_k}^{a_l}R_{a_k}(z),$$

$$R_{a_k}(z) = R_{a_l}(z) - R_{a_k}(z)\bar{V}_{a_k}^{a_l}R_{a_l}(z). \tag{2.11}$$

where the superscript a_l takes the value corresponding to the partitions following a_k, $a_l \subset a_k$. The perturbation $\bar{V}_{a_k}^{a_l}$ contains only those potentials from the partition a_k which do not appear in V_{a_l}, $\bar{V}_{a_k}^{a_l} = V_{a_k} - V_{a_l}$.

2.2.2 Operators $R_{a_k}(z)$

Now we describe the properties of the operators $R_{a_k}(z)$ which take part in kernels and free terms in equations.

Let $r_{a_k}^{(int)}$ and $r_{a_k}^{(ext)}$ be the resolvents of the intrinsic and extrinsic Hamiltonians for the partition a_k, $r_{a_k}^{(int)} = (h_{a_k}^{(int)} - z)^{-1}$, $r_{a_k}^{(ext)} = (h_{a_k}^{(ext)} - z)^{-1}$. It follows from (1.16) that the resolvent $R_{a_k}(z)$ can be explicitly given by means of the integral

$$R_{a_k}(z) = \frac{1}{2\pi}\oint_{\gamma} d\zeta\, r_{a_k}^{(int)}(\zeta) r_{a_k}^{(ext)}(z - \zeta) \tag{2.12}$$

where the integration is over a contour around points of spectrum of the operator $h_{a_k}^{(int)}$ in negative direction. Any resolvent $r_{a_k}^{(int)}(z)$ is determined by the resolvents $R_\omega(z) = (H_\omega - z)^{-1}$ of the subsystems ω_{l_i} $(i = 1, 2, \ldots, k)$ constituting the partition a_k, by means of the $(k-1)$-fold integral

$$r_{a_k}^{(int)}(z) = \frac{1}{(2\pi)^{k-1}} \oint d\zeta_1 \oint d\zeta_2 \ldots$$

$$\ldots \oint d\zeta_{k-1} R_{\omega_{l_1}}(\zeta_1) R_{\omega_{l_2}}(\zeta_2 - \zeta_1) R_{\omega_{l_k}}(\zeta_{k-1} - \zeta_{k-2});$$

the integration with respect to ζ_i, $(i = 1, 2, \ldots, l-1)$ being also extended over a contour around the spectrum of the operators $H_{\omega_{l_1}}$ in negative direction.

In momentum space, the resolvent $r_{a_k}^{(ext)}(z)$ reduces to the operator of multiplication by the function $(p_{a_k}^2 - z)^{-1}$ so that the integral (2.12) can be evaluated with the help of the residuum theorem. Performing the integration, we find the kernel of resolvent

$$R_{a_l}(P, P', z) = r_{a_l}^{(int)}(k_{a_l}, k'_{a_l}, z - p_{a_l}^2) \delta(p_{a_l} - p'_{a_l}). \tag{2.13}$$

Here $r_{a_k}^{(int)}(k, k', z)$ is the kernel of the internal Hamiltonian $h_{a_l}^{(int)}$.

In accordance with our assumptions concerning the structure of the point spectrum, the kernels of the energy operators $R_{a_l}(P, P', z)$ of subsystems have poles. Since the resolvent $r_{a_k}^{(int)}$ acts in the corresponding subspace as an operator of multiplication by the function $(z + \varkappa_A^2)^{-1}$, the operators $P_A R_{a_l}$ are given by the singular kernels

$$P_A R_{a_l}(P, P', z) = \frac{\psi_A(k_A)\overline{\psi_A(k'_A)}}{p_A^2 - \varkappa_A^2 - z} \delta(p_A - p'_A). \tag{2.14}$$

Here, the pole $(p_A^2 - \varkappa_A^2 - z)^{-1}$ is explicitly present. Note that the operator $R_0(z)$ acting as the operator of multiplication by the function $(P^2 - z)^{-1}$ can be also expressed in the form of an integral operator with singular kernel

$$R_0(P, P', z) = (P^2 - z)^{-1} \delta(P - P'). \tag{2.15}$$

We see that kernels of the operators $R_{a_k}(P, P', z)$ possess two types of singularities: poles and δ-functions. Since these kernels appears in equations of the perturbation theory (2.10) both as operator kernels and free terms, it seems reasonable to assume that resolvent kernels have similar singularities.

2.2.3 Non-connected parts

Now, we describe δ-like singularities of the kernel $R(P, P', z)$. First of all, we give some definitions.

The kernel $K(P, P')$ possessing δ-like singularities will be called disconnected. A disconnected kernel in which δ-like singularities do not depend on the intrinsic coordinates k_{A_l} of the partition a_l will be called a_l-connected kernel. In addition, an a_1-connected kernel is smooth. We call such kernels connected.

In the case of two bodies, the kernel of the resolvent can be decomposed into the sum of connected and disconnected parts by means of identity (2.9). In this identity, the kernel of the first term is disconnected and the kernel of the second term

$$(R_0VR)(p,p',z) = (p^2 - z)^{-1} \int V(p-p'')R(p'',p',z)\,dp'' \qquad (2.16)$$

is a smooth function for Im $z \neq 0$.

In order to perform a similar decomposition in the system of several particles we will make use of identities (2.9) and (2.10) to turn from the a_k-connected part to a_{k-1}-connected part ($k = N-1, N-2, \ldots, 2$). In the first step, we obtain the representation

$$R(z) = R_0(z) - \sum_{a_{N-1}} R_0(z)V_{a_{N-1}}R(z)$$

where the terms in the sum have a_{N-1} connected kernels. In the second step, we use identity (2.10) applied to the operators $R_{a_{N-1}}$ and $R(z)$ and then express the operator of perturbation $\bar{V}^{a_{N-1}}$ in the form of the sum

$$\bar{V}^{a_{N-1}} = \sum_{a_{N-2}\supset a_{N-1}} \bar{V}^{a_{N-1}}_{a_{N-2}}.$$

We find

$$R(z) = R_0(z) - \sum_{a_{N-1}} R_0(z)V_{a_{N-1}}R(z) +$$

$$+ \sum_{a_{N-1}} \sum_{a_{N-2}\supset a_{N-1}} R_0(z)V_{a_{N-1}}R_{a_{N-1}}\bar{V}^{a_{N-1}}_{a_{N-2}}R(z)$$

The components of the second and third groups are a_{N-1}- and a_{N-2}-connected, respectively.

Carrying on this procedure, we use at every step the identity (2.10) for the pair $R_{a_{N-l}}$ and we replace the perturbation $\bar{V}^{a_{N-l}}$ by the sum

$$\bar{V}^{a_{N-l}} = \sum_{a_{N-l-1} \supset a_{N-l}} \bar{V}^{a_{N-l}}_{a_{N-l-1}}.$$

After $N-1$ steps we obtain

$$R(z) = R_0(z) + \sum_{l=2}^{N-1} \sum_{a_l} [R_{a_l}(z)]_c^{(a_l)} + \sum_{a_2} [R_{a_2}(z)]_c^{(a_2)} \bar{V}^{a_2} R(z) \tag{2.17}$$

where $[R_{a_l}(z)]_c^{(a_l)}$ denotes a_l-connected part of the resolvent $R_{a_l}(z)$:

$$[R_{a_l}(z)]_c^{(a_l)} = \sum_{A_l} (-1)^N R_0(z) V_{a_{N-1}} R_{a_{N-1}} \bar{V}^{a_{N-1}}_{a_{N-2}} R_{a_{N-2}} \ldots \bar{V}^{a_{l+1}}_{a_l} R(z).$$

The symbol $\sum_{A_l}$ means the summation over all chains of partition A_l with fixed l.

It is clear that the kernels of the operators $[R_{a_2}(z)]_c^{(a_2)} \bar{V}^{a_2}$ are connected. Indeed, the operator $[R_{a_2}(z)]_c^{(a_2)}$ is constructed in such a way that it has just one δ-like singularity and the kernel of $\bar{V}^{a_2}$ is a smooth function of the corresponding variable. Consequently, the kernel of the product of these operators is a smooth function for $\operatorname{Im} z \neq 0$. For the same reason the kernels of the operators $[R_{a_2}(z)]_c^{(a_2)} \bar{V}^{a_2} R(z)$ are also smooth.

Finally, let us note that the first group of components in formula (2.17) which is called the disconnected part of the resolvent $[R(z)]_{nc}$ can be expressed by means of the operator $R_{a_l}(z)$.

The following relation holds

$$[R(z)]_{nc} =$$

$$= R_0(z) + \sum_{l=2}^{N-1} \sum_{a_l} (-1)^{\left(\sum n_i^{(a_l)} - 1\right)} \left(\sum n_i^{(a_l)} - 1\right)! \, (R_{a_l}(z) - R_0(z)). \tag{2.18}$$

Here $n_i^{(a_l)}$ denotes the number of subsystems of the partition a_l containing i particles so that $\sum i n_i^{(a_l)} = N$.

This relation can be proved if we follow in the opposite direction the considerations leading to the formula (2.17) and then reduce the similar terms.

Thus we succeeded in representing the operators $R(z)$ as the sum of connected and disconnected parts. The kernels of the disconnected parts have singularities of the δ-function type and the kernels of the connected parts are smooth functions for complex z.

2.2.4 Pole singularities

The second type of singularities results from pole singularities of the kernels R_0 and R_{a_k}. The existence of the singularities corresponding to the resolvent of the kinetic energy operator $(P^2 - z)^{-1}$ does not depend on the structure of discrete spectrum of the energy operators of subsystems, while the singularities connected with the energy of channels appear only if cluster states are present. For this reason, we will often call these last ones the cluster singularities.

We describe the pole singularities of the kernel of resolvent by means of the transition operator. This operator is determined by the relations

$$U_{a_l b_k}(z) = \bar{V}_{b_k} + \bar{V}_{a_l} R(z) \bar{V}_{b_k} \tag{2.19}$$

Using the operators R_{a_l} and R_{b_k} in these relations, we obtain the following representation of the resolvent $R(z)$ in terms of the transition operator:

$$R(z) = R_{b_k}(z) - R_{a_l}(z) U_{a_l b_k}(z) R_{b_k}(z) \tag{2.20}$$

Note that the transition operator for $l = k = N$ is called the T-matrix and is denoted by $T(z)$. Consequently,

$$T(z) = V - VR(z)V. \tag{2.21}$$

The resolvent $R(z)$ is expressed by means of the T-matrix by the relation

$$R(z) = R_0(z) - R_0(z)T(z)R_0(z). \tag{2.22}$$

In accordance with the last formula, the kernel of resolvent has pole singularities:

$$R(P, P', z) = \frac{\delta(P - P')}{P'^2 - z} - \frac{T(P, P', z)}{(P^2 - z)(P'^2 - z)} \tag{2.23}$$

Similarly, equation (2.20) permits to determine explicitly the cluster pole singularities $(E_A(p_A) - z)^{-1}$ and $(E_B(p_B) - z)^{-1}$. In order to do that, we multiply equation (2.20) by the operator P_B from the right. Because of (2.14), for the kernel $R(z)P_B$ we obtain the representation

$$(RP_B)(P,P',z) = \frac{\psi_B(k_B)\psi_B^*(k'_B)}{E_B(p_B)-z}\delta(p_B - p'_B) -$$

$$K_B(P,p'_B,z)\frac{\psi_B^*(k'_B)}{E_B(p'_B)-z} \tag{2.24}$$

where the singularities depending on the variable p'_B are explicitly shown. Here the kernel $K_B(P,p'_B,z)$ has singularities with respect to the variable P which are determined by the singularities (2.14) of the kernels R_{a_l} in (2.20):

$$K_B(P,p'_B,z) = \sum_A \frac{\psi_A(k_A)}{E_A(p_A)-z} K_{AB}(p_A,p'_B,z) + \tilde{K}_B(P,p'_B,z); \tag{2.25}$$

here the summation is over all detailed partitions including $A = 0$. In addition, we put here $\psi_A \equiv 1$, $\varkappa_A^2 \equiv 0$ for $A = 0$, as mentioned above. The kernel $\tilde{K}$ is free of pole singularities.

We associate an integral operator with the kernel $K_B(P,p'_B,z)$. This operator acts from the channel $\mathfrak{H}_B$ into the space $\mathfrak{H}$. It is expressed by resolvent by the equation

$$K_B(z) = R(z)\bar{V}_B L_B. \tag{2.26}$$

or in terms of the transition operator

$$K_B(z) = R_0(z)U_{0B}(z)L_B.$$

Similarly, the operators of transition K_{AB} acting from the channel B to the channel A can be associated with the kernels $K_{AB}(p_A,p'_B,z)$. For $A \neq 0$, the representation

$$K_{AB}(z) = L_A^{-1}U_{AB}L_B \tag{2.27}$$

can be used and for $A = 0$, these operators are determined by the equation

$$K_{0B}(z) = \left(1 - \sum_A P_A\right) K_B(z).$$

We will make use of these relations in the next section.

In many calculations the following separability property of pole singularities is important:

The singular denominators $P^2 - z$ and $E_A(p_A) - z$ do not vanish simultaneously so that the manifolds where the singular points are concentrated are disjoint.

To verify this property, we consider the difference $\delta I = (P^2 - z) - (E_A(p_A) - z)$. Because of the relation $P^2 = k_A^2 + p_A^2$, this equality can be written as $\delta I = k_A^2 + \varkappa_A^2$ and the strict inequality $\delta I \geq \epsilon > 0$ follows.

Further, the cluster singularities for the subsequent partitions a_l and a_{l+1} are also disjoint. Indeed, let a_{l+1} be the partition obtained from a_l by division of the subsystem Ω into two parts ω and ω'. Note that the relation $p_{a_{l+1}}^2 = p_{a_l}^2 + k_{\omega\omega'}^2$ is satisfied in this case. Consequently, the difference

$$\delta I_l = (p_{a_{l+1}}^2 - \varkappa_{I_{a_{l+1}}}^2 - z) - (p_{a_l}^2 - \varkappa_{I_{a_l}}^2 - z)$$

can be written as

$$\delta I_l = k_{\omega\omega'}^2 + (\varkappa_{I_{a_l}}^2 - \varkappa_{I_{a_{l+1}}}^2)$$

Note that from the assumption about the structure of the discrete spectra of Hamiltonians of subsystems it follows that $\varkappa_{I_{a_l}}^2 > \varkappa_{I_{a_{l+1}}}^2$, so that the positiveness of δI_l is guaranteed. It also follows from the proved theorem that the singularities of any cluster partitions a_l and a_k, $a_l \subset a_k$ are also disjoint.

Apart from the singularities which correspond to the scattering states, the kernel of resolvent has also poles on the real axis for $z = -E_i$. These poles correspond to the bound states ψ_i of the N-body system. The relation

$$R(P, P', z) = \sum_{i=1}^{n_0} \frac{\psi_i(P)\psi_i^*(P')}{z + E_i} + \tilde{R}(P, P', z)$$

holds; here $\tilde{R}(z)$ is a bounded operator in neighbourhood of the points of discrete spectrum. Let us remind that as a result of the assumption concerning the structure of the discrete spectrum of the many-body Hamiltonians, the eigenvalues $-\varkappa_A^2$ satisfy the inequalities $-E_i < \min_A \varkappa_A^2$.

In order to complete our discussion of singularities of the resolvent, we give a short description of the pole singularities of the operators which appear in the disconnected part of the resolvent and we establish the relation between the poles and δ-function singularities.

The operator H_{a-l} corresponds to N-body scattering problem with interactions between the subsystems belonging to the partition a_l switched off.

To analyse singularities of resolvent of this operator, identity (2.11) can be used for $k > l$. Repeating the procedure applied in the case of the operator H, we reach the conclusion that the singularities of R_{a_l} are described by very similar relations as the singularities of $R(z)$. The only difference is that the indices B, A in (2.24) and (2.25) assume now only the admissible values: $b_j \subset a_l$, $a_k \subset a_l$.

On the other hand, the scattering problem for the operator H_{a_l} is degenerate. The resolvent R_{a_l} is described explicitly in terms of the internal Hamiltonian $h_{a_l}^{(int)}$ and the resolvent of this Hamiltonian is given by the resolvents of the subsystems. Consequently, all the coefficients of pole singularities can be explicitly given by means of the transition operators of subsystems. Therefore, the computation of these coefficients reduces to solving a set of scattering problems for smaller number of particles; these problems must be solved before the N-body problem is considered.

2.3 Poles of Resolvent and Waves Operators

In this section, we will express wave operators in terms of transition operators.

2.3.1 Kernels of wave operators

In accordance with the representation (2.24), the kernel of the operator $R(z)L_A$ for $z = E_A(p'_A) \pm i\epsilon$ can be expressed by means of kernels of transition operators as

$$(R(z)L_A)(P, P'_A, E_A(p'_A) \pm i\epsilon) =$$

$$= \frac{1}{\mp i\epsilon}\left(\psi_A(k_A)\delta(p_A - p'_A) - K_A(P, p'_A, E_A(p'_A) \pm i\epsilon)\right). \tag{2.28}$$

Substituting this relation into (2.3) and taking the limit $\epsilon \downarrow 0$, we get the following representation:

$$U_A^{(\pm)}(P, p'_A) = \psi(k_A)\delta(p_A - p'_A) - K_A(P, p'_A, E_A(p'_A) \pm i\epsilon). \tag{2.29}$$

The other way is to extract the pole singularity $(P^2 - E_A(p'_A) \mp i\epsilon)^{-1}$ from relation (2.28); the relevant term corresponds to the resolvent $R_0(z)$. We get the relation

$$U_A^{(\pm)}(P, p'_A) = \psi(k_A)\delta(p_A - p'_A) - \frac{T_A(P, p'_A, E_A(p'_A) \pm i)}{P^2 - E_A(p'_A) \mp i0}$$

where $T_A(z)$ is the kernel of the operator acting from $\mathfrak{H}_A$ to $\mathfrak{H}$ and given by the equation $T_A(z) = U_{0A}(z)L_A$. For $A = 0$, the operator T_A reduces to the T-matrix and this representation takes a form

$$U_0^{(\pm)}(P, p'_A) = \delta(p_A - p'_A) - \frac{T(P, p', P'^2 \pm i0)}{P^2 - P'^2 \mp i0} \tag{2.30}$$

In accordance with (2.25), the kernel T_A possesses cluster singularities so that it can be expressed in the form of the sum

$$T_A(P, p'_A, z) = \sum_B \frac{F_B(k_B)}{E_B(p_B) - z} K_{BA}(p_B, p'_A, z) + \tilde{T}_A(P, p'_A, z).$$

Here Φ_A is a function which can be expressed in terms of ψ_A by the relation

$$\Phi_A(k_A) = (k_A^2 + \varkappa_A^2)\psi_A(k_A). \tag{2.31}$$

This function is called the form-factor. To obtain this relation we used the identity

$$\frac{-1}{P^2 - z}\frac{F_B(k_B)}{E_B(p_B) - z} = \psi_A(k_A)\left(\frac{1}{P^2 - z} - \frac{1}{E_B(p_B) - z}\right). \tag{2.32}$$

Along with the pole singularities the kernels $T_A(P, p'_A, z)$ have δ-function singularities which correspond to the disconnected part of the resolvent. However, these singularities exist only if the detailed partition consists of three or more clusters. In the case when these partition consists of two clusters the kernel T_A has no δ-function singularities. This can be proved with the aid of the explicit representations (2.18) for the disconnected part of the resolvent.

Note that the wave operators obey the integral equations of the perturbation theory, which differ from the equations for the resolvent only by free terms. In order to get such equations we multiply relation (2.10) by $\mp i\epsilon L_A$ from the right, then $z = E_A(p'_A) \pm i\epsilon$ put and take the limit $e \downarrow 0$. Taking equation (2.3) into account, we get the integral equation of second kind

$$U_A^{(\pm)}(P, p'_A) = L_A(P, p'_A) -$$
$$- \int (R_A\bar{V}_A)(P, P'', E_A(p'_A) \pm i0)U_A(P'', p'_A) \tag{2.33}$$

This can be written in the operator form as

$$U_A^{(\pm)} = L_A - R_A(E_A(p'_A) \pm i0)\bar{V}_A U_A^{(\pm)}.$$

It would be natural to take these equations as the basis for alternative determination of the wave operators and to investigate them using the properties of the kernels of these operators. As we will see later, in the case of the two-body system such approach indeed provides a possible formulation of the scattering problem. The principal property of these equations for $N = 2$ is that the Fredholm alternative can be applied. However, this property is lost in the case of three or more particles.

In order to see that, let us consider equation (2.10) given by a detailed partition A. We multiply this equation by the operator $\mp i\epsilon L_B$ for some detailed partition B which is different from A. In the limit $\epsilon \downarrow 0$. the non-homogeneous terms vanishes so that we get homogeneous equation

$$U_B^{(\pm)} = -R_A(E_B(p'_A) \pm i0)\bar{V}_A U_B^{(\pm)}.$$

for the operator $U_B^{(\pm)}$. Comparing this equation with (2.33) we reach the conclusion that the Fredholm alternative cannot be applied in this case. In the same time it is clear that the corresponding operator has a continuous spectrum. These difficulties make it impossible to use directly the equations of the theory of perturbations as a basis for a formulation of the N-body scattering problem for N ¿ 3. Consequently, one must formulate such integral equations for the wave operators in this case that the Fredholm alternative can be applied. We will consider this as the first problem in the next chapter.

2.3.2 The scattering operator

Now, we will express kernels of matrix elements of the scattering operator in terms of the transition operators. To this end, it is convenient to use the non-stationary definition (1.39) of these operators. Using equation (2.6) and taking relations (2.24) and (2.25) into account, we obtain an integral representation where the integration is computed along the contour surrounding the spectrum of the operator H. Further, note that the integrand has no singularities in the lower half-plane. We deform the contour of integration into the straight line $\lambda + ie$, $-\infty < \lambda < \infty$, parallel to the real axis.

As a result, we find the following representation:

$$\begin{aligned} S_{BA}(p_B, p'_A) = \\ = \frac{1}{2\pi i} \lim_{\substack{t_1 \to \infty \\ t_2 \to -\infty}} \lim_{\epsilon \downarrow 0} \int_{-\infty}^{\infty} d\lambda \left(\frac{\delta\psi(p_B, p'_A, \lambda)}{E_A(p'_A) - \lambda - i\epsilon} + \right. \\ \left. + \frac{K_{BA}(p_B, p'_A, \lambda + i\epsilon)}{(E_B(p_B) - \lambda - i\epsilon)(E_A(p'_A) - \lambda - i\epsilon)} \right) \times \\ \times \exp\{i(E_B(p_B) - \lambda - i\epsilon)t_1 - i(E_A(p'_A) - \lambda - i\epsilon)t_2\} \end{aligned} \tag{2.34}$$

where

$$\delta\psi(p_B, p'_A, \lambda) = \int dP'' L_B(P'', p_B) L_A(P'', p'_A) + \tilde{K}_A(P, p', \lambda + i\epsilon).$$

The result of the limiting procedure in this formula is determined by the relation between the location of pole singularities and the character of oscillations of the exponentials for $t_{1,2} \to \pm\infty$. The interconnection between these factors is the following: a finite limit exists only for those components of the resolvent for which the pole singularities are identical with zeros of the exponents. In order to find this limit, the following relations must be taken into account:

$$\lim_{|t|\to\infty} \frac{e^{\mp ixt}}{x \pm i0} = \begin{cases} \mp 2\pi i\delta(x), & t \to \pm\infty \\ 0, & t \to \mp\infty \end{cases} \tag{2.35}$$

It follows from this relation that the first term corresponding to the function $\delta\psi$ gives a nontrivial contribution only for $A = B$; in other cases the limit vanishes. The limit generated by the kernel K_{BA} is equal to the product of the δ-function $\delta(E_B(p_B) - E_A(p'_A))$ and the kernel K_{BA}. As a result, we obtain the equality

$$S_{AB}(p_B, p'_A) = \delta_{AB}\delta(p_B - p'_A) -$$

$$-2\pi i\delta(E_B(p_B) - E_A(p'_A))K_{AB}(p_B, p'_A, E_A(p'_A) + i0). \tag{2.36}$$

Here, the indices B and A run over all channels including $A = B = 0$.

Note that if we take only the limit $t_2 \to -\infty$ in (2.34), the representation

$$S_{AB} = \lim_{t\to\infty} \lim_{\epsilon\downarrow 0} i\epsilon e^{i(E_B - E_A)t} L_B^{-1} R(E_A + i\epsilon) L_A$$

follows from (2.24) and (2.25)

The preceding considerations are based on the formulae which have a very simple meaning. Let us regard the kernel of the resolvent $R(P, P', z)$ as a function of three variables. Then it follows from (2.29) and (2.30) that the kernels of the wave operators coincide with the residua of this function in the poles which correspond to the second argument $(P'2 - z)^{-1}$ and $(E_A(p'_A) - z)^{-1}$. Regarding the kernels of the wave operators as functions of two variables, the kernels of the scattering matrix can be identified with the residues of this function with respect to the first argument.

In this way, the wave operators and the scattering matrix can be determined as the coefficients of poles of the kernel of resolvent in momentum representation taken for the special values of the "energy" variables $E_A(p_A)$ and $E_B(p'_B)$. These values are determined by the laws of conservation of energy and momentum. The function $T(P, P', P'^2 \pm i0)$ is called T-matrix on half energy shell and the kernels are called the T-matrix components on half energy shell. The kernel $T(P, P', P'^2 + i0)$ for $P^2 = P'^2$ and the kernels $K_A(P, p'_A, E_A(p'_A) + i0)$, $K_A(p_A, p'_B, E_B(p'_B) + i0)$ for $E_A(p_A) = E_B(p'_B)$ are called the T-matrix and its components on energy shell.

2.3.3 Integral representations

The kernels of the scattering operator can be expressed directly in terms of wave functions by means of integral representations. In order to get such a representation we consider the relations (2.24), (2.10) and (2.30).

We take the kernels of operators in (2.20) for $z = E_B(p_B) + i\epsilon$, $E_A(p'_A) = E_B(p_B)$. Next we multiply this relation by $i\epsilon L_B^*$ from the left and by the operator L_A from the right. Finally we take a limit $\epsilon \downarrow 0$, taking into account the equality conjugated to (2.3),

$$\lim_{\epsilon \downarrow 0} i\epsilon L_B^* R(E_B + i\epsilon) = U_B^{(-)*}$$

We get the integral representation

$$K_{BA}(p_B, p'_A, E_A(p_A) + i0) = \int U_B^{(-)*}(X, p_B)\bar{V}_A(X) L_A(X, p'_A)\, dX \qquad (2.37)$$

where the momentum variables are on energy shell. As in (2.36), here the indices A and B can take all values including $A = B = 0$.

The physical meaning of the kernels K_{BA} is clear. In accordance with the rules of perturbation theory, these kernels are determined by the matrix elements of a perturbed potential computed between the initials L_A and final states $U_B^{(-)}$.

2.4 Singularities of Resolvent for Charged Particles

In this section, the principal singularities of a kernel of resolvent and wave operators for a system of charged particles are described. The expressions for the kernels of scattering operator in terms of the coefficients of these singularities are also deduced.

2.4.1 Resolvent and wave operators

In preceding section, the principal singularities of kernels of resolvent and of wave operators were described in the case of system of neutral particles. The term "principal" means that these singularities are responsible for the existence of nontrivial limits (2.3) which determine wave operators and the S-matrix. The character of these singularities were determined with the aid of equations (2.3) where some conjectures concerning the structure of kernels R_{a_l} were made. This approach, however, cannot be used in the case of long-range potentials. The reason is that the operators H_{a_l} do not describe the asymptotic dynamics of charged particles correctly . Consequently, we cannot perform our discussion in analogy with the case described above.

Despite this, we can obtain a definite information concerning the character of singularities if we analyse relations (2.7) only with the assumption that the limits exist, taking no account of equations of perturbation theory. More precisely, we can formulate the problem as to how change the pole singularities of the resolvent in order to ensure existence of the limits (2.7).

Let us analyse formulae (2.6) and (2.7). We saw in the preceding section that since the factor ϵ vanish in the limit only the components containing the poles $(E_A - z)^{-1}$ remain finite; these singularities correspond to the detailed partition A. In the case of charged particles the pole singularities do not ensure existence of the limit of modified expression (2.7) any more. However, assuming that the poles $(E_A - z)^{-1}$ transform into poles distorted by Coulomb interaction $(E_A - z)^{-1 \pm \eta_A}$, we get immediately finite result in the limit $\epsilon \downarrow 0$.

Indeed, consider relation (2.7). According to our hypothesis the relation is satisfied.

$$(RP_A)(P,P',z) = K_A(P,p'_A,z)\frac{\psi_A(k'_A)}{(E_A(p'_A)-z)^{1\mp\eta_A}} \tag{2.38}$$

where the kernel K_A is determined as a distribution on shell $E_A(p'_A) = z$. It is clear that the value of integral (2.6) is determined by the singular factor $(E_A(p'_A)-z)^{-1\pm\eta_A}$ and that the function $K_A(P,p'_A,z)$ can be taken outside the integral at the point $z = E_A(p'_A) \pm i0$. As a result, we get the integral

$$I_\pm = \oint dz(z - E_A \mp i\epsilon)^{-1\mp\eta_A}(E_A - z)^{-1\pm\eta_A}$$

where we integrate along the contour adjoining the real axis. The point $E_A \mp i\epsilon$ and the real axis are separated by the contour. The integral $I_\pm$ can be calculated explicitly and the result is $I_\pm = \dfrac{\pm 2\sinh\pi\eta}{i\epsilon\eta}$. As a finally result, we obtain the following representation of the modified wave operator:

$$U_A^{(\pm)}(P,p'_A) = \frac{\sinh\pi\eta_A(p'_A)}{\pi\eta_A(p'_A)}K_A(P,p'_A,E_A(p'_A)\pm i0) \tag{2.39}$$

A remarkable difference between this formula and the similar expression for neutral particles is the absence of the δ-function term $\delta(p_A - p'_A)\psi_A(k_A)$. This term originated from the resolvents of the operators H_{a_k} in the case of neutral particles. In the case of charged particles, the strong limits of these components vanish.

As we noted before, this formula is not too convenient because of the complex exponent of the resolvent in the integrand. From relations (2.39) it follows that the kernels of the wave operators depend on components of the kernels of the resolvent linearly so that the complex exponent can be transferred to the parameter ϵ. Comparing (2.7) and (2.39), we get the desired modified formula similar to (2.5):

$$U_A^{(\pm)}(P',p_A) == \frac{\sinh\pi\eta'_A}{\pi\eta'_A}\lim_{\epsilon\downarrow 0}(\mp i\epsilon)^{1\mp i\eta'_A}\times$$

$$\times\int dk'_A R(P,P',E_A(p'_A)\pm i\epsilon)\psi_A(k'_A), \quad \eta'_A = \eta_A(p'_A). \tag{2.40}$$

This formula can be also expressed in the form valid in any representation:

$$U_A^{(\pm)}(P',p_A) == \frac{\sinh\pi\eta_A}{\pi\eta_A}\lim_{\epsilon\downarrow 0}(\mp i\epsilon)^{1\mp i\eta'_A}R(E_A \pm i\epsilon)L_A. \tag{2.41}$$

2.4.2 The scattering operator

As in the case of neutral particles we will use the non-stationary definition to compute the kernels of the scattering operator. We assume that the relation

$$K_A(P,p_A',z) = \sum_B \frac{\psi_B(k_B)}{(E_B(p_B)-z)^{1-\eta_B(\sqrt{z})}} K_{BA}(p_B,p_A',z) + \tilde{K}_A(P,p_A',z) \tag{2.42}$$

holds. Correspondingly, we get a formula similar to (2.34) for the matrix element S_{BA}:

$$S_{BA}(p_B,p_A') =$$

$$= \frac{1}{2\pi i} \lim_{\substack{t_1\to\infty \\ t_2\to-\infty}} \lim_{\epsilon\downarrow 0} \int d\lambda \frac{\exp(i(E_B(p_B)-\lambda-i\epsilon)t_1 + i\eta_B \log t_1)}{(E_B(p_B)-\lambda-i\epsilon)^{1-\eta_B(p_B)}} \times$$

$$\times K_{BA}(p_B,p_A',\lambda+i\epsilon) \frac{\exp(i(E_A(p_{)}'-\lambda-i\epsilon)t_2 + i\eta_A \log(-t_2))}{(E_A(p_{)}'-\lambda-i\epsilon)^{1-\eta_A(p_A')}} \tag{2.43}$$

The result of the passage to the limit in (2.43) is determined by the relation between the fast oscillating exponents and the singular denominators expressed by the formulae

$$\lim_{|t|\to\infty} \frac{e^{\mp ixt+i\eta\log t}}{x\pm i0} = \begin{cases} \mp 2\pi i\delta(x)\frac{e^{-i\pi\eta/2}}{\Gamma(1-i\eta)}, & t\to\pm\infty \\ 0, & t\to\mp\infty \end{cases} \tag{2.44}$$

from which it follows that

$$S_{AB}(p_B,p_A') = 2\pi i \frac{\exp\left(-\frac{\pi}{2}(\eta_B(p_B)+\eta_A(p_A'))\right)}{\Gamma(1-i\eta_B(p_B))\Gamma(1-i\eta_A(p_A'))} \times$$

$$\times \delta(E_B(p_B)-E_A(p_A'))K_{AB}(p_B,p_A',E_A(p_A')+i0). \tag{2.45}$$

Note that as in the case of the generalized wave, operators the diagonal elements of the scattering operator do not contain δ-function terms. These kernels are proportional to the coefficients of the kernel of the resolvent with distorted poles. Thus we have described the principal singularities of the resolvent in the case of long-range potentials. According to our hypothesis, the kernel of resolvent can be expressed as

$$R(P,P',z) =$$

$$\sum_A \sum_B \frac{\psi_A(k_A)}{(E_A(p_A)-z)^{1-\eta_A(\sqrt{z})}} \tilde{T}(p_A,p_B',z) \frac{\psi_B^*(k_B')}{(E_B(p_B')-z)^{1-\eta_B(\sqrt{z})}} \tag{2.46}$$

where the summation run over all detailed partitions including $A, B = 0$. Unfortunately, nothing can be said about the properties of the functions $\tilde{T}(p_A, p'_B, z)$. In particular, we cannot say what type of singularities replaces the δ-functions in the disconnected part of the resolvent.

However, the example of the two-body problem, where the solution is known explicitly, shows that a new type of singularities appears. We will see later that in the neighbourhood of energy shell $p^2 = p'^2 = z$ the kernel of the Coulomb T-matrix can be written in the form

$$T_c(p, p'_B, z) \sim \frac{\gamma q_1 q_2}{\pi} \frac{\eta(z)(4z)^{-i\eta(z)}}{(e^{\pi\eta(z)} - 1)|p - p'|^{2(1+i\eta(z))}}$$

where $\eta(z) = \frac{\gamma q_1 q_2 \sqrt{\mu_{12}}}{\sqrt{2z}}$. It follows from this relation that on energy shell $p^2 = p'^2 = z$ the kernel $T_c(p, p', z)$ and, consequently, also the kernel $\tilde{T}(p, p', z)$ has a strong singularity given by $(1 - \cos\theta)^{1+i\eta}$, where $\cos\theta = (\hat{p}, \hat{p}')$. This singularity is non-integrable on the unit sphere. Consequently, the kernel $T_c(p, p', z)$ must be taken as a generalized function. We will determine such a function in the Chapter 5.

CHAPTER 3

The Method of Integral Equation

This chapter will be devoted to the description of the stationary scattering method based on the use of linear integral equations.

As we have seen previously, the problem of solving the scattering problem in the stationary formalism reduces to the study of singularities of resolvent kernel. It will be shown that the integral equations are reliable means for determination of the resolvent and for description of its singularities. These equations also serve as a basis for numerical calculations.

In this chapter, we will work in momentum representation. For that reason, it is more appropriate to study the T-matrix (2.22), because the T-matrix is less singular than the resolvent kernel. Problems related to the transformation to coordinate representation will be deferred to the next chapter.

This chapter will be organized as follows. The two-body system will be discussed in Section 1 where the basic technical procedures used for the study of integral equations will be described. This section will serve as a standard for the discussion of three- and more-particle systems carried out in Sections 2 and 3.

The justification of the asymptotic completeness hypotheses, however, will not be discussed here. These and related problems will be deferred to the Chapter 4.

3.1 Integral Equations for the Two-Body T-matrix

As we already noticed, it is more appropriate to discuss the T-matrix $t(z)$ determining the resolvent $r(z)$, in momentum representation

$$r(z) = r_0(z) - r_0(z)t(z)r_0(z) \tag{3.1}$$

where $r(z) = (h - z)^{-1}$ and h is the energy operator of the two-particle system. From the equation of perturbation theory for the resolvent

$$r(z) = r_0(z) - r_0(z)vr(z) \tag{3.2}$$

we get the following equation for $t(z)$

$$t(z) = v - vr_0(z)t(z). \tag{3.3}$$

To obtain this equation, it suffices just to insert (3.1) into (3.2) and cancel all unnecessary terms.

The study of the equation (3.3) for the integral kernel $t(k, k', z)$

$$t(k, k', z) = v(k - k') - \int v(k - q)\frac{1}{q^2 - z}t(q, k', z)\, dq \tag{3.4}$$

will be the subject of this section.

The variables k, k' and z play different roles in the kernel $t(k, k', z)$. The first argument k is the integration variable and, therefore, the kernel solves the integral equation in this variable. Variables k' and z are parameters; k' enters the leading term $v(k - k')$ z does the kernel $v(k - q)(q^2 - z)^{-1}$. Consequently, equation (3.4) represents a family of integral equations of second kind of the form

$$f = f_0(k') + a(z)f \tag{3.5}$$

where the solution f is a function of k and the integral kernel $a(z)$ is of the form

$$a(z)f(k) = \int \frac{v(k - q)}{q^2 - z} f(q)dq.$$

The leading term $f_0(k')$ depends on k'

$$f_0(k') = v(k - k').$$

Equation (3.5) as compared to equation (3.2) is more convenient because the former contains only smooth functions whereas the leading term in the latter is a distribution $\delta(k-k')(k'^2-z)^{-1}$.

Equation (3.4) will be used to determine the operator t(z) and to describe its properties. Equation (3.4) is useful for such purposes if it satisfies the Fredholm alternative: the equation possesses a unique solution in the given functional space if the conjugate homogeneous equation has no nontrivial solution.

In our case, the conjugate homogeneous equation takes the form

$$\psi(k) = \frac{1}{k^2 - z^*} \int v(k-q)\psi(q)\, dq \tag{3.6}$$

with $\psi(k)$ belonging to the class of functions satisfying

$$\left| \int \psi(k) f(k)\, dk \right| < \infty \tag{3.6$'$}$$

In the derivation we used the fact that the kernel $v(k-k')$ is self-adjoint. This follows from its reality

$$\overline{v(k)} = v(-k).$$

Equation (3.6) can be rewritten in the form of the Schrödinger equation

$$(k^2 - z^*)\psi(k) = -\int v(k-q)\psi(q)\, dq$$

for the eigenvalue z^*.

Let us show that the Fredholm alternative can be applied to equation (3.4). We choose a class of functions $f(k)$ in such a way that the solutions of the homogeneous equation (3.6) belong to the Hilbert space $\mathfrak{H}$. Then the eigenvalue z^*, for which these equations have solutions becomes discrete eigenvalue of the operator h. This information allows a detailed characterization of the kernel $t(k, k', z)$.

Let us first discuss the Fredholm character of equation (3.4). For this equation to be Fredholm it is suffices that $a(z)$ is a completely continuous operator. This means, roughly speaking, that this operator "improves" the properties of the functions f(k) to which it is applied; i.e., it makes them smoother and more rapidly decreasing at infinity. The precise meaning of this statement can be found in many textbooks and will not be repeated here. Equation (3.5) with $a(z)$ completely continuous will be sometimes called a compact equation.

To select a convenient class of functions $f(k)$, we will first discuss (in a more detailed way) the properties of the integral $I(z)$ defined as

$$I(z) = \int \frac{\varphi(q)dq}{q^2 - z} \tag{3.7}$$

which appears in equation (3.4).

If the function $\varphi(k)$ is bounded and decreases at infinity as

$$\varphi(q) \leq C(1 + |q|)^{-1-\theta}, \quad \theta > 0, \tag{3.8}$$

the integral exists for all z except for those belonging to the real positive semiaxis. Let Π_0 denote the complex z-plane with the real positive semiaxis cut off. The integral $I(z)$ is then defined for all z from Π_0. However, to ensure that the integral remains finite even when z approaches the cut, we must add some additional restrictions on the function $f(k)$. A widely used restriction, the so called Hölder condition, is

$$|\varphi(q + h) - \varphi(q)| \leq C(1 + |q|)^{-1-\theta}|h|^{\alpha} \tag{3.9}$$

If both conditions (3.8) and (3.9) are satisfied, the integral $I(z)$ is defined for all z from Π_0 and satisfies the Hölder condition in z

$$|I(q + \Delta) - I(z)| \leq C|\Delta|^{\alpha} \tag{3.9 }$$

for all z except a vicinity of the point $z = 0$ where α is replaced by $\min(1/2, \alpha)$. The constant C is uniform in z for z from Π_0 and decreases when $|z| \to \infty$.

The aforementioned statement will be called the singular integral lemma to stress the fact that the integral $I(z)$ becomes singular at the real positive semiaxis. The fact that the point $z = 0$ must be treated separately can be easily explained by writing the singular kernel (3.7) in spherical coordinates

$$\frac{x^{1/2}dx}{x - z}, \quad x = |q|^2;$$

the term $x^{1/2}$ satisfies the Hölder condition with $\alpha = 1/2$ in vicinity of the point $x = 0$.

The singular integral lemma defines a class of functions for which equation (3.4) is of Fredholm type. Let us assume that $v(k)$ satisfies the conditions (3.8) and (3.9), i.e.,

$$|v(k)| \leq C(1+|k|)^{-1-\theta},$$
$$|v(k+h)-v(k)| \leq C(1+|k|)^{-1-\theta}|h|^{\alpha} \tag{3.10}$$

and that the exponent α exceeds $1/2$. We will see soon that this is a natural condition. Let the functions $f(k)$ satisfy analogous conditions with different exponents θ' and α' and let us discuss the integral

$$g(k,z)=\int \frac{v(k-q)g(q)}{q^2-z}dq. \tag{3.11}$$

We are interested in the properties of the function $g(k,z)$ as a function of k for fixed z and we require the properties of smoothness and asymptotic behaviour of this function to be better than those of $f(k)$.

Moreover, it follows from the singular integral lemma that

$$|g(k,z)| \leq C(z)(1+|k|)^{-1-\theta},$$
$$|g(k+h,z)-g(k,z)| \leq C(z)(1+|k|)^{-1-\theta_1}|h|^{\alpha_1}, \tag{3.12}$$

where α_1 can be taken smaller than, but arbitrarily close to α.

The exponents θ' and α' are not present in these estimates. They determine the properties of $g(k,z)$ as a function of variable z which plays a role of parameter in the integral equation. This is the difference between our integral equation and the so-called singular integral equations which have kernels with moving singularities $\frac{1}{x-x'}$ and which are not of Fredholm type.

Consequently, if the smoothness parameter α' and the parameter θ' determining the asymptotic rate of decreasing of the function $f(k)$ are smaller than α and θ respectively, then the integral operator $a(z)$ improves them and becomes completely continuous.

Let us now turn to homogeneous equation (3.6) which we will investigate in the class of functions $\psi(k)$ satisfying condition (3.6') for all $f(k)$ from our class. For these functions ψ, the integral

$$\varphi(k)=\int v(k-q)\psi(q)\,dq \tag{3.13}$$

on the right hand side of (3.6) determines Hölder decreasing functions with exponents $\theta - \theta'$ and $\alpha - \alpha'$ because we can assume that $v(k-q)$ as a function of q and k has exponents θ' and α' and $\theta - \theta'$ and $\alpha - \alpha'$, respectively. Equation for $\varphi(k)$ reads

$$\varphi(k) = \int \frac{v(k-q)}{q^2 - z^*} \varphi(q)\, dq \tag{3.14}$$

and it follows from this equation that $\varphi(k)$ indeed has the exponents θ_1 and α_1. As a result, for complex z the function $\psi(k)$

$$\psi(k) = \frac{\varphi(k)}{k^2 - z^*} \tag{3.15}$$

is square integrable and therefore is an eigenfunction of the discrete spectrum. Consequently, for complex z equation (3.6) has no nontrivial solutions.

For z approaching the cut, i.e., for $z = \lambda + i0$, $\lambda > 0$, the function $\psi(k)$ remains square integrable. Indeed, it follows from equation (3.14) and from the fact that the kernel $v(k - k')$ is a self-adjoint that

$$\int \overline{\varphi(k)} \frac{1}{k^2 - \lambda - i0} \varphi(k)\, dk$$

is real. This is possible only when $|\varphi(k)|^2 = 0$ for $k^2 = \lambda$. Because $\varphi(k)$ is Hölder, the singularity of $\psi(k)$ $k^2 = \lambda$ in (3.15) is square integrable. More precisely, in vicinity of this singularity we have

$$|\psi(k)| = \frac{\varphi(k) - \varphi(\sqrt{\lambda}\hat{k})}{k^2 - \lambda} \leq C \frac{1}{|k^2 - \lambda|^{1-\alpha_1}}$$

and at $\alpha_1 > \frac{1}{2}$, this singularity is sufficiently weak. It is precisely the point where the condition $\alpha_1 > \frac{1}{2}$ is used to deduce the exponent of the function $v(k)$ which guarantee smoothness.

The point $z = 0$ where all arguments given above lose their power must be treated separately. This point may be singular or regular. If it is singular, it does not need to appear in the discrete spectrum of h. In order to avoid complications related to this problem, we will assume here that equation (3.14) has no solution for $z = 0$. The case when the point $z = 0$ is singular will be treated briefly at the end of Section 4.

Therefore, the homogeneous equation (3.14) has no nontrivial solutions for all z from Π_0 except the points belonging to the discrete spectrum of the operator h.

On the other hand, the discrete spectrum of the operator h is well known. It was verified, for example, that for the potential which is an almost arbitrary function on sphere of finite radius and which behaves outside the sphere as

$$|v(x)| \leq C(1+|x|)^{-\epsilon} \tag{3.16}$$

with $\epsilon > 0$, there is no positive discrete spectrum of the operator h. There are also restrictions under which the negative spectrum is finite. For example if $\epsilon > 2$, the non-positive spectrum of the operator h consists of a finite number of eigenvalues with finite multiplicity.

We will assume that these restrictions on the potential are satisfied, so that the discrete spectrum of the operator h is negative and finite. The related eigenfunctions are given by (3.15) with $z^* = -\varkappa_i$, $i = 1, 2, \ldots, l$.

Now we can apply our formalism to equation (3.4). It remains to prove that the leading term of this equation belongs to the class of Hölder decreasing functions. This follows from the estimates for $v(k)$; however, the dependence on k' makes these estimates non-uniform. This difficulty can be easily overcome. It suffices to subtract first few iterations

$$t^{(n)}(k,k',z) =$$
$$(-1)^n \int dk_1 dk_2 \ldots dk_n \frac{v(k-k_1)}{k_1^2 - z} \frac{v(k_1-k_2)}{k_2^2 - z} \cdots v(k_n - k').$$

By generalizing the singular integral lemma, it is possible to prove that

$$|t^{(n)}(k,k',z)| \leq C(1+|k|)^{\theta_1}(1+|k'|)^{\theta_2}, \quad \theta_1 + \theta_2 < 1 + \theta$$

for sufficiently large n. For the difference

$$\tilde{t}^{(n)}(k,k',z) = t(k,k',z) - \sum_{i=1}^{n} t^{(i)}(k,k',z)$$

the following equation holds

$$\tilde{t}^{(n)}(k,k',z) = \tilde{t}^{(n+1)}(k,k',z) + \int \frac{v(k-q)}{q^2 - z} \tilde{t}^{(n)}(q,k',z)\, dq \tag{3.17}$$

where estimates of the leading term are already uniform in k' and z.

This makes it possible to apply the Fredholm alternative to equation (3.4). Hence, once the position of the singular points is known, we can characterize the properties of the kernel of T-matrix. The following statement holds:

The kernel $t(k,k',z)$ is a decreasing Hölder function of k, k', and z for all z from the cut complex plane Π_0 except the singular points $\varkappa_i^2$, $i = 1,,2,\dots,l$, which belong to the discrete spectrum of the operator h. In vicinity of any singular point the following representation holds:

$$t(k,k',z) = \sum_{i=1}^{l} \frac{\varphi_i(k)\varphi_i^*(k)}{z+\varkappa_i^2} + \tilde{t}^{(n)}(k,k',z) \tag{3.18}$$

where $\tilde{t}^{(n)}(k,k',z)$ is a decreasing Hölder function of z away from the real axis.

The functions $\varphi_i(k)$ are given by equation (3.13) or, if one takes the Schrödinger equation $\varphi_i(k)+\varkappa_i^2)\psi_i(k)$ into account, these functions represent form factors of the two-body system.

Let us justify the representation (3.18). For this purpose, we write the resolvent $r(z)$ as

$$r(z) = \sum_i \frac{p_i}{z+\varkappa_i^2} + \tilde{r}(z)$$

where p_i is a projector on the eigensubspace corresponding to the eigenvalue $z = -\varkappa_i 2$. $\tilde{r}(z)$ is an operator bounded in vicinity of the singular points $z_i = -\varkappa_i^2$, $i = 1,,2,\dots,l$ analytic in z. The T-matrix $t(z)$ can be equivalently expressed as

$$t(z) = \sum_i (z+\varkappa_i^2)^{-1} c_i + \tilde{t}(z) \tag{3.19}$$

where $c_i = vp_iv$, $\tilde{t}(z) = v - v\tilde{r}(z)v$.

The operator p_i is an integral operator with kernel $\psi_i(k)\psi_i^*(k)$. Therefore, because of (3.13), the kernel of the operator c_i has the form of an operator of multiplication

$$c_i(k,k') = \varphi_i(k)\varphi_i^*(k')$$

and equation (3.18) holds.

Finally, we must show that the kernels $\tilde{t}(k,k',z)$ are decreasing Hölder functions. To prove this, we make use of the following reasoning. We note that by virtue of the Hilbert identity

$$r(z_1) - r(z_2) = (z_1 - z_2) r(z_1) r(z_2),$$

the operator $t(z)$ satisfies the equation

$$t(z_1) - t(z_2) = (z_1 - z_2) t(z_1) r_0(z_1) r_0(z_2) t(z_2). \tag{3.20}$$

This expression will be called the Hilbert identity for T-matrix. In particular. it follows from this identity that the operator $t(z)$ is an analytic function of z for all z from Π_0 except the singular points $-\varkappa_i^2$ and

$$\frac{\mathrm{d}}{\mathrm{d}\,z} t(z) = t(z) r_0^2(z) t(z).$$

This function can be represented as a contour integral

$$t(z) = \frac{1}{2\pi i} \oint \frac{t(\xi)}{\xi - z} d\xi$$

where the integration contour consists of a circle of radius R, the segment $[0, R]$ adjoining the real axis, and a circle C_ϵ of radius ϵ centered at the points $-\varkappa_i^2$. The integral can be written in a form of the sum (3.19) with the kernel $\tilde{t}(k, k', z)$ given by

$$\tilde{t}(k, k', z) =$$
$$= v(k - k') + \frac{1}{2\pi i} \int_0^\infty d\lambda (t(k, k', \lambda + i0) - t(k, k', \lambda - i0))).$$

Transforming the integrand by means of (3.20) we get the representation

$$\tilde{t}(k, k', z) =$$
$$= v(k - k') + \frac{1}{2\pi i} \int_0^\infty dq\, t(k, q, q^2 + i0)(q^2 - z)^{-1} t(q, k', q^2 - i0))). \quad (3.21)$$

It follows that for z close to the singular points $-\varkappa_i^2$, the kernel $\tilde{t}(k, k', z)$ can be estimated by the kernels $t(k, k', z)$ evaluated at real positive energies. These estimates can be obtained by using the singular integral lemma.

At this point we stop our description of the properties of the two-body T-matrix kernel.

Finally, let us discuss general conclusions which can be drawn from our reasoning. It is possible to distinguish three basic steps in investigating integral equations of second kind. Namely,

1. we have to obtain estimates of iterations of integral equations;
2. starting from these estimates, we have to choose a space of functions into which these iterations belong (even starting at some order of iterations) so that it is possible to describe the integral equation in the form of a compact equation in terms of elements and operators of this space;

3. we have to study the related homogeneous equation.

In the next section we will investigate how this scheme works in the three-body case.

To conclude this section, we present a series of expressions for wave and scattering operators. According to (3.20), the wave operator kernel $u^{(\pm)}(k,k')$ has the form

$$u^{(\pm)}(k,k') = \delta(k-k') - \frac{t(k,k',k'^2 \pm 0)}{k^2 - k'^2 \mp 0}. \tag{3.22}$$

This kernel solves the integral equation of second kind

$$u^{(\pm)}(k,k') = \delta(k-k') - \frac{1}{k^2 - k'^2 \mp 0} \int v(k-q) u^{(\pm)}(q,k')\, dq$$

which can be obtained analogously to the more general equation (2.33). If we select out the δ-function $\delta(k-k')$ and the term $(k^2 - k'^2 \mp 0)^{-1}$, we arrive at equation (3.4) for T-matrix where z should be replaced by $z = k'^2 \pm i0$. As we have proved, this equation has, similarly to the equation for $u^{(\pm)}(k,k')$, a unique solution.

The kernel of the scattering operator is expressed in terms of the T-matrix as

$$s(k,k') = \delta(k-k') - 2\pi i \delta(k^2 - k'^2) t(k,k',k'^2 + i0) \tag{3.22'}$$

which follows from (2.36) for $N = 2$. According to (3.18), the kernel $t(k,k',z)$ appearing in this representation is a decreasing Hölder function. We have already used this fact in Chapter 1.

At this point we stop our discussion of general properties of the T-matrix and wave operators for two-body systems.

3.2 Compact Integral Equations for Three-Particle Systems

In this section we derive integral equations for the three-body T-matrix components and we discuss properties of their kernels.

3.2.1 Derivation of compact equations

To start with, we will describe notations for relative coordinates and momenta which are convenient in the three-particle problem. We will use three pairs of coordinates and momenta $\{x_\alpha, y_\alpha\}$, $\{k_\alpha, p_\alpha\}$ which have been introduced in Chapter 1 (equations (1.9) and (1.12')). The subscript α takes the values 1, 2, and 3 corresponding to the label denoting the free particle not entering the two-particle subsystem. For instance, the pair of coordinates (1.9) induced by partition (31)(2) is denoted by $\{x_2, y_2\}$. Various relative coordinates and momenta are related to each other by the relations

$$k_1 = c_{12}k_2 + s_{12}p_2, \quad c_{12} = \left(\frac{m_1 m_2}{(m_1+m_3)(m_2+m_3)}\right)^{1/2},$$

$$p_1 = s_{12}k_2 - c_{12}p_2, \quad c_{12}^2 + s_{12}^2 = 1, \quad s_{12} = -s_{21}$$

for $\alpha = 1$, $\beta = 2$; other values of α and β can be obtained by exchanging particles labels in equations above.

Let us turn to the derivation of compact equations. We will discuss the equation for the T-matrix:

$$T(z) = V - VR_0(z)T(z) \tag{3.23}$$

which can be obtained exactly as in the two-particle case. The perturbation in this case equals the sum of two-particle potentials

$$V = V_1 + V_2 + V_3$$

and, analogously, VR_0 is equal to the sum

$$VR_0(z) = \sum_\alpha V_\alpha R_0(z).$$

Let us discuss, for example, the operator $V_1 R_0$ which enters this equation. In the momentum representation, its kernel reads

$$v_1(k_1 - k_1')\delta(p_1 - p_1')(P'^2 - z)^{-1}.$$

This kernel contains δ-functions; thus the singularity does not disappear after iterating equation (3.23). This can easily be seen in diagrams corresponding to iterations of $(V_1 R_0)^n$ (see Figure 7). The line corresponding to the third particle remains unconnected any order of perturbation theory. Therefore, contrary to the two-particle case, there is no possibility to reduce equation (3.23) to an equation with completely continuous kernel in any functional space.

Let us note however that in the product of the operators $V_1 R_0(z)$ and $V_2 R_0(z)$, the δ-functions disappear after iterating equation (3.23). Indeed, when computing the kernel of the operator $V_1 R_0(z) V_2 R_0(z)$, we can use p_1 and p_2 as the integration variables and the δ-functions disappear after integration. The analogous situation arises when we multiply two operators of type $V_1 R_0(z)$ with different subscripts. All such iterations correspond to connected diagrams shown in Figure 8.

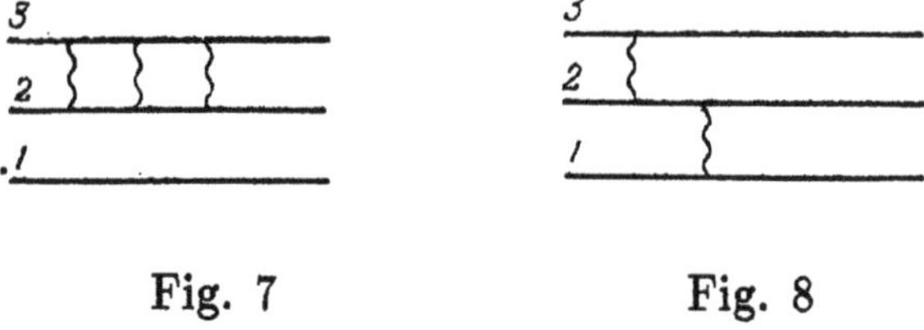

Fig. 7 Fig. 8

These considerations suggest the following procedure for constructing convenient equations for T-matrix.

Let us introduce the operators

$$M_{\alpha\beta} = \delta_{\alpha\beta} V_\alpha - V_\alpha R_0(z) V_\beta$$

with $\alpha, \beta = 1, 2, 3$. It is easy to prove that the T-matrix is expressed in terms of $M_{\alpha\beta}$ as

$$T(z) = \sum_{\alpha,\beta} M_{\alpha\beta}(z). \tag{3.24}$$

Using equation (2.9), it can be shown that the operators $M_{\alpha\beta}$ satisfy the equations

$$M_{\alpha\beta} = \delta_{\alpha\beta} V_\alpha - V_\alpha R_0(z) \sum_\gamma M_{\gamma\beta}(z). \tag{3.25}$$

The system of equations (3.25) has no advantages over equation (3.23). However, we can perform the following transformation. We divide the sum on the right hand side into two terms

$$V_\alpha R_0(z) M_{\alpha\beta}(z) + V_\alpha R_0(z) \sum_{\gamma\neq\alpha} M_{\gamma\beta}(z)$$

and move the first term to the left hand side of equation (3.25). We obtain

$$(1 + V_\alpha R_0(z)) M_{\alpha\beta} = \delta_{\alpha\beta} V_\alpha R_0(z) \sum_{\gamma\neq\alpha} M_{\gamma\beta}(z).$$

Finally, we invert the operator $1 + V_\alpha R_0(z)$ on the left hand side. To evaluate the result of the inversion, we will discuss the operator $T_\alpha(z)$ defined as

$$T_\alpha(z) f(P) = \int t_\alpha(k_\alpha, k'_\alpha, z - p_\alpha^2) f(k'_\alpha, p_\alpha)\, dk'_\alpha.$$

Above, $t_\alpha(k, k', z)$ denotes the solution of equation (3.4) obtained with $v(k)$ replaced by $v_\alpha(k)$. Using this equation, we can easily show that $T_\alpha(z)$ satisfies the equations

$$T_\alpha(z) = V_\alpha - V_\alpha R_0(z) T_\alpha(z), \quad T_\alpha(z) = V_\alpha - T_\alpha(z) R_0(z) V_\alpha \tag{3.26}$$

Hence

$$(1 + V_\alpha R_0(z))^{-1} V_\alpha = T_\alpha, \quad (1 + V_\alpha R_0(z))^{-1} = 1 - T_\alpha R_0. \tag{3.27}$$

Taking these equalities into account, we obtain a set of equations for $M_{\alpha\beta}(z)$:

$$M_{\alpha\beta}(z) = \delta_{\alpha\beta} T_\alpha(z) - T_\alpha(z) R_0(z) \sum_{\gamma\neq\alpha} M_{\gamma\beta}(z). \tag{3.28}$$

This set represents the main result of this section. In what follows, this set will serve as the basic tool to study the T-matrix.

Equation (3.28) determines the T-matrix uniquely. Namely, let $M_{\alpha\beta}(z)$ be nine operators satisfying set of equations (3.28). Then, the operator

$$\tilde{T}(z) = \sum_{\alpha,\beta} M_{\alpha\beta}(z)$$

coincides with the operator $T(z)$. To prove this statement, one has to reverse the considerations leading to (3.28). Multiplying equation (3.28) by the bounded operator $I + V_\alpha R_0(z)$ and taking into account equation (3.27), we find that the operator $M_{\alpha\beta}$ satisfies (3.25) and, consequently, the operator $\tilde{T}(z)$ satisfies equation (3.23) of perturbation theory. By virtue of uniqueness of solutions of this equation for complex z we obtain the required statement.

It is convenient to rewrite equations for the components in the matrix form. Let ω_β be a vector function with three components $\omega_\beta = \{M_{1\beta}(z), M_{2\beta}(z), M_{3\beta}(z)\}$, and let $A(z)$ be the operator defined by the matrix elements

$$\begin{pmatrix} 0 & T_1(z)R_0(z) & T_1(z)R_0(z) \\ T_2(z)R_0(z) & 0 & T_2(z)R_0(z) \\ T_3(z)R_0(z) & T_3(z)R_0(z) & 0 \end{pmatrix}$$

Then, for any fixed β equations (3.28) can be written in the form of equation for the vector function ω_β

$$\omega_\beta = \omega_\beta^{(0)} + A(z)\omega_\beta$$

where the leading term $\omega_\beta^{(0)} = \delta_{\alpha\beta} T_\alpha(z)$, $\alpha = 1, 2, 3$.

Later it will be shown that the equations obtained are free from drawbacks from which equation (3.23) suffer. Let us first note that kernels of the leading terms in equation (3.28) contain δ-functions. Therefore, instead of the operators $M_{\alpha\beta}$, we introduce operators $W_{\alpha\beta}$

$$W_{\alpha\beta}(z) = M_{\alpha\beta}(z) - \delta_{\alpha\beta} T_\alpha(z) \tag{3.28$'$}$$

for which a set of equations analogous to equation (3.28) holds

$$W_{\alpha\beta}(z) = Q_{\alpha\beta}^{(0)}(z) - T_\alpha R_0(z) \sum_{\gamma\neq\alpha} W_{\gamma\beta}(z). \tag{3.29}$$

The leading terms of this set of equations are different, however

$$Q^{(0)}_{\alpha\alpha}(z) = 0, \quad Q^{(0)}_{\alpha\beta}(z) = -T_\alpha(z)R_0(z)T_\beta(z), \ \ \alpha \neq \beta.$$

We used the notation

$$Q^{()n)}_{\alpha\gamma_1\gamma_2\ldots\gamma_n\beta} =$$
$$= (-1)^{n+1}T_\alpha(z)R_0(z)T_{\gamma_1}(z)\ldots T_{\gamma_n}(z)R_0(z)T_\beta(z) \tag{3.30}$$

where the subscripts $\gamma_1, \gamma_2, \ldots \gamma_n$ run through the values constrained by the condition

$$\gamma_i \neq \gamma_{i+1}, \ \ i = 1, 2, \ldots, n-1, \quad \gamma_1 \neq \alpha, \quad \gamma_n \neq \beta.$$

It is easy to verify that the kernels of the leading terms of equations (3.29) are decreasing Hölder functions. Indeed, let us take for example

$$Q^{(0)}_{1,2}(P, P', z) =$$
$$= \frac{t_1(k_1, k_1(p_1, p_2'), z - p_1^2)t_2(k_2(p_1, p_2'), k_2', z - p_2^2)}{|s_{12}|^2(k_1^2(p_1, p_2') + p_1^2 - z)} \tag{3.31}$$

where the variable k_α with arguments $k_\alpha(p_\alpha, p_\beta')$ for $\alpha = 2$ denotes the functions

$$k_2(p_2, p_2') = \frac{c_{21}}{s_{21}}p_2 + \frac{1}{s_{21}}p_1', \quad \beta = 1,$$
$$k_3(p_2, p_3') = \frac{c_{32}}{s_{32}}p_2 - \frac{1}{s_{32}}p_3', \quad \beta = 3; \tag{3.32}$$

for other values of α and β the functions can be obtained from (3.32) by cyclic permutation of the subscripts. The prime in the argument p_β' corresponds to an independent configuration of the system. The relative momenta are related to each other such that $k_\alpha(p_\alpha, p_\beta) = k_\alpha$.

Thus, contrary to the equation of perturbation theory, δ-functions in equation (3.28) disappear after one iteration. Later, we will show that this feature guarantees compactness of the equations (3.28).

Apart from compactness, this set of equations possesses one more feature which distinguishes it from other possible modifications of the equations of perturbation theory (3.23): the homogeneous equation (3.28) is equivalent to the Schrödinger equation. More precisely, knowing a solution of the homogeneous set

$$\Phi_\alpha = -T_\alpha(z)R_0(z)\sum_{\gamma\neq\alpha}\Phi_\gamma \tag{3.33}$$

it is possible to construct a square integrable solution of the Schrödinger equation

$$\left(H_0 + \sum_\alpha V_\alpha\right)\Psi = \lambda\Psi$$

of the form

$$\Psi = \sum_\alpha \Psi_\alpha$$

where

$$\Psi_\alpha = R_0\Phi_\alpha.$$

Indeed, multiplying equation (3.33) by $I + R_0 V_\alpha$ and taking (3.27) into account, we obtain

$$\Psi_\alpha = -R_0 V_\alpha \Psi. \tag{3.34}$$

Summing over α, we get the formula

$$\Psi = -R_0(V_1 + V_2 + V_3)\Psi.$$

By multiplying this formula by $H_0 - z$, it can be shown that Ψ satisfies the Schrödinger equation. The condition of square integrability of the function Ψ will be verified later after discussing properties of solutions of equation (3.28).

3.2.2 Properties of solutions of compact equations

Let us now discuss equations (3.28) in more details. According to the scheme drafted at the end of the previous section, we must start with the study of kernels of iterations which are expressed in terms of the operators $Q^{(n)}_{\alpha\gamma_1\ldots\beta}(z)$ as

$$Q^{(n)}_{\alpha\beta}(z) = \sum_{\gamma_1,\gamma_2,\ldots,\gamma_n} Q^{(n)}_{\alpha\gamma_1\gamma_2\ldots\gamma_n\beta}(z) \tag{3.35}$$

The prime in summation symbol $\sum'$ indicates that the summation runs only over admissible values of $\gamma_1, \gamma_2, \ldots, \gamma_n$, i.e.,

$$\gamma_1 \neq \alpha, \quad \gamma_1 \neq \gamma_2, \quad \ldots \quad \gamma_n \neq \beta.$$

Let us first consider the kernel of the operator $Q^{(0)}_{\alpha\beta}(z)$ determining the leading term in equation (3.29). For $\alpha = 1$ and $\beta = 2$, $Q^{(0)}_{\alpha\beta}(z)$ is given explicitly by equation (3.31). It can be seen that this kernel is well-behaving for complex z with $\operatorname{Im} z = 0$ and becomes of singular for $\operatorname{Im} z \to 0$. Some singularities result from singularities of the kernels $t_1(k_1, k_1', z)$ and $t_2(k_2, k_2', z)$ present in this equation. For example, in equation (3.29) we have terms of the form

$$\frac{\varphi_A(k_\alpha}{p_\alpha^2 - \varkappa_A^2 - z} J^{(0)}_{\alpha\beta}(p_\alpha, P', z)$$

where $J^{(0)}_{\alpha\beta}$ is continuous in vicinity of the singularity. The symbol of detailed partition A contains in this case the index of the pair of particles α ($\alpha = 1, 2, 3$) and the index of the eigenvalue j ($j = 1, 2, ..., N_\alpha$), i.e., $A = \{\alpha, j\}$. Apart from singular term of the form $(p_\alpha^2 - \varkappa_A^2 - z)^{-1}$, there exist terms as singular as $(p_\beta^2 - \varkappa_B^2 - z)^{-1}$ and a term in which both these singularities are multiplied. It is important that the only term which is k_α dependent is the form-factor $\varphi_A(k_\alpha)$. Analogously, the term with singularity of the type $(p_\beta'^2 - \varkappa_B^2 - z)^{-1}$ depends on k_β' only through $\varphi_B^*(k_\beta')$, and the term with the singularity $(p_\alpha^2 - \varkappa_A^2 - z)^{-1}(p_\beta'^2 - \varkappa_B^2 - z)^{-1}$ depends on k_α and k_β' only through $\varphi_A(k_\alpha)$ and $\varphi_B(k_\beta')$. These singularities will be called the main and two-particle singularities, respectively. These singularities exist only in the presence of a two-body bound state and appear in the same form in all iterations of equations (3.29).

Besides these singularities, other singularities appear in equation (3.31) as the result of vanishing of the term $k_\alpha^2(p_\alpha, p_\beta') + p_\alpha^2 - z$. It is important that the position of this singularity depends on both variables p_α and p_β'. Moreover, it does not only depend on their magnitudes but on their directions as well. Such singularities which do exist even if there is no two-body bound state, will be called secondary or three-particle singularities. It turns out that the secondary singularities become weaker and completely disappear with an increasing order of iterations.

It should be noted that two- and three-particle singularities do not intersect. Indeed, the secondary singularities arise when the expression $k_\alpha^2(p_\alpha, p_\beta') + p_\alpha^2 - z$ in the denominator, equals zero. If we put, for example, $z = -\varkappa_A^2 + p_\alpha^2$ which is equivalent to the vanishing of the two-body denominator, this expression becomes $k_\alpha^2(p_\alpha, p_\beta') + \varkappa_A^2$ and is positive at $\varkappa_A^2 > 0$. An analogous feature of separability of singularities was described in Chapter 2.

Before describing the properties of the kernels $Q^{(n)}_{\alpha\gamma_1\ldots\gamma_n\beta}$ for $n > 1$ we introduce some definitions. The kernel $Q(P,P',z)$ is called a kernel of type $\mathcal{D}_{\alpha\beta}$ if it can be represented in the form

$$Q(P'P',z) = F(P,P',z)+$$

$$+\sum_j \frac{\varphi_A(k_\alpha)}{p_\alpha^2 - \varkappa_A^2 - z} J(p_\alpha, P', z)+$$

$$+\sum_l G(P, p'_\beta, z) \frac{\varphi_B^*(k'_\beta)}{{p'_\beta}^2 - \varkappa_B^2 - z}+$$

$$++\sum_{j,l} \frac{\varphi_A(k_\alpha)}{p_\alpha^2 - \varkappa_A^2 - z} H(p_\alpha, p'_\beta, z) \frac{\varphi_B^*(k'_\beta)}{{p'_\beta}^2 - \varkappa_B^2 - z} \tag{3.36}$$

where the summation is over all eigenvalues of the energy operators h_α and h_β of the two-body subsystems. The kernels F, J, G, and H will be called components of the kernel $Q(P,P',z)$.

Now, let us consider the kernels $Q^{(n)}_{\alpha\gamma_1\ldots\gamma_n\beta}(P,P',z)$ of the operator $Q^{(n)}_{\alpha\gamma_1\ldots\gamma_n\beta}$ for $n > 1$. According to definition (3.30), these kernels are represented in the form of the $(n+1)$-dimensional singular integrals

$$Q^{(n)}_{\alpha\gamma_1\ldots\gamma_n\beta}(P,P',z) = (-1)^{n+1} \int t_\alpha(k_\alpha, k_\alpha^(1), z - p_\alpha^2) \times$$

$$\times \frac{\delta(p_\alpha - p_\alpha^{(1)})}{P^{(1)2} - z} \cdots \frac{\delta(p_\beta^{(n+1)} - p'_\beta)}{P^{(n+1)2} - z} \times$$

$$\times t_\beta(k_\alpha^(n+1), k'_\beta, z - {p'_\beta}^2)\, dP^{(1)} dP^{(2)} \ldots dP^{(n+1)}. \tag{3.37}$$

For

$$n = 0,$$

the δ-functions remove all integrations in (3.37) and the kernels $Q^{(0)}_{\alpha\beta}(P,P',z)$ can be explicitly expressed in terms of the kernels $t_\alpha(k,k',z)$ by equations (3.31). These kernels were discussed already.

Let us start our investigation of the kernels $Q^{(n)}_{\alpha\gamma_1\ldots\gamma_n\beta}$ with the kernel $Q^{(1)}_{\alpha\gamma_1\beta}$. As an example we take $\alpha = 1$, $\gamma_1 = 2$, and $\beta = 3$ and in order not to overload the formulae by subscripts we will temporarily denote the operator $Q^{(1)}_{123}(z)$ as $Q(z)$ and its kernel as $Q(P,P',z)$. If, for $n = 1$, $\alpha = 1$, $\gamma_1 = 2$, and $\beta = 3$ we take $p_1^{(1)}$, $p_2^{(1)} = q$, $p_2^{(2)}$, and $p_3^{(2)}$ as the integration variables, for the kernel $Q(P, P', z)$ we obtain

$$Q(P, P', z) = \frac{1}{|s_{21}s_{23}|^3} \int t_1(k_1, k_1(p_1, q), z - p_1^2) \times$$

$$\times (k_1^2(p_1, q) + p_1^2 - z)^{-1} t_2(k_2(q, p_1), k_1(q, p_3'), z - q^2) \times$$

$$\times (k_3^2(p_3', q) + p_3' - z)^{-1} t_3(k_3(p_3', q), k_3', z - p_3'^2)\, dq. \tag{3.38}$$

If we replace the kernels t_1 and t_3 by the corresponding expression of the form (3.18), we find that the kernel $Q(P, P', z)$ is of the $\mathcal{D}_{\alpha\beta}$ type. We denote its components by F, G, J, and H. Representation (3.37) for the kernel $F(P, P', z)$ can be obtained by replacing the kernels t_1 and t_2 by $\tilde{t}_1$ and $\tilde{t}_2$, respectively. The kernels G, J, and H can be obtained by exchanging in (3.38) t_3 by $\varphi(k_3(q, p_3'))$ or t_1 by $\varphi_1^*(k_1(q, p_1))$, or by making both substitutions simultaneously.

According to representation (3.18) of the kernel $t_2(k_2(q, p_1), k_1(q, p_3'), z - q^2)$, the integral (3.38) defining Q consists of two terms, the first of whose contains a product of two and the second a product of three singular denominators in the integrand. The denominators corresponding to the two- and three-body singularities are singular in various regions of the integration range. Therefore, the integral representation of the components F, G, J, and H can be split into parts in such a way that in each part only two-body or only three-body singularities are present. The integrals containing the two-body singularities are of the same form as analogous two-body integrals (3.11). They are decreasing Hölder functions of z outside the real axis on the complex plane with a cut from $-\varkappa_A^2$ to ∞.

The integral generated by the three-particle singularities is more complicated. If we denote

$$a = c_{21}p_1, \quad b = c_{23}p_3', \quad \xi = s_{12}^2(z - p_1^2), \quad \eta = s_{23}^2(z - p_3'^2),$$

then such integral takes the form

$$I_{13} = \int \frac{dq f(q, P, P')}{((q-a)^2 - \xi)((q-b)^2 - \eta)} \tag{3.39}$$

where f is a smooth function vanishing outside the sphere of a sufficiently large radius $q^2 = 1 + |z|$. It is clear that the strongest singularities of the function I_{13} coincide with the singularities of a simpler expression

$$I_{13}^{(0)} = \int \frac{dq}{((q-a)^2 - \xi)((q-b)^2 - \eta)}$$

which follows from equation (3.39) when the smooth function f is taken outside the integral.

It is possible to determine the function $I_{13}^{(0)}$ explicitly by means of the identity

$$\frac{1}{uv} = \frac{1}{2}\int_{-1}^{1} dt \left((u\frac{1+t}{2} + v\frac{1-t}{2} \right)^{-2}$$

The following formula holds

$$I_{13}^{(0)} = \frac{i\pi^2}{\sqrt{A}} \log \frac{\sqrt{\xi} + \sqrt{\eta} + A}{\sqrt{\xi} + \sqrt{\eta} - A} \tag{3.40}$$

where $A = (a-b)^2 = (c_{21}p_1 - c_{23}p_3')^2$. It is clear that the kernel is a square integrable function in each variable. We emphasize that the integral $I_{31}(p_1, p_3')$ is non-singular in vicinity of the two-particle energy shell. More precisely, if the variable z takes values in a neighbourhood of the point $z = p_1^2 - \varkappa_A^2$ or $z = p_3'^2 - \varkappa_B^2$ then the kernels F, G, J, and H are smooth bounded functions. This completes our discussion of the kernel $Q_{123}^{(1)}$. Any other kernel $Q_{\alpha\beta\gamma}^{(1)}$ can be investigated similarly.

Now, let us consider the kernel $Q_{\alpha\beta}^{(1)}$ of the first iteration given by the sum (3.35). Using the kernel $Q_{123}^{(1)}$ as an example, we have convinced ourselves that the kernels $Q_{\alpha\beta}^{(1)}$ are of the type $\mathcal{D}_{\alpha\beta}$. Moreover, if

$$p_\alpha^2 \geq |z| + 1 \quad \text{or} \quad p_\beta'^2 > |z| + 1,$$

$$k_\alpha^2(p_\alpha, p_\beta' + p_\alpha^2 - z > 1 \quad \text{and} \quad k_\beta'^2(p_\alpha, p_\beta' + p_\beta'^2 - z > 1 \tag{3.41}$$

or the variables p_α, p_β are located in the neighbourhood of the two-particle energy shell:

$$p_\alpha^2 = \operatorname{Re} z + \varkappa_A^2 \quad \text{or} \quad p_\beta'^2 = \operatorname{Re} z + \varkappa_B^2 \tag{3.42}$$

then the secondary singularities do not appear and the components of such kernels are smooth functions.

Let us now proceed to the investigation of the kernels $Q_{\alpha\gamma_1\ldots\gamma_n\beta}^{(n)}$ for $n \geq 2$. These kernels can be expressed in terms of the kernels $Q_{\alpha\gamma_1\ldots\gamma_n\beta}^{(n-1)}$ by

$$Q_{\alpha\gamma_1\ldots\gamma_n\beta}^{(n)}(P, P', z) =$$

$$= \int dP'' t_\alpha(k_\alpha, k''_\alpha, z - p_\alpha^2) \frac{\delta(p_\alpha - p''_\alpha)}{P''^2 - z} Q^{(n-1)}_{\alpha\gamma_1\ldots\gamma_n\beta}(P'', P', z) \qquad (3.43)$$

If we take p''_{γ_1} and p''_α as the integration variables, then the integration over this last variable disappears because of the presence of the δ-function. Let us set $n = 2$. Contributions of the two- and three-particle singularities may be studied separately as it was done previously. Integrals containing the two-particle singularities are estimated with the help of the singular integral lemma. The difficulties are related to the integrals containing intersecting three-particle singularities. It is possible to show, as it was done above that the integrals have no singularities if the variables p_α and p'_β satisfy one of the conditions (3.41) or (3.42). If these conditions are not satisfied, then the components $Q^{(2)}_{\alpha\gamma_1\gamma_2\beta}$ have secondary singularities. We will not discuss these singularities here. We only note, that the singularities of $Q^{(2)}_{\alpha\gamma_1\gamma_2\beta}$ are weaker than those of the representation (3.40).

It is possible to show that for $n = 3$ the secondary singularities do not arise. As an example, we will prove this claim for the kernel $Q^{(n)}_{\alpha\gamma_1\ldots\beta}$ with $\alpha = 1$, $\gamma_1 = 2$, $\gamma_3 = 2$, $\beta = 1$. Analysing the reasoning above, we can verify that the integral which may introduce the secondary singularity has the form

$$\int dq I_{13}(p_1, q) I_{31}(q, p'_1) f(q)$$

where the integrals I_{13} and I_{31} are defined by the formulae of the type (3.39) in which the variables p_1, p''_3 and p''_3, p'_1 are replaced by p_1, q and q, p'_1, respectively. The function $f(g)$ is smooth and can be explicitly expressed in terms of the two-particle T-matrices. By using the representation (3.40) for the integrals I_{13} and I_{31}, we find that these integrals are square integrable functions of the variable q. Hence, the secondary singularities in these integrals disappear in the course of integration over q.

The kernels $Q^{(n)}_{\alpha\gamma_1\ldots\beta}$ can be successively expressed in terms of ordinary singular integrals (3.43) containing the kernels $Q^{(n-1)}_{\gamma_1\ldots\beta}$ in the integrands. These kernels are of the type $\mathcal{D}_{\alpha\beta}$ and their components are decreasing Hölder functions. The same holds for the kernels $Q^{(n)}_{\alpha\beta}$. More precisely, their properties may be specified as follows.

Let us denote by $N(P, \theta)$ an estimator defined as

$$N(P, \theta) = \sum_{\alpha \neq \beta} (1 + |p_\alpha)^{-1-\theta} (1 + |p_\beta)^{-1-\theta}$$

and by $\Pi^2_{-k^2}$ the complex plane with the cut $[-\varkappa^2, \infty)$ where $\varkappa^2 = ds \max_A \varkappa_A^2$. Then the following statement holds: The F components of the iterated kernels $Q_{\alpha\beta}^{(n)}$ satisfy the estimate

$$|F(P, P', z)| \leq CN(P, \theta)(1 + p_\beta'^2)^{-1},$$

$$|F(P, P' + \Delta P, z) - F(P, P', z)| \leq CN(P, \theta)(1 + p_\beta'^2)^{-1} |\Delta P|^\mu \qquad (3.44)$$

uniformly with reference to P' and z from $\Pi^2_{-k^2}$. The same estimates hold for F as a function of P'. The components G, J, and H satisfy the estimates which can be obtained from (3.44) by setting $k_\beta' = 0$ or $k_\alpha = 0$ or both of them.

To prove these estimates, it is necessary to apply the singular integral lemma and to use at the same time the estimates (3.12) for the kernels $t_\alpha(k, k', z)$. Here, we will only present the formulation of the final result.

We will call the kernels of the type $\mathcal{D}_{\alpha\beta}$ which satisfy the estimate (3.44) the kernels of the class $\tilde{\mathcal{D}}_{\alpha\beta}$. Evidently, the iterated kernels $Q_{\alpha\beta}^{(n)}$ belong to the class $\tilde{\mathcal{D}}_{\alpha\beta}$ for any n starting from $n = 4$.

The considerations presented above make it possible to find the class of functions in which the integral equation (3.29) should be considered. Namely, one should consider a set of vector functions

$$\omega = (\rho_1(P), \rho_2(P), \rho_3(P), \sigma_{11}(p_1), \ldots$$

$$\ldots, \sigma_{1N_1}(P), \ldots, \sigma_{31}(p_3), \ldots, \sigma_{3N_3}(p_3))$$

where the components $\rho_\alpha(P)$ as functions of P satisfy the estimates (3.44). Estimates for $\sigma_A(p_\alpha)$ can be obtained from (3.44) by setting $k_\alpha = 0$. This set of functions will be denoted $\mathfrak{M}$.

With the help of the elements ω we will discuss the operator $A(z)$ defined as follows: $o' = A(z)\omega$ means that

$$\rho_\alpha'(P) = -\int \tilde{t}_\alpha(k_\alpha, k_\alpha', z - p_\alpha^2) \frac{\delta(p_\alpha - p_\alpha')}{P'^2 - z} \sum_{\beta \neq \alpha} \chi_\beta P' \, dP', \qquad (3.45)$$

$$\sigma_A'(p_\alpha) = -\int \varphi_A^*(k_\alpha') \frac{\delta(p_\alpha - p_\alpha')}{P'^2 - z} \sum_{\beta \neq \alpha} \chi_\beta P' \, dP' \qquad (3.46)$$

where $\chi_\alpha(P)$ denotes the function

$$\chi_\alpha(P) = \rho_\alpha(P) + \sum_i \frac{\varphi_A(k_\alpha)\sigma_A(p_\alpha)}{p_\alpha^2 - \varkappa_A^2 - z}, \quad A = \{\alpha, i\} \tag{3.47}$$

Let us note that if we add to equation (3.45) equation (3.46) multiplied by $\varphi_A(k_\alpha)(p_\alpha^2 - \varkappa_A^2 - z)^{-1}$ and denote by $\chi'_\alpha(P)$ the function constructed from $\rho'_\alpha(P)$ and $\sigma'_A(p_\alpha)$ and according to equation (3.47), we obtain

$$\chi'_\alpha(P) = -\int t_\alpha(k_\alpha, k'_\alpha, z - p_\alpha^2)\frac{\delta(p_\alpha - p'_\alpha)}{P'^2 - z}\sum_{\beta\neq\alpha}\chi_\beta P'\,dP'$$

or, in the operator form

$$\chi'_\alpha = -T_\alpha(z)R_0(z)\sum_{\beta\neq\alpha}\chi_\beta.$$

Therefore, the operator $A(z)$ acting on elements of the set $\mathfrak{M}$ is closely related to the set of equations (3.29). The very complicated construction discussed above was necessary for systematic investigations of the main singularities of the kernels $W_{\alpha\beta}(P, P', z)$.

Finally, we reduce the analysis of the system (3.29) to the study a of equation of second kind defined on $\mathfrak{M}$ with the operator $A(z)$. Instead of the operators $W_{\alpha\beta}(z)$, we will discuss operators $\tilde{W}^{(4)}_{\alpha\beta}(z)$ obtained from $W_{\alpha\beta}(z)$ by subtracting the first three iterations of the system (3.29). These operators satisfy the following set of equations

$$\tilde{W}^{(4)}_{\alpha\beta}(z) = Q^{(4)}_{\alpha\beta}(z) - T_\alpha(z)R_0(z)\sum_{\gamma\neq\alpha}\tilde{W}^{(4)}_{\gamma\beta}(z) \tag{3.48}$$

It follows from the statement formulated above that the kernels $Q^{(4)}_{\alpha\beta}(z)$ belong to the classes $\tilde{\mathcal{D}}_{\alpha\beta}$. Let us denote their components by $F^{(4)}_{\alpha\beta}$, $G^{(4)}_{\alpha B}$, $J^{(4)}_{A\beta}$, $H^{(4)}_{AB}$. For fixed β, P', and z, the set of the kernels $F^{(4)}_{\alpha\beta}(P, P', z)$, $J^{(4)}_{A\beta}(p_\alpha, P', z)$ can be treated as an element of $\mathfrak{M}$. We denote this element $\omega^{(4)}_\beta(P', z)$. Analogously, with the set of kernels $G^{(4)}_{\alpha B}(P, p'_\beta, z)$, $H^{(4)}_{AB}(p_\alpha, p'_\beta, z)$, for fixed B, p'_β, and z, we associate the element $\omega^{(4)}_B(p'_\beta, z)$.

Consider the equations

$$\omega_\beta(P', z) = \omega^{(4)}_\beta(P', z) + A(z)\omega_\beta(P', z) \tag{3.49}$$

$$\omega_B(p'_\beta, z) = \omega^{(4)}_B(p'_\beta, z) + A(z)\omega_B(p'_\beta, z) \tag{3.50}$$

Let these equations have a solution for some z. By $F_{\alpha\beta}(P,P',z)$, $J_{A\beta}(p_\alpha,P',z)$, $G_{\alpha B}(P,p'_\beta,z)$, $H_{AB}(z)(p_\alpha,p'_\beta,z)$ we denote components of $\omega_\beta(P',z)$ and $\omega_B(p'_\beta,z)$ and we will regard these kernels as components of the kernels $\tilde{W}^{(4)}_{\alpha\beta}(P,P',z)$ from the class $\tilde{\mathcal{D}}_{\alpha\beta}$. It is easy to verify that these kernels satisfy equations (3.48). Therefore, we reduced the problem investigation of the kernels $W_{\alpha\beta}(P,P',z)$ to the problem of solvability of the equation

$$\omega = \omega^{(0)} + A(z)\omega \tag{3.51}$$

on the set $\mathfrak{M}$.

Now we have to prove that equation (3.51) is of Fredholm type and we must consider the corresponding homogeneous equation. However, we cannot confine ourselves to the study of this equation because the matrix elements of the operator $A(z)$ contain δ-functions and therefore this operator is not completely continuous. Therefore, it is necessary to discuss the integrated equations and the powers of the operator $A^n(z)$. These operators can be expressed in terms of $Q^{(n)}_{\alpha\beta}$. Indeed, the equation(3.48) integrated n times reads

$$\tilde{W}^{(4)}_{\alpha\beta}(z) = \sum_{i=0}^{n} Q^{(4+i)}_{\alpha\beta}(z) + \sum_{\gamma\neq\alpha} Q^{(n)}_{\alpha\beta}(z)R_0(z) \sum_{\gamma_2\neq\gamma_1} \tilde{W}^{(4)}_{\gamma_2\beta}(z) \tag{3.52}$$

This equation can be reduced to the equation

$$\omega = \omega^{(n)} + A^n(z)\omega, \quad \omega^{(n)} = \sum_{i=1}^{n} A^i(z)\omega^{(0)}$$

in the same way as the equation (3.48) has been reduced to the equation (3.51). As a result, we obtain explicit representations of kernels of the operators $A^n(z)$ in terms of the components $Q^{(n)}_{\alpha\beta}$. It is possible to verify by means of the obtained estimates for the components that the operator $A^n(z)$ improves properties of the components of the elements ω and, consequently, is a compact operator. We will not present a detailed proof of this statement here. The reasoning is based on the singular integral lemma and have been discussed in detail in the two-body case.

To conclude, we will describe a set of singular points at which the homogeneous equation

$$\omega = A(z)\omega \tag{3.53}$$

has nontrivial solution. We denote by $\Psi_\alpha(P)$ the function

$$\Psi_\alpha(P) = (P^2 - z)^{-1}\chi_\alpha(P)$$

where the function χ_α is expressed in terms of the components ρ_α and σ_A of the element ω by equation(3.47). Equation (3.35) can be written in terms of $\Psi_\alpha(P)$ as

$$\Psi_\alpha = -R_0(z)T_\alpha(z)\sum_{\beta\neq\alpha}\Psi_\beta$$

and, correspondingly, the function $\Psi = \sum_\alpha \Psi^{-1}$ satisfies the Schrödinger equation $H\Psi = z\Psi$. Note that according to the estimate (3.44), this function is square integrable for all z from $\Pi_{-\varkappa^2}$ not touching the cut $[-\varkappa^2, \infty)$. It is more complex to verify that for z on the cut the solution remains square integrable. For this purposes, a reasoning similar to that discussed in the two-body case can be used. The corresponding constructions are very complicated and will not be discussed here.

Thus we arrived at the conclusion that the set of singular points of equation (3.51) coincides with the set of eigenvalues of the operator $A(z)$. We denote this set by Σ_H.

Therefore, way we concluded the study of equation (3.29) and now we can characterize the properties of the kernels $W_{\alpha\beta}(z)$. Let $\Pi'_{-\varkappa^2}$ be the complex plane with a cut $[-\varkappa^2, \infty)$ and with neighbourhoods of the points from Σ_H eliminated. The following statement holds.

For all z from $\Pi'_{-\varkappa^2}$ the kernel $W_{\alpha\beta}(P, P', z)$ can be represented in the form of the sum

$$W_{\alpha\beta}(P, P', z) = \sum_{j=1}^{4} Q_{\alpha\beta}^{(j)}(P, P', z) + \tilde{W}_{\alpha\beta}^{(4)}(P, P', z).$$

The kernels $Q_{\alpha\beta}^{(j)}$ are of the type $\mathcal{D}_{\alpha\beta}$ and their properties have been discussed earlier. The kernels $\tilde{W}_{\alpha\beta}^{(4)}$ belong to the class $\tilde{\mathcal{D}}_{\alpha\beta}$, so that their components $F_{\alpha\beta}$, $G_{\alpha\beta}$, $J_{\alpha\beta}$, and $H_{\alpha\beta}$ satisfy the estimate (3.44).

Let us now examine what these results yield as applied to the resolvent. By $R_\alpha(z)$ we denote the resolvent of the energy operator $H_\alpha = H_0 + V_\alpha$. This operator is expressed in terms of the T-matrix T_α as

$$R_\alpha(z) = R_0(z) - R_0(z)T_\alpha(z)R_\alpha(z). \tag{3.54}$$

The analogous relation holds for the operators $R(z)$ and $T(z)$:

$$R(z) = R_0(z) - R_0(z)T(z)R(z). \tag{3.55}$$

Comparing formulae (3.24) and (3.28') with equality (3.55), we obtain the representation

$$R(z) = R_0(z) + \\ + \sum_\alpha (R_\alpha(z) - R_0(z)) - R_0(z)\sum_{\alpha\beta} W_{\alpha\beta}(z)R_0(z) \tag{3.56}$$

which makes it possible to describe all singularities of the resolvent kernel in terms of singularities of the operator $W_{\alpha\beta}(z)$. In particular, equalities (2.24) and (2.25) which have been discussed in Chapter 2 as hypotheses, follow from the representations (3.36). We note that the last term on the right hand side of (3.56) coincides with the connected part of the resolvent

$$[R(z)]_c = -R_0(z)\sum_{\alpha\beta} W_{\alpha\beta}(z)R_0(z).$$

At this point we conclude our discussion of the properties of the resolvent kernels for three-particle systems. In this section we obtained compact equations, on the basis of whose we discussed main singularities of the kernel $T(z)$ of the T-matrix and gave grounds for the representations of the resolvent kernel (2.24) and (2.25).

3.3 Integral Equations for Resolvent and Wave Operators

In this section, we describe compact integral equations for components of the resolvent and the wave operators. Such equations are useful in numerical calculations. We also describe here integral representations for the components $W_{\alpha\beta}$ and express the scattering operator in terms of these components.

3.3.1 Components of the resolvent

We define components of the resolvent as

$$R_{\alpha\beta}(z) = -R_0(z)M_{\alpha\beta}(z)R_0(z) \tag{3.57}$$

The resolvent can be expressed in terms of these components

$$R(z) = R_0(z) + \sum_{\alpha,\beta} R_{\alpha\beta}(z).$$

Let us multiply both sides of the compact equations for the T-matrix by $R_0(z)$. Taking into account the identity

$$R_0(z)T_\alpha(z) = R_\alpha(z)V_\alpha \tag{3.58}$$

which follows from definition of the T-matrix (2.31) and from equality (2.29), we arrive at the following set of integral equations for the components

$$R_{\alpha\beta}(z) = (R_\alpha(z) - R_0(z))\delta_{\alpha\beta} - R_\alpha(z)V_\alpha \sum_{\gamma\neq\alpha} R_{\gamma\beta}(z). \tag{3.59}$$

It is also possible to introduce components of the resolvent in correspondence with the partition of the potentials in equation of the perturbation theory (2.9):

$$R_\alpha^{(\beta)}(z) = R_0\delta_{\alpha\beta} - R_0(z)V_\alpha R(z). \tag{3.60}$$

Equations for the operators $R_\alpha^{(\beta)}(z)$ can be obtained in the same way as those for the T-matrix.

Indeed, inserting into (3.60) the obvious identity $R(z) = \sum_\alpha R_\alpha^{(\beta)}(z)$ following from equation(3.9), we obtain

$$R_\alpha^{(\beta)}(z) = R_0(z)\delta_{\alpha\beta} - R_0(z)V_\alpha \sum_\gamma R_\gamma^{(\beta)}(z).$$

Then, by moving the diagonal term $R_\alpha^{(\beta)}(z)$ to the left hand side, we get

$$(I + R_0(z)V_\alpha)R_\alpha^{(\beta)}(z) = R_0(z)\delta_{\alpha\beta} - R_0(z)V_\alpha \sum_{\gamma\neq\alpha} R_\gamma^{(\beta)}(z);$$

inverting the operator $I + R_0(z)V_\alpha$ and taking the identity $(I + R_0(z)V_\alpha)^{-1}R_0(z)$
$= R_\alpha(z)$ into account, we obtain the following system

$$R_\alpha^{(\beta)}(z) = R_\alpha(z)\delta_{\alpha\beta} - R_\alpha(z)V_\alpha \sum_{\gamma\neq\alpha} R_\gamma^{(\beta)}(z).$$

This system differs from (3.59) only in the free term. We note that the two types of the components are related by the equation

$$R_\alpha^{(\beta)}(z) = \sum_\gamma R_{\alpha\gamma}(z) + \delta_{\alpha\beta}R_0(z).$$

Subsequent discussion will be based on the partition (3.57).

3.3.2 Components of wave operators

In accordance with representation (2.3), we define components of the wave operators U_B for $B = 0$ as

$$U_{\alpha B}^{(\pm)} = \lim_{\epsilon\downarrow 0} \mp i\epsilon R_{\alpha\beta}(E_{AB}(p'_A) \pm i\epsilon)L_B(p'_B). \tag{3.61}$$

The components of the wave operator $U_{\alpha 0}^{(\pm)}$ describing processes with three free particle in the initial state, are defined by

$$U_{\alpha 0}^{(\pm)} = \lim_{\epsilon\downarrow 0} \mp i\epsilon(\delta_{\alpha 1}R_0(P'^2 \pm i\epsilon)+$$

$$+\sum_\beta R_{\alpha\beta}(P'^2 \pm i\epsilon)L(P')). \tag{3.61'}$$

The components of the operators $U_{\alpha B}$ and $U_{\alpha 0}$ can be expressed in terms of components of the T-matrix. Let us denote by $T_{\alpha B}$ a kernel defined as the sum

$$T_{\alpha B}(P, p'_\beta, z) =$$

$$= G_{\alpha B}(P, p'_\beta, z) + \sum_j \frac{\varphi_A(k_\alpha) H_{AB}(p_\alpha, p'_\beta, z)}{p_\alpha^2 - \varkappa_A^2 - z} \tag{3.62}$$

and by $T_{\alpha 0}$ the kernel

$$T_{\alpha 0}(P, P', z) = \sum_\beta M_{\alpha\beta} P, P', z).$$

Repeating the reasoning which leads us to formulae (2.29) and (2.30), we obtain the following expressions for the kernels $U_{\alpha B}$ and $U_{\alpha 0}$:

$$U_{\alpha B}(P, p'_\beta) = \delta_{\alpha\beta} \psi_B(k_\beta \delta(p_\beta - p'_\beta) -$$

$$- \frac{T_{\alpha B}(P, p'_\beta, E_B(p'_\beta) \pm i0)}{P^2 - E_B(p'_\beta) \mp i0} \tag{3.63}$$

$$U_{\alpha 0}(P, P') = \delta_{\alpha 1} \delta(P - P') - \frac{T_{\alpha 0}(P, P', P'^2 \pm i0)}{P^2 - P'^2 \mp i0} \tag{3.63$'$}$$

The wave operators $U_B^{(\pm)}$ and $U_0^{(\pm)}$ can be restored from their components form

$$U_B^{(\pm)} = \sum_\alpha U_{\alpha B}^{(\pm)}, \quad U_0^{(\pm)} = \sum_\alpha U_{\alpha 0}^{(\pm)}. \tag{3.63$''$}$$

Note, that the kernel $T_{\alpha B}$ coincides, up to the term φ_A^*, with residuum of the kernel $W_{\alpha\beta}(P, P', z)$ taken at the two-particle pole $(E_B(p'_\beta) - z)^{-1}$. Since the two-particle singularities do not intersect with the three-particle ones, this leads to the following important property of the kernel $T_{\alpha B}$: the kernels $T_{\alpha B}$ have only two-particle singularities explicitly singled out in the representation (3.62). On the other hand, the kernel T_{0B} contains three-particle singularities and, moreover, has two-particle polar singularities of the type $(E_A(p_A) - P'^2 \mp i0)^{-1}$ which, however, do not intersect with the three-particle ones.

In order to derive compact integral equations for kernels of the wave operators, we multiply both sides of equation(3.59) by $\mp i\epsilon L_B$ or $\mp i\epsilon L_0$. Then we set $z = E_B(p'_\beta) \pm i\epsilon$, $B \neq 0$, or $z = P'^2 \pm i\epsilon$, $B = 0$, respectively and taking into account equations (3.60) and (3.61) we pass to the limit $\epsilon \downarrow 0$. As a result we obtain

$$U_{\alpha B}^{(\pm)} = \delta_{\alpha\beta} L_B - R_0(E_B \pm i0) T_\alpha(E_B \pm i0) \sum_{\gamma \neq \alpha} U_{\alpha B}, \tag{3.64}$$

$$U_{\alpha 0}^{(\pm)} = \delta_{\alpha 1} L_0 - R_0(P'2 \pm i0) T_\alpha(P'2 \pm i0) \sum_{\gamma \neq \alpha} U_{\alpha 0}.$$

It should be stressed that integral equations of second type for the wave operators $U_{\alpha B}^{(\pm)}$ and $U_{\alpha 0}^{(\pm)}$ differ only in free terms. At the same time, the homogeneous equations are equivalent to the Schrödinger equation for eigenvalues for the operator H.

3.3.3 Integral equations for components

We will show that the components $W_{\alpha\beta}(z)$ can be defined by means of the following integral representations

$$T_{\alpha\beta}(z) \tilde{R}_B(z) = W_{\alpha\beta}(z) R_0(z) L_B \tag{3.65}$$

$$\tilde{R}_B(z) V_{B\alpha}(z) = L_B^* R_0(z) W_{\beta\alpha}(z); \tag{3.66}$$

where $T_{\alpha B}$ denotes an operator mapping from $\mathfrak{H}_B$ to $\mathfrak{H}$ defined by the kernel $T_{\alpha B}(P, p'_\beta, z)$. The symbol $V_{B\alpha}(z)$ denotes an operator mapping from $\mathfrak{H}$ to $\mathfrak{H}_B$ with the kernel

$$V_{B\alpha}(p_\beta, P', z) =$$
$$= J_{B\alpha}(p_\beta, P', z) + \sum_j \frac{H_{AB}(p_\beta, p'_\alpha, z) \varphi_A^*(k'_\alpha)}{E_A(p'_\alpha) - z}$$

and $\tilde{R}_B(z)$ denotes the resolvent of the energy operator in channel B

$$\tilde{R}_B(z) = (\tilde{H}_B - z)^{-1}.$$

First we prove the validity of the relations

$$T_\alpha(z) R_0(z) L_A = -F_A \tilde{R}_A(z) \tag{3.67}$$

where Φ_A denotes an operator mapping from $\mathfrak{H}_A$ to $\mathfrak{H}$ defined by the kernel

$$\Phi_A(P, p'_\alpha) = \varphi_A(k_\alpha) \delta(p_\alpha - p'_\alpha).$$

Equation (2.14) written in terms of the operator $T(z)$ reads

$$R_0(z)L_A - R_0(z)T_\alpha(z)R_0(z)L_A = L_A\tilde{R}_A(z).$$

Multiplying this equation by $H_0 - z$ and collecting similar terms, we obtain equation(3.67).

Further, with the help of these relations we show that the iterations $Q^{(n)}_{\alpha\beta}$, whose kernels are of the type $\mathcal{D}_{\alpha\beta}$, satisfy equations (3.65) and (3.66) in which the kernels $T_{\alpha\beta}$ are to be replaced by corresponding components $\tilde{T}^{(n)}_{\alpha B}$ and $\tilde{V}^{(n)}_{B\alpha}$ of the iterations $Q^{(n)}_{\alpha\beta}$. Now, to prove the relation (3.65) for the operators $W_{\alpha\beta}(z)$ we must employ equations (3.29). On one side, the operators

$$N_{\alpha B}(z) = -\sum_{\beta} W^{(n)}_{\alpha\beta}(z)R_0(z)L_B(\tilde{H}_B - z)$$

satisfy the system of equations

$$N_{\alpha B}(z) = \tilde{T}^{(n)}_{\alpha B} - T_\alpha(z)R_0(z)\sum_{\gamma\neq\alpha} N_{\gamma B}(z) \tag{3.68}$$

which differs from the compact equations (3.29) only in the free term. It follows from these properties of the kernels $Q^{(n)}_{\alpha\beta}(z)$ that the kernels of the operators $N_{\alpha B}$ are of the form

$$N_{\alpha B}(P, p'_\beta, z) = \rho_{\alpha B}(P', p'_\beta, z) +$$

$$+\sum_j \frac{\varphi_A(k_\alpha)}{E_A(k_\alpha) - z}\sigma_{AB}(p_\alpha, p'_\beta, z)$$

where the components $\rho_{\alpha B}$ and σ_{AB} are decreasing Hölder functions. On the other hand, it follows from the definition of the operators $T_{\alpha B}$ and from equation (3.48) that the operator $\tilde{W}^{(n)}_{\alpha\beta}(z)$ constructed in terms of $\tilde{T}^{(n)}_{\alpha\beta}$ satisfy the same equations (3.68). Using the theorem of uniqueness of solutions of equation (3.68) in the considered class of functions, we obtain relation (3.65).

It also follows from the statement demonstrated above that the components $T_{\alpha B}$ in terms of whose the wave operators are expressed, satisfy equations (3.68) with $n = 0$. These equations are more convenient for the investigation of properties of the kernels in momentum representation than the equations (3.46) are, because, in the former, the singular factor $(P^2 - E_B(p'_\beta) \mp i0)^{-1}$ has already been subtracted.

We now turn to proving equation (3.66). We note that from definitions of the kernels $J_{\alpha B}$ and H_{AB} it follows that

$$\tilde{R}_B(z)V_{B\alpha}^{(n)} = \tilde{R}_B(z)\tilde{V}_{B\alpha}^{(n)} - \tilde{R}_B(z)\Phi_B^* \sum_{\gamma\neq\beta} W_{\gamma\alpha}^{(n)}(z). \tag{3.69}$$

On the other hand, by multiplying equation (3.48) from the left by the operator $L_B^* R_0(z)$ and by taking relations (3.66) for the kernels $Q_{\alpha\beta}^{(n)}$ and $T_{\alpha\beta}^{(n)}$ into account, we find that the operators $\tilde{N}_{B\alpha} = L_B^*(z)R_0(z)W_{\beta\alpha}^{(n)}(z)$ satisfy the analogous relations

$$\tilde{N}_{B\alpha} = \tilde{R}_B(z)\tilde{V}_{B\alpha}^{(n)} - \tilde{R}_B(z)\Phi_B^* \sum_{\gamma\neq\beta} W_{\gamma\alpha}^{(n)}(z). \tag{3.70}$$

Comparing equations (3.69) and (3.70), we conclude that equations (3.66) are satisfied for the kernels $W_{\alpha\beta}^{(n)}$ and hence also for $W_{\alpha\beta}(z)$.

3.3.4 Kernels of the scattering operator

We express the kernels of the scattering operator in terms of the components of $W_{\alpha\beta}(z)$.

As it was shown in Chapter 2, kernels of matrix elements of the scattering operator S_{AB} coincide with the residua of the resolvent kernel taken at the poles corresponding to the physical processes$B \to A$ under consideration. Since we have already described the polar singularities of the T-matrix kernel, it remains only to use the relations (2.36). The kernels of S_{AB} are then expressed in terms of the components

$$\begin{gathered}
S_{AB}(p_\alpha, p'_\beta) = \delta(p_\alpha - p'_\alpha)\delta_{AB} - \\
-2\pi i\delta(E_A(p_\alpha) - E_B(p'_\beta)H_{AB}(p_\alpha, p'_\beta, E_B(p'_\beta) + i0), \\
S_{0B}(P, p'_\beta) = \\
= 2\pi i\delta(P^2 - E_B(p'_\beta)\sum_\alpha T_{\alpha B}(P, p'_\beta, E_B(p'_\beta) + i0), \\
S_{A0}(p_\alpha, P') = \\
= 2\pi i\delta(E_A(p'_\alpha - P'^2)\sum_\beta V_{A\beta}(p_\alpha, P'^2, P'^2 + i0),
\end{gathered} \tag{3.71}$$

$$S_{00}(P,P') = \delta(P-P')-$$
$$-2\pi i\delta(P^2 - P'^2)\sum_{\alpha,\beta} M_{\alpha\beta}(P,P',P'^2+i0).$$

In particular, it follows from these that although the operators $\sum_\alpha T_{\alpha B}$, $\sum_\beta V_{A\beta}$, and H_{AB} do not coincide with the transition operators K_{0B}, K_{A0}, and K_{AB} defining the kernel of S_{AB} (2.36), the kernels coincide on the energy shell. Note that the kernels K_{0B}, K_{A0} and K_{AB} are smooth bounded functions. On the contrary, the kernel $T(P,P',P'^2+i0)$ contains numerous singularities corresponding to the kernels $T_\alpha(P,P',z)$ and $Q^{(0)}_{\alpha\beta}(P,P',z)$. The following relation holds

$$T(P,P',P'^2+i0) =$$
$$= \sum_\alpha t_\alpha(k_\alpha,k'_\alpha,{k'_\alpha}^2+i0)\delta(p_\alpha - p'_\alpha)+$$
$$+\sum_{\alpha\neq\beta} Q^{(0)}_{\alpha\beta}(P,P',P'^2+i0) + \sum_{\alpha,\beta} W^{(1)}_{\alpha\beta}(P,P',P'^2+i0). \tag{3.72}$$

where the factors T_α and $Q^{(0)}_{\alpha\beta}$ with the strongest singularities, namely, δ-functions and poles have been isolated.

At this point we conclude our discussion of general consequences of integral equations (3.29). The results obtained will be used in Chapter 4 to prove the asymptotic completeness of wave operators.

3.4 Examples

In this section we illustrate the technique of working with the compact equations on two examples.

3.4.1 Scattering on rigid centre

We will consider a system of three particles, two of them are light non-interacting particles and the third particle is infinitely heavy. For definiteness we consider the case $m_3 = \infty$, $v_{12} = 0$. Then, it follows from (1.9) that $k_1 = p_2$ and $k_2 = p_1$. In this case the variables in the Schrödinger equation separate and the corresponding wave functions and wave operators can be written explicitly in two-body terms. Let us suppose for simplicity that the two-body subsystems (13) and (23) possess no bound state. Then only one operator U_0 exists such that its kernel is a product of the two-body wave operators

$$U_0^{(\pm)}(P,P') = u_1^{(\pm)}(k_1,k_1')u_2^{(\pm)}(k_2,k_2'), \tag{3.73}$$

where

$$u_\alpha^{(\pm)}(k_\alpha,k_\alpha') = \delta(k_\alpha - k_\alpha') - \frac{t_\alpha(k_\alpha,k_\alpha',k_\alpha'^2 \pm i0)}{k_\alpha^2 - k_\alpha'^2 \mp i0}. \tag{3.73'}$$

The scattering operator factorizes as well:

$$S_{00}(P,P') = s_1(k_1,k_1')s_2(k_2,k_2'). \tag{3.74}$$

Here the scattering operators of the subsystems are

$$s_\alpha(k_\alpha,k_\alpha') = \delta(k_\alpha - k_\alpha') -$$

$$-2\pi i\delta(k_\alpha^2 - k_\alpha'^2)t_\alpha(k_\alpha,k_\alpha',k_\alpha'^2 \pm i0). \tag{3.74'}$$

Substituting equations (3.73) and (3.74) into equations (3.73') and (3.74') we obtain

$$U_0^{(\pm)}(P,P') =$$

$$= \delta(P-P') - \sum_\alpha \frac{t_\alpha(k_\alpha,k_\alpha',k_\alpha'^2 \pm i0)}{k_\alpha^2 - k_\alpha'^2 \mp i0}\delta(p_\alpha - p_\alpha') +$$

$$+\sum_{\substack{\alpha,\beta \\ \alpha\neq\beta}} \frac{t_\alpha(k_\alpha, k'_\alpha, {k'_\alpha}^2 \pm i0) t_\beta(k_\beta, k'_\beta, {k'_\beta}^2 \pm i0)}{(k_\alpha^2 - {k'_\alpha}^2 \mp i0)(k_\beta^2 - {k'_\beta}^2 \mp i0)} \tag{3.75}$$

and analogous expressions for the scattering operator

$$S_{00}(P, P') = \delta(P - P') -$$

$$-2\pi i \sum_\alpha t_\alpha(k_\alpha, k'_\alpha, {k'_\alpha}^2 \pm i0)\delta(p_\alpha - p'_\alpha)\delta(k_\alpha^2 - {k'_\alpha}^2) -$$

$$-\sum_{\alpha\beta} 4\pi^2 \delta(k_\beta^2 - {k'_\beta}^2)(\delta(P - P') \times$$

$$\times t_\alpha(k_\alpha, k'_\alpha, {k'_\alpha}^2 + i0) t_\beta(k_\beta, k'_\beta, {k'_\beta}^2 + i0) \tag{3.76}$$

where, by virtue of the relations $p_2^2 = k_1^2$ and ${}_1'^2 = {k'_2}^2$, $P^2 = k_1^2 + k_2^2$, $P'^2 = {k'_1}^2 + {k'_2}^2$.

Let us now compare these expressions with that obtained on the basis of the compact equations (3.29). The first three terms can easily be identified with the analogous terms in equations (3.63') and (3.72) which correspond to the kernels $T_\alpha(P, P', z)$. However, the terms quadratic in the T-matrix terms (3.75) and (3.76) differ in form from the analogous terms in (3.63') and (3.72). At first sight, the appearance of the δ-function $\delta(k_\beta^2 - {k'_\beta}^2)$ in the quadratic terms (3.76) seems to contradict the compactness of the integral equations (3.29), since, as we have shown, their iterations have no δ-functions type singularities and consequently, they cannot compensate for the appearing singularity. However, a more detailed analysis shows that the δ-function type singularity appears when the terms $Q_{12}^{(0)}$ and $Q_{12}^{(0)}$ with different order of the subscripts are added.

Indeed, the assumption that the third particle is infinitely heavy simplifies the dependence of the variables in the singular denominator and in arguments of the T-matrix in (3.31). The following relations hold

$$k_1(p_1, p'_2) = p'_2, \quad k'_1 = p'_2, \quad P^2 = k_1^2 + k_2^2,$$

$$k_2(p_1, p'_2) = p_1, \quad k_2 = p_1, \quad P'^2 = {k'_1}^2 + {k'_2}^2. \tag{3.77}$$

Product of the T-matrices in the numerator of the kernel Q_{12} equals that of the kernel Q_{21}, their denominators are complex conjugate and of different sign. Taking into account the identity

$$\frac{1}{k_2^2 - {k'_2}^2 - i0} + \frac{1}{{k'_2}^2 - k_2^2 - i0} = 2\pi i \varkappa_2^2 - {k'_2}^2) \tag{3.78}$$

we find that the sum of these kernels coincides with the last term in (3.76).

Analogous reasoning lead to the conclusion that the quadratic terms in (3.63) and (3.75) are identical on energy shell. At the same time, the smooth terms which remain after subtracting products of two-particle wave functions (3.73') from (3.63) are compensated by a remainder of the perturbation theory series. This is a rather important result allowing us to find the sum of formal polynomial series in the two-particle T-matrices. This result cannot be obtained by any other means.

Note that the singularities of the scattering operator corresponding to the first iteration $Q^{(0)}_{\alpha\beta}$ can be expressed in terms of only the two-particle T-matrices. This result is valid for any masses of the particles.

Let us denote by $T_{\alpha\beta}$ an operator with the kernel

$$T_{\alpha\beta}(P, P') =$$

$$= -\frac{1}{|s_{\alpha\beta}|^3} \frac{t_\alpha(\hat{k}_\alpha, \hat{k}_\alpha(p_\alpha, p'_\alpha), E_\alpha) t_\beta(\hat{k}_\beta, \hat{k}_\beta(p_\beta, p'_\beta), E_\beta)}{E_{\beta\alpha} - E'_\beta - i0} \tag{3.79}$$

where

$$E_{\beta\alpha} = k^2_\beta(p_\alpha, p'_\alpha), \quad E'_\beta = {k'_\beta}^2, \quad E_\alpha = k^2_\alpha,$$

$$k^2_\alpha + p^2_\alpha = {k'_\beta}^2 + {p'_\beta}^2 = E$$

and by $t_\alpha(\hat{k}, \hat{k}', E)$ we denoted kernel of the two-particle T-matrix on energy shell:

$$t_\alpha(\hat{k}, \hat{k}', E) = t_\alpha(E^{1/2}\hat{k}, E^{1/2}\hat{k}', E + i0).$$

It is easily to see that the numerator in (3.31) equals that in (3.79) on energy shell $k^2_\alpha + p^2_\alpha = {k'_\beta}^2 + {p'_\beta}^2 = z$ if the denominators, which are identical in both cases, vanish. Consequently, the kernel of the difference $Q^{(0)}_{\alpha\beta} - T_{\alpha\beta}$ has no singularities on energy shell and the operator $T_{\alpha\beta}$ contains all three-particle singularities of $Q^{(0)}_{\alpha\beta}$. Physically, this operator describes processes of subsequent two-particle rescattering of the particles from the subsystems β and α.

3.4.2 Discrete spectrum in neighbourhood of zero

Here, we will discuss the character of discrete spectrum of the energy operator of three-particle system in the case when the two-particle subsystems possess virtual levels at zero energy, i.e., in case when equation (3.4) has a nontrivial solution for $z = 0$. It is also assumed that the negative discrete spectrum of these operators is absent.

It is possible to show that under the above stated conditions, the kernels of the two-particle T-matrices are singular at $z = 0$. The following representation holds

$$t(k, k', z) = \frac{\varphi(k)\varphi^*(k')}{i\sqrt{z}} + \tilde{t}(k, k', z) \tag{3.80}$$

where $\tilde{t}$ and $\varphi(k)$ are smooth functions decreasing at infinity. It is assumed that the particles are identical and we set $m = 1$. In this case $|\varphi(0)|^2 = (2\pi^2)^{-1}$.

The second term in (3.80) represents the Fredholm contribution to the three-particle equation. Let us now consider the contributions of the first term in more details. Equation (3.28) with this term reduces to a one dimensional equation. More precisely, we will look for its solution $\Phi_\alpha(P)$ in the form

$$\Phi_\alpha(P) = \frac{\varphi(k)\varphi^*(k')}{\sqrt{z-p^2}} F(p) + \tilde{\Phi}_\alpha(P).$$

The first term is leading for z in vicinity of zero. $F(p)$ satisfies the equation

$$F(p) = f_0(p) + \int \frac{\varphi(k_1(p, p_2''))\varphi^*(k_2(p_2'', p))}{z - k_2^2(p, p_2'') - p_2''^2} \frac{F(p_2'')}{\sqrt{z - p_2''^2}} dp_2''.$$

This equation can be simplified in vicinity of zero by noticing that the contributions to the singular part result from values of $\varphi(p)$ at $p = 0$. As a result, for $p^2 \leq p_0$ and for small negative z we have

$$F(p) = f_0(p) + \frac{1}{2\pi 62} \int_{|q|<p_0} \frac{F(q)dq}{\sqrt{\lambda + \frac{3}{4}q^2}(\lambda + p^2 + (p, q) + q^2)},$$

$$\lambda = -\frac{4}{3}z.$$

This equation can be significantly simplified by separating angular variables. For example, the equation for the spherically symmetric solution reads

$$F_0(p) = f_0(p) + \frac{2}{\pi^3\sqrt{3}} \int_0^{p_0/\lambda} \frac{F_0(q)dq}{\sqrt{1 + \frac{3}{4}q62}} \log \frac{1 + p^2 + pq + q^2}{1 + p^2 - pq + q^2}. \tag{3.81}$$

In this equation we changed the integration variables as follows: $q \to \sqrt{\lambda}q$, $p \to \sqrt{\lambda}p$. For $z \to 0$ the integration range becomes infinite. Since the kernel decreases slowly, the corresponding operator in not of the Fredholm type. To discuss the character of the spectrum at the boundary, we make additional simplifications and keep only the leading terms at large values of p and q. We obtain the following integral operator

$$K_\Lambda h(p) = \int_0^{p_0/\lambda} \log\frac{p^2+pq+q^2}{p^2-pq+q^2} h(q)\frac{dq}{q}.$$

This operator has a discrete spectrum for finite λ and for $\lambda \to \infty$ (i.e., $z \to 0$) the eigenvalues λ_n behave asymptotically as

$$K_n = \tilde{K}\left(\frac{\pi}{\Lambda}n\right), \quad n = 0, 1, 2, \ldots$$

where $\Lambda = \frac{1}{2}\log\lambda$ and

$$\tilde{K}(\omega) = \int_0^\infty p^{i\omega-1}\log\frac{1+p+p^2}{1-p+p^2}dp = -\frac{2\pi}{\omega}\frac{\sinh\frac{\pi\omega}{2}}{\sinh\frac{\pi\omega}{3}}.$$

Thus, solutions of equation (3.81) become infinite for the values of z for which the equation

$$1 + \frac{4}{3\pi^3}\tilde{K}\left(\frac{\pi k}{\Lambda}\right) = 0$$

has a solution. We are interested in large positive ω. In this case, the equation reduces to

$$\frac{\omega}{P}2\pi = \frac{4}{3\pi^3}e^{\pi\omega/6}$$

and has a unique solution. We denote this solution by ω_0. Then for $\Lambda = \pi n/\omega_0$, the solution of equation (3.18) $F(p)$ becomes infinite. Using the relation between Λ and λ, we find that the eigenvalues λ_n have the following asymptotic distribution

$$\lambda_n = \lambda_0 \exp\{-2\pi n/\omega_0\} \tag{3.82}$$

where λ_0 contains all undefined constants that appeared in the course of the calculations. Equation(3.82) is the main result of our discussion. It is possible to prove that our approximations do not change the final result and the asymptotics of the discrete spectrum for $\lambda \to 0$ is indeed described by this expression. We note that for the above effect to appear, the virtual levels must exist for at last two pairs of particles.

Therefore, if two-particle subsystems are free of bound states but possess virtual states, then an infinite number of bound states appears in the spectrum of energy operator of the three-particle system near the point $z=0$. The physical origin of this effect lies in the fact that the loosely bounded two-particle subsystems create a slowly decreasing attractive interaction. The exponential series of the eigenvalues (3.82) is a consequence of this interaction.

3.5 Compact Integral Equations for N-particle Systems

3.5.1 Difficulty of the problem

Equations of the type (3.28) can be written also for N-particle systems. However, such equations are not convenient for treatment of N-particle systems at $N \geq 4$ for the same reason as equations of perturbation theory (3.23). We will discuss this point in detail.

According to the rule described in Section 2, the T-matrix $T(z)$ must be split into components

$$M_{a_{N-1}b_{N-1}} = V_{a_{N-1}}\delta(a_{N-1}, b_{N-1}) - V_{a_{N-1}}R(z)V_{b_{N-1}} \tag{3.83}$$

Here, instead of the subscripts α and β denoting $N(N-1)/2$ pairs of the particles 12, 13, ..., $1N$, ..., $(N-1)/N$, we use symbols a_{N-1} and b_{N-1} corresponding to their partition. The operator $T(z)$ can be expressed in terms of the components $M_{a_{N-1}b_{N-1}}(z)$ by the equation

$$T(z) = \sum_{a_{N-1}b_{N-1}} M_{a_{N-1}b_{N-1}}(z), \tag{3.84}$$

where the summation runs over all $N(N-1)/2$ partitions a_{N-1} and b_{N-1}. Using the same reasoning as in the three-particle case, we can show that the components $M_{a_{N-1}b_{N-1}}(z)$ satisfy equations of the type (3.28):

$$M_{a_{N-1}b_{N-1}}(z) =$$

$$= \delta(a_{N-1}, b_{N-1})T_{a_{N-1}} - T_{a_{N-1}}R_0 \sum_{a_{N-1}\neq b_{N-1}} M_{a_{N-1}b_{N-1}}(z). \tag{3.85}$$

Here and in what follows, we often omit the variable z in notation of the operators. The operator $T_{a_{N-1}}$ corresponds to the N-particle system in which all potentials except $V_{a_{N-1}}$ are equal to zero. The kernel of this operator can be explicitly expressed in terms of two-particle T-matrix as

$$T_{a_{N-1}}(P, P', z) =$$

$$= t_{a_{N-1}}(k_{a_{N-1}}, k'_{a_{N-1}}, z - p^2_{a_{N-1}})\delta(p_{a_{N-1}} - p'_{a_{N-1}}).$$

The kernels of integral equations (3.35) contain, similarly to that of equation (3.28) δ-functions $\delta(p_{a_{N-1}} - p'_{a_{N-1}})$. These δ-functions, as we have seen above, disappear in the three-body problem after one iteration and this fact makes equations (3.28) compact. For $N \geq 4$, however, this property is lost. Indeed, in the course of iteration of equation(3.35), terms of the type

$$T_{a_{N-1}} R_0 T_{b_{N-1}} R_0 T_{a_{N-1}} R_0 \ldots R_0 T_{b_{N-1}}$$

do appear. In equation above, the pairs α and β corresponding to the subscripts a_{N-1} and b_{N-1}, share one particle. These iterations describe rescattering of particles of the pair α and β constituting the three-particle subsystem. Their kernels contain δ-functions depending on momenta of the remaining $N-3$ particles not entering the subsystem. This situation can be exemplified by diagrams of a five-particle system (see Figure 9). Here the disconnected lines pertain to the non-interacting particles 1 and 2. It is possible to prove, by continuing the analysis of equation (3.35) that the δ-functions appear in all iterations which follow from multiplication of the two-particle T-matrices pertaining to the same subsystem of an arbitrary number of particles. These δ-functions disappear, for example, when the product

$$T_{a^{(1)}_{N-1}} R_0 T_{a^{(2)}_{N-1}} R_0 \ldots R_0 T_{b_{N-1}}$$

contains no identical terms, i.e., if $a^{(i)}_{N-1} \neq a^{(k)}_{N-1}$, $i, k = 1, 2, \ldots, N-2$ and when each partition of the full chain

$$\{a_{N-1}, a_{N-2}, \ldots, a_2\}$$

is represented by at least one of the operators $T_{a^{(i)}_{N-1}}$, and the last operator$T_{b_{N-1}}$ corresponds to the partition b_2 different from a_2. Such a situation is illustrated on Figure 10.

Thus, to obtain equations with connected kernels, the system (3.35) must be rearranged. To do that, it is necessary to incorporate as many features of the three-particle compact equations (3.28) as possible. Namely, the rearranged equations must be compact and the corresponding homogeneous equations must be equivalent to the Schrödinger equation in the space of square integrable functions.

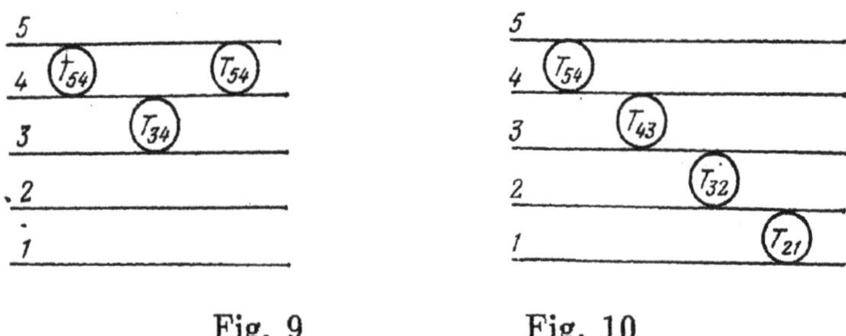

Fig. 9 Fig. 10

In the case of an N-particle system, we are faced with the problem of choosing the optimal rearrangement of equation (3.35), the problem caused by the ambiguity in subsequent partition of operators $T(z)$ or $M_{a_{N-1}b_{N-1}}$ into components. It is possible, for example, to split $T(z)$ into N components with respect to the number of particles, into $\frac{1}{2}N(N-1)$ components with respect to the number of two-particle potentials, and into $\frac{1}{6}N(N-1)N-2)$ components with respect to the number of three-particle subsystems, etc. In the three-particle case, only the first two ways are possible and the number of components with respect to the number of particles or two-particle potentials coincides being equal to $N = \frac{1}{2}N(N-1) = 3$. For $N \geq 4$, however, the number of equations obtained by different possibilities of inclusion of the potentials differs from the number of particles. As a result, we can introduce various numbers of components on the basis of any idea and obtain equations with compact kernels. It is possible to derive equations with compact kernels even without partitioning the T-matrix into components by separating the non-connected part from the T-matrix or from the resolvent. Such equations, however, have a serious defect: it is impossible to prove that their homogeneous equations are equivalent to the Schrödinger equation. They can have, in principle, spurious solutions not satisfying the Schrödinger equation.

Such solutions arise as a result of factorization of operators in integral equations. To illustrate their appearance, we consider the simplest type of equations resulting from separation of the non-connected part.

In the case of a three-particle system, we can start with equation (2.17) written in the form

$$R(z) - R_0(z) =$$

$$= \sum_{\alpha}(R_\alpha(z) - R_0(z)) + \sum_{\alpha\neq\beta} R_\alpha(z)V_\alpha R_0(z)V_\beta R(z) \qquad (3.86)$$

The kernel of this equation has the same structure as the kernel of the operator $Q^{(0)}_{\alpha\beta}$ discussed in Section 2. Hence, this kernel is compact.

Note that by virtue of the identity

$$I - \sum_{\substack{\alpha,\beta \\ \alpha\neq\beta}} R_\alpha(z)V_\alpha R_0(z)V_\beta = \left(I - \sum_{\alpha}\rho_\alpha V_\alpha\right)\left(I + \sum_{\beta} R_0 V_\beta\right)$$

the homogeneous part of equation (3.86) can be factorized as

$$\left(I - \sum_{\alpha}\rho_\alpha V_\alpha\right)\left(I + \sum_{\beta} R_0 V_\beta\right)\Psi = 0. \qquad (3.87)$$

Evidently, there are two types of solutions of this equation. The first, which is of interest, satisfies the Schrödinger equation resulting from the multiplication of the identity

$$\left(I + \sum_{\beta} R_0 V_\beta\right)\Psi = 0 \qquad (3.88)$$

from the left by the operator H_0. The second solution corresponds to zeros of the term $I - \sum_\alpha \rho_\alpha V_\alpha$

$$\left(I - \sum_{\alpha}\rho_\alpha V_\alpha\right)\chi = 0$$

and, in general, does not satisfy equation (3.88). The N-particle equations, in which the non-connected part has been subtracted,

$$R(z) = [R(z)]_{nc} + \sum_{a_2}[R_{a_2}]^{(a_2)}_{nc}\bar{V}_{a_2}R(z) \qquad (3.89)$$

have similar defects. These equations can be factorized as well.

3.5.2 Compact equations for four-particle systems

Analysing the difficulties discussed above, we found that to get rid of the spurious solutions, specification of the components of T-matrix must presuppose subsequent inclusion of two-particle interactions corresponding ot the scattering processes of the simplest two-particle subsystems as well as of more complex configurations comprising various partitions. Below, we demonstrate derivation of compact equations for the components in the four-particle case.

We start with rearrangement of the set (3.85) which, in the present case, consists of six equations for each fixed b_{N-1}. The components corresponding to chains of partitions arise from rewriting the operators

$$W_{a_{N-1}b_{N-1}} = M_{a_{N-1}b_{N-1}} - \delta(a_{N-1}, b_{N-1})T_{a_{N-1}}$$

as a sum of few terms corresponding to all possible ways of associating the interactions to the pair defining the partition a_{N-1}. This process corresponds to transition to the subsequent horizontal level of the "tree" diagram on Figures 2 and 3. The operator W_{12}, for example, is written as the sum

$$W_{12} = M_{12,123} + M_{12,124} + M_{12,34} \tag{3.90}$$

where the subscript b_{N-1} was omitted. Terms of this sum are defined to be

$$\begin{aligned} M_{12,123} &= -T_{12}R_0(T_{12} + T_{23}) + T_{12}R_0(W_{12} + W_{23}), \\ M_{12,124} &= -T_{12}R_0(T_{14} + T_{24}) + T_{12}R_0(W_{14} + W_{24}), \\ M_{12,34} &= -T_{12}R_0T_{34} + T_{12}R_0W_{34} \end{aligned} \tag{3.91}$$

As we have already stated, there exist 18 chains of the partition and, hence, we have the same number of components. Representations analogous to equation (3.90) and (3.91) hold also for the remaining 15 components. Note that the components $M_{12,123}$ and $M_{12,124}$ correspond to the switching on the interaction which leads to appearance of three-particle subsystems (123) and (124). The component $M_{12,34}$ corresponds to formation of two two-particle subsystems of particles 12 and 34.

The set of 18 equations for the components follows after insertion of equation (3.90) into the right hand side of equation(3.91). However, this is not the final set of equations. To obtain compact equations, the diagonal part of these equations must be inverted. Evidently, these diagonal terms create the δ singularities discussed above. The result of inversion can be described in terms of a solution of the three-particle problem.

In order to make inversion, we must collect all the components corresponding to the same subsystems. There are four groups of components for three-particle subsystems and three groups for two-particle subsystems. For example, components $M_{12,123}$, $M_{13,123}$, $M_{23,123}$ belong to the group of the first type, whereas the components $M_{12,34}$, $M_{34,12}$ to the group of the second type.

Let us now discuss how inversion is carried out in these groups.

The structure of the diagonal block of the first type is as follows.

$$\begin{aligned} M_{12,123} &= \ldots - T_{12}R_0(M_{13,123} + M_{23,123} + \ldots) \ldots, \\ M_{13,123} &= \ldots - T_{13}R_0(M_{13,123} + M_{23,123} + \ldots) \ldots, \\ M_{23,123} &= \ldots - T_{23}R_0(M_{13,123} + M_{23,123} + \ldots) \ldots, \end{aligned} \tag{3.92}$$

and that of the second type,

$$\begin{aligned} M_{12,34} &= \ldots - T_{12}R_0 M_{34,12} - \ldots, \\ M_{34,12} &= \ldots - T_{12}R_0 M_{12,34} - ||, \ldots. \end{aligned} \tag{3.93}$$

The non-diagonal and non-homogeneous terms are omitted here.

Now, we write our equations in the form of an operator equation

$$M = M_0 + AM.$$

The "matrix operator" A, defined by a 18×18 matrix, can be decomposed into two terms

$$A = A_0 + \tilde{A}$$

with A_0 being a quasidiagonal matrix consisting of 2×2 and 3×3 blocks. A typical block of first type which has already been discussed in the three-particle case reads

$$C_{123} = \begin{pmatrix} 0 & T_{23}R_0 & T_{23}R_0 \\ T_{13}R_0 & 0 & T_{13}R_0 \\ T_{12}R_0 & T_{12}R_0 & 0 \end{pmatrix}$$

whereas the blocks of second type

$$C_{12,34} = \begin{pmatrix} 0 & T_{12}R_0 \\ T_{34}R_0 & 0 \end{pmatrix}$$

correspond to the case of two non-interacting pairs of particles. The matrix $(I - A_0)^{-1}$ can be expressed in terms of matrices $(I - C_{123})^{-1}$ and $(I - C_{12,34})^{-1}$ and analogous matrices comprising the remaining 2×2 and 3×3 blocks. Matrices of the type $M^{(123)} = (I - C_{123})^{-1}$, the elements of whose we denote by $M^{(123)}_{\alpha\beta}$ can be explicitly expressed through solutions of the set of three-particle equations

$$\begin{pmatrix} M^{(123)}_{23,\beta} \\ M^{(123)}_{31,\beta} \\ M^{(123)}_{12,\beta} \end{pmatrix} = \begin{pmatrix} T_{23} & 0 & 0 \\ 0 & T_{31} & 0 \\ 0 & 0 & T_{12} \end{pmatrix} - \begin{pmatrix} 0 & T_{23}R_0 & T_{23}R_0 \\ T_{13}R_0 & 0 & T_{13}R_0 \\ T_{12}R_0 & T_{12}R_0 & 0 \end{pmatrix} \begin{pmatrix} M^{(123)}_{23,\beta} \\ M^{(123)}_{31,\beta} \\ M^{(123)}_{12,\beta} \end{pmatrix},$$

$$\beta = 23, 31, 12$$

and matrices $M^{(12,34)} = (I - C_{12,34})^{-1}$ through solutions of

$$\begin{pmatrix} M^{(12,34)}_{12,\beta} \\ M^{(12,34)}_{34,\beta} \end{pmatrix} = \begin{pmatrix} T_{12} & 0 \\ 0 & T_{34} \end{pmatrix} - \begin{pmatrix} 0 & T_{12}R_0 \\ T_{34}R_0 & 0 \end{pmatrix} \begin{pmatrix} M^{(12,34)}_{12,\beta} \\ M^{(12,34)}_{34,\beta} \end{pmatrix}$$

$$\beta = 12, 34.$$

The solution of the last equation is given explicitly in terms of two-particle T-matrices T_{12} and T_{34}.

As a result of inversion of the diagonal terms described above we obtain a set of compact equations. Kernels and non-homogeneous terms of these equations are explicitly given in terms of the operators $M^{(a_k)}_{A_2B_2}$. Let us write down two typical equations in details

$$\begin{aligned} M_{123,12} = {} & M^{(123)}_{12,12} + M^{(123)}_{12,13} + M^{(123)}_{12,23} - T_{12}\delta(12, b_{N-1}) - \\ & - M^{(123)}_{12,12} R_0 (M_{234,23} + M_{134,13} + M_{14,23} + M_{23,13}) - \\ & - M^{(123)}_{12,13} R_0 (M_{124,12} + M_{234,23} + M_{34,13} + M_{24,23}) - \\ & - M^{(123)}_{12,23} R_0 (M_{124,12} + M_{134,13} + M_{34,12} + M_{24,13}) \end{aligned} \tag{3.94}$$

$$M_{34,12} = M^{(12,34)}_{12,12} + M^{(12,34)}_{12,34} - T_{12}\delta(12, b_{N-1}) +$$

$$+ M^{(12,343)}_{12,12} R_0 (M_{134,34} + M_{234,34}) +$$

$$+ M^{(12,34)}_{12,34} R_0 (M_{123,12} + M_{134,12}) \tag{3.94'}$$

The remaining 16 equations are can be obtained from these equations by permutation of indices. These equations represent the desired generalization of the compact equations (3.28) to the case of four particles.

3.5.3 N-particle problem

Let us now formulate compact equations for an arbitrary number N of particles. First we introduce a new notation. We will denote by T_{a_l} the T-matrix of the operator H_{a_l}

$$T_{a_l} = V_{a_l} - V_{a_l} R_{a_l} V_{a_l}. \tag{3.95}$$

Resolvent R_{a_l} is expressed in terms of the operator T_{a_l} as

$$R_{a_l} = R_0 - R_0 T_{a_l} R_0. \tag{3.96}$$

By virtue of equation (3.15), the kernel of the T-matrix T_{a_l} in coordinate representation reads

$$T_{a_l}(P, P', z) = t^{(int)}_{a_l}(k_{a_l}, k'_{a_l}, z - p^2_{a_l}) \delta(p_{a_l} - p'_{a_l})$$

where $t^{(int)}_{a_l}$ denotes the kernel of T-matrix of the operator $h^{(int)}_{a_l}$. By $L_{A_i B_k}$ we denote matrix operators, rows and columns of whose are labelled by chains of partitions $A_i = (a_i, a_{i+1}, \ldots, a_{N-1})$ and $B_k = (b_k, b_{k+1}, \ldots, b_{N-1})$. With any partition a_k we associate components T_{a_k} of the T-matrix

$$M^{(a_k)}_{a_{N-1} b_{N-1}} \equiv M^{(a_k)}_{A_{N-1} B_{N-1}} = \delta(a_{N-1} b_{N-1}) V_{a_{N-1}} - V_{a_{N-1}} R_{a_k} V_{b_{N-1}}. \tag{3.97}$$

For $k = 1$, i.e., when the interaction is completely switched on, these components coincide with the components $M_{a_{N-1} b_{N-1}}$ defined by equations (3.85). For $k \geq 2$, the components $M^{(a_k)}_{A_{N-1} B_{N-1}}$ can be explicitly written in terms of the T-matrices of the subsystems forming the partition a_k. It follows from equation (3.19) that the T-matrix T_{a_k} equals the sum of components $M^{(a_k)}_{A_{N-1}B_{N-1}}$ over all chains A_{N-1}, B_{N-1}:

$$T_{a_k} = \sum_{A_{N-1}B_{N-1}} M^{(a_k)}_{A_{N-1}B_{N-1}}.$$

We define operators $M^{(a_k)}_{A_iB_i}$ for $2 \le i \le N-1$ using recurrent relations

$$M^{(a_k)}_{A_iB_i} = M^{(a_{i+1})}_{A_{i+2}B_{i+2}}\delta(a_{i+1}, b_{i+1})\delta(a_i, b_i) -$$

$$- \sum_{(d_{i+1}\neq d_{i+1})\subset a_i} \sum'_{(D_{i+2}\neq C_{i+2})\subset a_{i+1}} M^{(a_{i+1})}_{A_{i+2}C_{i+2}} R_0 M^{(a_k)}_{D_{i+1}B_{i+1}} \tag{3.98}$$

where the first summation runs over the partitions d_{i+1} following a_i and different from a_{i+1}. The prime above the symbol of the second sum indicates that the summation runs only over linked chains C_{i+2}, D_{i+2} for which the conditions $c_{N-1} \neq d_{N-1}, \ldots c_{i+2} \neq d_{i+2}$, $c_{N-2} \supset d_{N-1}, \ldots c_{i+2} \supset d_{i+3}$ are satisfied. For $i = N-2$, the operators $M^{(a_{N-1})}_{A_NB_N}$ appear on the right hand side and are, by definition, equal to $T_{a_{N-1}}$. If the components $M^{(a_k)}_{A_iB_i}$ are known, then the components $M^{(a_k)}_{A_{i+1}B_{i+1}}$ and the T-matrix T_{a_k} can be computed by using the formula

$$M^{(a_k)}_{A_{i+1}B_{i+1}} =\sim_{a_i} M^{(a_k)}_{A_{i+1}a_i, B_{i+1}b_i} \tag{3.99}$$

where b_i is any partition preceding b_{i+1} and

$$T_{a_k} = \sum_{b_{N-1}} \sum_{A_i} M^{(a_k)}_{A_iB_i}. \tag{3.100}$$

The sum in (3.100) runs over all A_i and b_{N-1}, moreover, for each b_{N-1} the chain B_i, $B_i = (b_i, b_{i+1}, \ldots, b_{N-1})$ must be selected such that $b_{N-2} \supset b_{N-1}$. Derivation of equations (3.99) and (3.100) is based on the obvious relation

$$\sum_{a_{k-1}} \sum_{(d_k\neq a_k)\subset a_{k-1}} = \sum_{d_k \neq a_k} \tag{3.101}$$

which is known as the sum rule.

In this notation, the operators $M^{(a_{k-1})}_{A_kB_k}$ satisfy the equations

$$M^{(a_{k-1})}_{A_kB_k} = M^{(a_k)}_{A_{k+1}B_{k+1}}\delta(a_k, b_k) -$$

$$- \sum_{d_k\neq a_k} \sum'_{(D_{k+1}\neq C_{k+1})\subset a_k} M^{(a_k)}_{A_{k+1}C_{k+1}} R_0 M^{(a_{k-1})}_{D_kB_k}. \tag{3.102}$$

The simplest from the system (3.102) is the set of equations for $M^{(a_{N-2})}_{A_{N-1}B_{N-1}}$ which coincides with equation (3.85). The most detailed partition of the T-matrix is obtained for $k = 2$ and represents the desired generalization of the compact equations

$$M_{A_2B_2} = M^{(a_2)}_{A_3B_3}\delta(a_2, b_2)-$$

$$-\sum_{d_2\neq a_2}\ \sum_{(D_3\neq C_3)\subset a_2}{}' \ M^{(a_2)}_{A_3C_3}R_0M_{D_2B_2}. \tag{3.103}$$

Here in the operators $M_{A_2B_2}$ the subscript a_1 was omitted.

It is possible to show that after $N-2$ iterations, the δ-functions in kernels of these equations disappear. This allows us to prove that for $\operatorname{Im} z \neq 0$, equations (3.103) are compact. The case when z reaches the real axis is more difficult from the technical point of view and there exists no complete proof of compactness for $\operatorname{Im} z = 0$. However, the results obtained so far leave little space to doubt the compactness of these equations even for $\operatorname{Im} z = 0$.

Note, that the derivation of equations (3.103) was based on inversion of diagonal singular terms in the chain of equations of the type (3.102) for $k = N-1, N-2, \ldots, 3$. Above, we described the details of such an inversion for $N = 4$. The general case can be obtained by induction $a_2^{(N-1)} \to a_3^{(N)}$ with the help of very tedious constructions. We will not discuss this procedure here.

It is sometimes convenient to consider instead of equations for the T-matrix, the compact equations for components of the wave operators defined for $A \neq 0$ as

$$U^{(\pm)}_{B_2,A} = \lim_{\epsilon\downarrow 0} \mp i\epsilon R_0(E_A \pm i\epsilon)\times$$

$$\times \sum_{\sigma_{N-1}} M_{B_2A_2}(E_A \pm i\epsilon)R_0(E_A \pm i\epsilon)L_A(E_A) \tag{3.98$'$}$$

and for $A = 0$, as

$$U^{(\pm)}_{B_2,0} = \lim_{\epsilon\downarrow 0} \mp i\epsilon R_0(E_0 \pm i\epsilon)\times$$

$$\times \sum_{\sigma_{N-1}} M_{B_2A_2}(E_0 \pm i\epsilon)R_0(E_0 \pm i\epsilon)L_0(E_0)$$

Equations for these components follow from multiplication of equation (3.103) by operators $\mp i\epsilon R_0(E_A \pm i\epsilon)$ and $R_0(E_A \pm i\epsilon)L_A$ and by subsequent passage to the limit $\epsilon \downarrow 0$.

Further, let us consider the homogeneous part of equation (3.103)

$$\Phi_{A_2} = -\sum_{d_2 \neq a_2} \sum_{(D_3 \neq C_3) \subset a_2}' M^{(a_2)}_{A_3 C_3} R_0 \Phi_{D_2}. \tag{3.104}$$

Performing the operations inverse to those which led us to equations (3.102) or (3.103), we return to the original homogeneous set (3.85). It is important to note that each of the operations carried out in deriving equation (3.105) was invertible so that no spurious solutions have been acquired by multiplication of the inverse operators. We denote by $\Phi_{a_{N-1}}$ the components corresponding to the homogeneous equation (3.85). These components are expressed in terms of the components Φ_{A_2} as

$$\Phi_{a_{N-1}} = \sum_{a_2, a_3, \ldots, a_{N-2}} \Phi_{A_2}.$$

Now, we multiply the homogeneous equation (3.85) by R_0 and denote

$$\Psi_{a_{N-1}} = R_0 \Phi_{a_{N-1}}$$

Note, that from equations (3.95) and (3.96) it follows that

$$T_{a_{N-1}} R_0 = R_{a_{N-1}} V_{a_{N-1}}. \tag{3.105}$$

From this we find that these functions satisfy the set of equations

$$\Psi_{a_{N-1}} = \sum_{b_{N-1} \neq a_{N-1}} R_{a_{N-1}} V_{a_{N-1}} \Psi_{b_{N-1}}. \tag{3.106}$$

Multiplying equation (3.106) by the operator $H_{a_N} - z$ and summing the equations obtained in this way, we find that the function

$$\Psi = R_0 \sum_{A_2} \Phi_{A_2} \tag{3.107}$$

satisfies the Schrödinger equation $(H - z)\Psi = 0$.

To complete our discussion of the set of homogeneous equations (3.104) we must verify square integrability of the functions (3.107). It is possible to proceed along the scheme used in the two- and three-particle cases. Of course, from the technical point of view, the N-particle problem in much more complex. The reasoning has, however,a standard character and will not be described here.

Hence, the set of equations (3.103) possesses the required properties: the Fredholm alternative is applicable and the corresponding homogeneous equation is equivalent to the Schrödinger equation for eigenfunctions.

Let us now discuss properties of solutions of this set of equations.

3.5.4 Singularities of the kernels $M_{A_2B_2}$

The properties of the kernels $M_{A_2B_2}$ can be studied along the same lines as those of the kernels $M_{\alpha\beta}$ in case of three-particle systems. However, we are not going to describe all arguments necessary for the discussion of the set (3.103). Instead, we will collect here the final results only.

The operators $M_{A_2B_2}$ can be represented in the form of the sum

$$M_{A_2B_2} = \sum_{k=1}^{n} Q^{(k)}_{A_2B_2} + M^{(n)}_{A_2B_2}$$

where by $Q^{(k)}_{A_2B_2}$ we denoted iterations of the compact equations (3.103) of the order k. The operators $M^{(n)}_{A_2B_2}$ satisfy a set of equations which is analogous to (3.103) but having a different free terms $Q^{(n+1)}_{A_2B_2}$. The kernels $M_{A_2B_2}$ possess, as the resolvent kernel does, the singularities of two types: the polar and the δ-type ones.

The δ-type singularities appears only in the kernels $Q^{(k)}_{A_2B_2}$ corresponding to the first $N-3$ iterations, i.e., for $k \leq N-3$. These kernels correspond to the disconnected parts of the T-matrix and can be expressed explicitly in terms of components $M^{(a_k)}_{A_iB_i}$ of the T-matrix of the partitions a_k, $k \geq 2$.

The cluster polar singularities of the kernels $M_{A_2B_2}$ arise as a consequence of the representation (2.14) for the resolvents of the subsystems energy operators. These singularities appear in the same form in the kernels of all iterations $Q^{(k)}_{A_2B_2}$ starting with $k = N-2$ and in the kernels $M^{(k)}_{A_2B_2}$ for $k \geq N-2$. To describe these singularities, we introduce some notation.

By $\tilde{M}^{(k)}_{A_{l+1}B_{l+1}}$ with $l \leq k$ we denote components of the T-matrix for the operator $h^{(int)}_{a_k}$. These components are defined by the recurrent relations (3.97) and (3.98) in which the resolvent R_0 should be replaced by the resolvent of internal kinetic energy operator $r^{(int)}_{a_k}(z) = (h^{(int)}_{a_k} - z)^{-1}$.

Further, let A be a detailed partition consisting of k clusters. The components of the form-factor Φ_A of this partition will be denoted by $\Phi^{A_{k+1}}_{A}$. It is possible to express these components by means of the form-factor Φ_A either with the help of the recurrent relations of the type (3.98) or as solutions of the homogeneous equations

$$\Phi_A^{A_{k+1}} = - \sum_{d_{k+1} \neq a_{k+1}} \sum'_{(D_{k+2} \neq C_{k+2}) \subset a_{k+1}} \tilde{M}^{a_{k+1}}_{A_{k+2} C_{k+2}} \Phi_A^{D_{k+1}}.$$

In this notation the cluster polar singularities can be represented as

$$\begin{aligned} Q_{A_2 B_2}(P, P', z) &= F_{A_2 B_2}(P, P', z) + \\ &+ \sum_{i=2}^{N-1} \sum_C \frac{\Phi_C^{A_{i+1}}(k_C)}{E_C(p_C) - z} G_{CB_2}^{a_2, a_3, \ldots, a_{i-1}}(p_C, P', z) + \\ &+ \sum_{j=2}^{N-1} \sum_D \frac{\Phi_D^{b_{N-1}}(k'_D)}{E_D(p'_D) - z} J_{CB_2}^{b_2, b_3, \ldots, b_{j-1}}(P, p'_D, z) + \\ &+ \sum_{i=2}^{N-1} \sum_{j=2}^{N-1} \sum_C \sum_D \frac{\Phi_C^{A_{i+1}}(k_C)}{E_C(p_C) - z} \frac{\Phi_D^{b_{N-1}}(k'_D)}{E_D(p'_D) - z} \times \\ &\times H_{CD}^{a_2, a_3, \ldots, a_{i-1}, b_2, b_3, \ldots, b_{j-1}}(p_C, p'_D, z). \end{aligned} \tag{3.108}$$

Here, for $i(j) = 2, 3, \ldots, N-2$, the summation runs over all detailed partitions $C(D)$ present in the partitions $a_i(b_j)$.

Components F, G, J, and H corresponding to the kernels $Q^{(k)}_{A_2 B_2}$ of not too high order can possess additional polar singularities depending not only on the modulus but also on directions of momentum variables. Such, so called, secondary singularities have been treated in detail in the three-particle case. In the N-particle case, however, the form of these singularities has not been studied in general yet. Also the proof that the secondary singularities disappear after sufficiently large number of iterations, as they did in the case of three-particle subsystems, is still missing. Nevertheless, the hypothesis of disappearing of the secondary singularities seems to be natural. Let $\Pi'_{-\varkappa^2}$ be a complex plane with the cut $[-\varkappa^2, \infty)$, $\varkappa^2 = \max_A \varkappa_A^2$ with neighbourhoods of singular points in equation(3.103) excluded. The hypothesis consist of the following:

Components F, G, J, and H of the kernels $M^{(n)}_{A_2 B_2}$ are, for sufficiently large n, decreasing Hölder functions of their arguments for all z from $\Pi'_{-\varkappa^2}$ up to the real axis.

This statement concludes our discussion of properties of the compact equations and their solutions.

To conclude this section, we stress that compact equations allow for a complete solution of the problem of description of singularities of the resolvent kernel. This resolvent kernel has the form of the sum

$$R(z) = R_0(z) - R_0(z) \sum_{A_2, b_{N-1}} M_{A_2 B_2}(z) R_0(z). \tag{3.109}$$

The disconnect part of the resolvent (2.13) can be expressed explicitly in terms of the operators $Q^{(k)}_{A_2 B_2}$ for $k \leq N-3$ and the cluster polar singularities (2.23) - (2.25) are given by the representation (3.108), respectively.

3.6 Charged Particles

In this section, we will construct compact integral equations for systems of charged particles. We will mainly consider the three-particle case and will describe the ways of modifications of integral equations to the case of N particles only briefly.

As it was shown in Chapter 2, the long-range forces change substantially the asymptotic behaviour of particles. This leads, on one hand, to appearance of new singularities in the kernel of the resolvent $R(z)$ and, on the other hand, it is reflected in the fact that equations (3.4) and (3.28) are not compact anymore and that the spectrum of their kernels acquires a continuous part. This is explained from the point of view of local properties of the kernels by the fact that the kernels of charged particles acquire new singularities which do not disappear in the course of iterations. It is interesting that equations (3.4) and (3.28) are non-compact although their kernels are connected. Here we again face the problem of separating the Coulomb singularities which are not weaker than the δ ones. The existence of such singularities has been discussed firstly time in the part devoted to discussion of the Coulomb S-matrix.

Hence, we again have to find a modification of the integral equations leading to Fredholm equations. In this section, we describe one such modification based on inversion of singular Coulomb contributions in the framework of the integral equation method.

3.6.1 Two-particle system

First we consider a system of two charged particles and illustrate the difficulties which appear in the presence of long-range forces. The results obtained will be used for derivation of modified compact equations for three-particle systems.

In momentum representation the Schrödinger equation for the Coulomb wave function reads

$$p^2 u_c(p,p') + \int v_c(p-q) u_c(q,p')\, dq = p'^2 uc(p,p') \tag{3.110}$$

where $v_c(p-q)$ is the Fourier transform of the Coulomb potential

$$v_c(p-q) = \frac{n}{2\pi^2 |p-q|^2}, \quad n = \sqrt{2\mu_{12}}\gamma q_1 q_2.$$

The function $u_c(p,p')$, is explicitly known

$$u_c(p,p') = -\frac{1}{2\pi^2} \lim_{\epsilon \downarrow 0} \frac{\mathrm{d}}{\mathrm{d}\,\epsilon} \frac{(p^2 - (|P'| + i\epsilon)^2)^{i\eta}}{((p-p')^2 + \epsilon^2)^{1+i\eta'}},$$

$$\eta = \frac{n}{2|p|}, \quad \eta' = \frac{n}{2|p'|} \tag{3.111}$$

satisfies also the following homogeneous equation of perturbation theory

$$u_c(p,p') = -\frac{n}{2\pi^2} \int \frac{u_c(q,p') dq}{|p-q|^2 (p^2 - p'^2 - i0)}. \tag{3.112}$$

At the same time, the kernel of resolvent $r_c(z)$ satisfies the non-homogeneous equation of perturbation theory

$$r_c(p,p',z) = \frac{\delta(p-p')}{p'^2 - z)} - \frac{n}{2\pi^2} \int \frac{r_c(q,p',z) dq}{|p-q|^2 (p^2 - z)}. \tag{3.113}$$

If we take $z = p'^2 + i0$, both equations, the homogeneous as well as the non-homogeneous one, have a solution for the same value of the variable z. Hence, equation (3.113) cannot be of the Fredholm type for real and positive z and has a continuous spectrum. This fact is not of a fundamental importance in the two-particle case because the function u_c is explicitly known. However, it becomes important for a system of three or more particles since in this case, the explicit form of a solution of the Schrödinger equation is not known.

To avoid difficulties arising as a consequence of non-compactness of equations of perturbation theory in the two-particle case, we can employ the screened Coulomb potential

$$v_\mu(r) = n \frac{e^{-\mu r}}{r} \tag{3.114}$$

and then to pass to the limit $\mu \to 0$ in the final formulae. When realizing this procedure, one must take into account that the wave function for the potential (3.104) satisfies the non-homogeneous equation

$$u_\mu(p,p') = \delta(p-p') -$$

$$-\frac{n}{2\pi^2}\frac{1}{p^2-p'^2-i0}\int\frac{u_\mu(q,p')}{|p-q|^2(p^2-p'^2-i0)}dq \qquad (3.115)$$

and does not reach a definite limit when $\mu \to 0$. However, the part of wave function having no limit can be explicitly separated out in the form of the multiplier $Ze^{i\eta'\log\mu}$ which does not depend on p. As a result, the Coulomb wave function u_c can be obtained as a limit of renormalized function

$$u_c(p,p') = \lim_{\mu\downarrow 0} Z^{-1}(\mu,p')u_\mu(p,p').$$

3.6.2 Three charged particles

Let us now return to the three-particle problem. By analogy with two-particle systems, the wave functions of three charged particles satisfy the same homogeneous set of equations (3.28). Thus, this system does not have a unique solution and must be rearranged in accordance with the scheme used to transform equation (3.4) to compact equations (3.28). Namely, singular parts of kernels of equations (3.28) giving rise to continuous spectrum must be found and explicitly inverted.

The simplest way of reconstruction of the set (3.28) consists of introduction of new components $M^{(c)}_{\alpha\beta}$ to the Hamiltonian H_c which contain already all Coulomb potentials $v^{(c)}_\alpha(x_\alpha)$, $\alpha = 1,2,3$, $v^{(c)}_\alpha(x_\alpha) = n_\alpha/|x_\alpha|$, i.e.,

$$H_c = \sum_\alpha V^{(c)}_\alpha.$$

Thus we set

$$M^{(c)}_{\alpha\beta} = -\delta_{\alpha\beta}V^{(c)}_\alpha - V^{(s)}_\alpha R(z)V^{(s)}_\beta$$

where $V^{(s)}_\alpha$ denotes the short-range part of two-particle potentials so that

$$v_\alpha(x) = v^{(s)}_\alpha(x) + \frac{n_\alpha}{|x_\alpha|}.$$

By the same reasoning which led us to equation (3.28), we obtain the modified equations

$$M^{(c)}_{\alpha\beta} = T^{(c)}_\alpha - T^{(c)}_\alpha R_c \sum_{\gamma\neq\alpha} M^{(c)}_{\alpha\gamma} \qquad (3.116)$$

where

$$T^{(c)}_\alpha = V^{(s)}_\alpha R_c V^{(s)}_\alpha.$$

After perestroika (reconstruction), the pure Coulomb operators became singled out from the components $M^{(c)}_{\alpha\beta}$ and are transferred to the kernel $T^{(c)}_{\alpha}$. It is possible to prove that equations (3.116) are compact. Now it remains to study the properties of kernels containing information about the long-range parts of the potentials. However, unlike equation(3.28), the kernels $T^{(c)}_{\alpha}$ are not known explicitly and we must study them with the help of some independent equations. This difficulty remains even if only two particles are charged and it is possible to construct a T-matrix in which the short-range potential is for the charged pair is switched on. However, besides such an operator, we must known the T-matrix for a system in which the short-range potential acts between charged and neutral particles. It is not possible to find such an operator explicitly.

Thus, the set of equations (3.116) cannot help us to overcome the difficulties of the Coulomb problem because to study the properties of the operators $T^{(c)}_{\alpha}$ is as complicated as the original problem itself. Nevertheless, as we will see in Chapter 7, equations (3.116) are in many cases convenient for numerical calculation.

There exists another yet possibility of rearranging the equations. To this end, we consider the Coulomb singularities appearing in equation (3.28) and we will try to invert the corresponding operators without making use of properties of the intermediate Coulomb three-particle problems.

Consider equation (3.68) for components $T_{\alpha B}$ of wave operators U_B, $\beta = 1$ describing scattering of particle 1 by a bound state of particles 2 and 3. Let us assume, for simplicity, that only particles 1 and 2 are charged and that the pair (23) has only one bound state ψ_1.

To determine Coulomb singularities of kernels of equations (3.68), we replace, by analogy with the two-particle system, the Coulomb potential $n_\alpha(|x_\alpha|)^{-1}$ by the screened potential $n_\alpha(|x_\alpha|)^{-1}e^{-\mu|x_\alpha|}$ and then pass to the limit $\mu \downarrow 0$.

With the help of equations of perturbation theory, it can be shown that the T-matrix for the potential v_μ becomes singular in the limit $\mu \downarrow 0$:

$$t^{(\mu)}(p,p',z) = \frac{n}{2\pi^2}\frac{1}{|p-p'|+\mu^2} + \tilde{t}^{(\mu)}(p,p',z). \tag{3.117}$$

We show that it is this singularity which by interacting with the singularity

$(z + \varkappa_1^2)^{-1}$ of two-particle T-matrix for the pair (23), is a source of the main singularity in the integral equation for three-particle wave function at negative energies $p'^2 - \varkappa_1^2 < 0$.

Indeed, consider equations (3.68) with $n = 0$ integrated once

$$N_{\alpha B}(z) = \tilde{T}^{(0)}_{\alpha B}(z) - \tilde{T}^{(1)}_{\alpha B}(z) - \sum_{\gamma_1 \neq \gamma_2} Q^{(0)}_{\alpha\gamma_1}(z) R_0(z) N_{\gamma_2 B}(z).$$

Solutions of these equations for $z = E_B(p'_\beta)$ define the components of the wave operators U_B in equation (3.64).

It can be verified with the help of representation (3.117) for the kernels t_3, that all operators $Q^{(0)}_{\alpha\beta}$, have sufficiently smooth kernels for $z < 0$ and for any $\mu \geq 0$. Moreover, these kernels can acquire weak singularities in passing to the limit $\mu \downarrow 0$ disappearing in the subsequent integration of equations (3.68). Consider the most dangerous operator $Q^{(0)}_{31}$. Its kernel can be represented in the form

$$Q^{(0)}_{31}(P, P', E_1(p') + i0) = Q^{(c)}(P, P') + \tilde{Q}(P, P')$$

where the kernel $Q^{(c)}$, which contains the leading Coulomb singularities, is given by the formula

$$Q^{(c)}(P, P') = \\ = \frac{\tilde{n}}{2\pi^2(|p_1 - p'_1|^2 + \mu^2)} \frac{\varphi_1(k_1(p_3, p'_1))\varphi_1^*(k'_1)}{k_1^2(p'_1, p_3) + \varkappa_1^2)(p_1^2 - p_1{}^2 - i0)}. \tag{3.118}$$

Here $\tilde{n} = n_1(|s_{12}|)^{-1}$. Further, note that at the point $p_1 = p'_1$, where the Coulomb singularity is located,

$$\varphi_1(k_1(p_3, p'_1)) = \varphi_1(k_1), \quad k_1(p_3, p'_1) = k_1$$

so that the singular part of (3.118) takes a simple form

$$Q^{(c)}(P, P')|_{p_1 = p'_1} \equiv Q^{(0)}(P, P') = \\ = \frac{\tilde{n}}{2\pi^2(|p_1 - p'_1|^2 + \mu^2)} \frac{\varphi_1(k_1)\varphi_1^*(k'_1)}{k_1^2(p'_1, p_3) + \varkappa_1^2)(p_1^2 - p_1{}^2 - i0)}. \tag{3.119}$$

The difference $Q^{(c)} - Q^{(0)}$ has now weaker singularities which are getting more smooth by subsequent integration, but the main singularity (3.119) remains. Let us write the set of iterated equations (3.68) in matrix form

$$\omega = \omega_0 + A\omega$$

where the vector function ω is defined by its components $(\omega_1, \omega_2, \omega_3)$ and the matrix operator A by components $A_{\alpha\gamma} = \sum_{\gamma_1}(1 - \delta_{\gamma\gamma_1})Q^{(0)}_{\alpha\gamma_1}$. On the basis of the results presented above we can write the operator A as a sum of two terms

$$A = A_0 + \tilde{A}$$

where the operator A_0 is generated by the singular kernel $Q^{(0)}$ and the operator $\tilde{A}$ contains all smoother terms. Following the scheme used for derivation of compact equations (3.28), we move to the left hand side the term $A_0\omega$ responsible for the appearance of Coulomb singularities in the limit $\mu \downarrow 0$ and then we invert the operator $(I - A_0)$.

Inverting the operator $I - A_0$ is analogous to solving an integral equation with separable kernel and can be carried out explicitly. Let us denote

$$\psi = (I - A_0)^{-1}\omega.$$

The component $\psi^{(0)}$ of the vector function ψ takes the form

$$\psi^{(0)}(P, p_1') = \psi_1(k_1) \int K(p_1, p_1')\varphi_1(k_1')\omega^{(1)}(P')\, d\varkappa_1' dp_1'$$

where the unknown kernel $K(p_1, p_1')$ satisfies the equation

$$K(p, p') = \delta(p - p') - \frac{\tilde{n}}{2\pi^2} \frac{1}{p^2 - p'^2 - i0} \int \frac{K(q, p')dq}{|p - q|^2 + \mu^2}.$$

This equation is equivalent to equation (3.115) of perturbation theory for screened Coulomb potential. Therefore, after renormalization

$$K(p, p') \to Z^{-1}(\pi', \mu)K(p, p')$$

we can pass to the limit $\mu \downarrow 0$ in equation for the vector ω and obtain the equation

$$\omega = \varphi_0 + (I - A_0)^{-1}\tilde{A}\omega, \quad \varphi_0 = (I - A_0)^{-1}\omega_0 \tag{3.120}$$

This equation represents the desired modification of the compact integral equations (3.28) for $\mathrm{Re}\, z < 0$.

Instead of screening Coulomb potential in order to be able to invert the singular Coulomb operators, we can construct the inverse operator directly. This is possible because the solution (3.111) of equation (3.112) is known.

This makes it possible to determine the kernel of the operator $(I - K_c)^{-1}$ explicitly, where K_c is the Coulomb operator from (3.110) with the kernel

$$K_c(p,p') = -\frac{\tilde{n}}{2\pi^2} \frac{1}{p^2 - E - i0} \frac{1}{|p-p'|^2}.$$

Namely,

$$(I - K_c)^{-1} = I - N_c$$

with N_c being an integral operator whose kernel written in terms of the two-particle Coulomb T-matrix reads

$$N_c(p,p') = t_c(p,p',p'^2 + i0)(p^2 - E - i0)^{-1}.$$

and $t_c(p,p',z)$ is defined to be

$$t_c(p,p',z) = \frac{n}{2\pi^2|p-p'|^2} -$$

$$- \frac{i\sqrt{z}n^2}{\pi^2(p^2 - = z)(p'^2 - z)} \int_0^1 dt\, t^{i\eta}(1-ty)^{-1}\left(1 - \frac{t}{y}\right)^{-1} \tag{3.121}$$

where

$$\eta = \frac{n}{2\sqrt{z}} \quad \text{and } x^2 = 1 + \frac{(p^2 - z)(p'^2 - z)}{z(|p-p'|^2}, \quad y = \frac{x+1}{x-1}.$$

To sketch the scheme of the direct inversion of the Coulomb singular operator, we use the above mentioned model with one charged particle. Now we must rearrange the integral equation (3.28) just for pure Coulomb interactions. As in the case $\mu \neq 0$, it is possible to verify that the most singular part of integral kernels is of the form (3.119) with $\mu = 0$. Proceeding along the lines which led us to the derivation of the modified equation (3.120), we obtain a compact integral equations in which the operator $(I - A_0)^{-1}$ is replaced by the operator $I - N_c$ which is known explicitly.

The case of three charged particles with an arbitrary number of bound states in two-particle subsystems can be treated similarly. The most dangerous singularities arise in kernels $Q^{(0)}_{\alpha\beta}$ $(\alpha, \beta = 1,2,3)$ when Coulomb singularities $|p_\beta - p'_\alpha|^{-2}$, contained in the kernel of the T-matrix $t_\alpha(k_\alpha(p_\alpha, p'_\alpha), k'_\alpha, z - p^2_\alpha)$ intersect polar singularities $(E_B(p'_\beta) - E - i0)^{-1}$ of the kernel

$$t_\beta(k_\beta(p_\alpha, p'_\alpha), k'_\beta, p'^2_\beta + i0).$$

Inversion of the operator $(I - A_0)^{-1}$ can be then reduced to solving of a set of independent problems of the type (3.110) or (3.104) and the resulting modified equation can be written in the form (3.120).

Singularities of integral equations (3.28) have a simple physical meaning. Consider, for example, the scattering of particle 1 on a bound state of particles 2 and 3. At large separations between particle 1 and the target, only Coulomb interaction remains which asymptotically has the form

$$\tilde{v}_\alpha^{(c)}(y_\alpha) = \frac{n_{\alpha\alpha}}{|y_\alpha|}, \qquad n_{\alpha\alpha} = \sum_{\beta\neq\alpha} \frac{n_\beta}{|s_{\alpha\beta}|}.$$

On the other hand, the Fourier transform of the potential $\tilde{v}_\alpha^{(c)}$ coincides with the Coulomb singularity $\frac{n_{\alpha\alpha}}{2\pi^2|p_\alpha - p'_\alpha|^2}$ which is to be inverted in integral equations (3.28). Thereby, the singularities (3.119) have a two-particle origin and their inversion reduces to solving two-particle problems with the effective potential $\tilde{v}_\alpha^{(c)}$. Indeed, this circumstances made it possible to construct compact equations at negative energies where all processes are essentially two-particle ones.

Significantly more serious difficulties arise at positive energies where the channel of decay into three free particles opens. In this case a new type of singularities appears related to the crossing of the free-particle resolvent pole $(P^2 - z)^{-1}$ with Coulomb singularities $|p_\beta - p'_\beta|^{-2}$ of the two-particle T-matrices. As a consequence, we face the problem of inverting the singular part of the kernels $Q_{\alpha\beta}$ which does not reduce to any effective two-particle problem. Such direct inversion has not yet been achieved. Nevertheless, elimination of the three-particle Coulomb singularities can be carried out in configuration space with the help of certain auxiliary constructions based on the locality of the Schrödinger equation. Such approach will be discussed in the next Chapter.

Now we briefly describe methods of modifications of integral equations in the N-particle case. We can introduce, as in the three-particle case, a new "unperturbed" Hamiltonian H_c containing all pure Coulomb potentials

$$H_c = H_0 + \sum_{a_{N-1}} V_{a_{N-1}}^{(c)}$$

where

$$v_{a_{N-1}}^{(c)}(x_{a_{N-1}}) = \frac{n_{a_{N-1}}}{|x_{a_{N-1}}|}$$

and define, with respect to this Hamiltonian, new components $M^{(c)}_{A_k B_k}$ of the T-matrix by equations (3.97) - (3.102) in which the operators R_0 are replaced by $R_c(z) = (H_c - z)^{-1}$ and the potentials $V_{a_{N-1}}$ by short-range parts $V^{(s)}_{a_{N-1}}$: $V_{a_{N-1}} = V^{(c)}_{a_{N-1}} + V^{(s)}_{a_{N-1}}$. Such modified equations are compact but suffer from the shortcomings discussed above. Most importantly, the properties of the operators $T^{(c)}_{a_k}$ are not known for the pure Coulomb problem.

For energies below the nearest three-cluster threshold the rearrangement of equations (3.103) can be carried out by inverting singular Coulomb terms having a two-cluster nature. As it was noted previously, the kernels of equations (3.103) become compact after $N-2$ iterations. It is possible to prove that in the presence of long-range forces the most dangerous singularity of these kernels arises by crossing the cluster singularity $(E_A(p_A) - z)^{-1}$ with the effective Coulomb interaction $|p_{a_2} - p'_{a_2}|^{-2}$. The inversion procedure for such singularities is completely analogous to that used in the three-particle problem. The construction of the inverse operator $(I - A_0)^{-1}$ is then reduced to solving a set of two-particle Coulomb problems of the type (3.110).

At this point we conclude description of compact equations in momentum space. The next two chapters will be devoted to the study of properties of resolvent and wave operator kernels in configuration space.

CHAPTER 4

Configuration Space. Neutral Particles

This chapter deals with wave functions and resolvent kernels in configuration space. In particular, we show that these wave functions can be found by solving either the Schrödinger equation or the differential equations for components with some asymptotics boundary conditions. The main advantage of this so-called differential formalism is that it is very convenient for effective computational methods based on boundary value problems for wave functions.

4.1 Two-particle System

Following the plan outlined in the previous chapter, we will deal with the two-particle scattering problem first. The considerations of this section will serve as a model for the few-body problem.

4.1.1 Wave functions

Let us consider the Schrödinger equation for system of two particles

$$(-\Delta + v(x) - k^2)\psi(x, k) = 0.$$

In order to determine the wave functions, we must add to this equation some asymptotics boundary conditions. These can be determined from the asymptotics of kernels of the Fourier transform of the wave operators

$$\psi^{(\pm)}(x, k) = \int e^{i(x,k')} u^{(\pm)}(k, k')\, dk' \tag{4.1}$$

for $|x| \to \infty$. In contrast to (1.13), the normalizing factor here is chosen equal to one, as it is customary in physical literature.

For $|x| \to \infty$, the integral (4.1) rapidly oscillates and its asymptotics can be found by the stationary phase method. Since the integrand is singular, the resulting expression is determined not only by the critical points but also by the position of singularities. This sort of integrals will not appear in following sections. The results concerning asymptotics of these integrals will be summarized in the last section of this chapter. Everywhere in this chapter, the asymptotics formulae will be used without detailed proofs.

Let us consider the asymptotics of the function $\psi(x, k)$. We note that the δ-function part of the kernel $u(k.k')$ (3.22) generates a plane wave $e^{i(k,x)}$. Separating this wave out, we write the function $\psi(x, k)$ as a sum

$$\psi^{(\pm)}(x, k) = e^{i(k,x)} + \varphi^{(\pm)}(x, k) \tag{4.2}$$

where the term $\varphi(x, k)$ is given by the singular integral

$$\varphi^{(\pm)}(x, k) = \int \delta, dq e^{i(x,q)} \frac{t(q, k, k^2 \pm 0)}{q^2 - k^2 \mp i0}. \tag{4.3}$$

For x large, the asymptotics of this integral is determined by two factors: the critical point of the angle variables $\hat{q}_0 = \hat{x}$ and the pole singularity $(q^2 - k^2 \mp i0)^{-1}$. As a result, we find a spherical wave:

$$\varphi^{(\pm)}(x, k) = f^{(\pm)}(\hat{x}, k) \frac{e^{(\pm i|k||x|)}}{|x|}. \tag{4.4}$$

The amplitude of this wave is given by the T-matrix on energy shell:

$$f^{(\pm)}(\hat{x}, k) = -2\pi^2 t(\pm|k|\hat{x}, k, k^2 \pm i0).$$

Let us discuss the physical meaning of these expressions. The plane wave which remains after switched off the interaction between particles describes the free motion. This wave corresponds with to two different fluxes of particles. Particles approaching one another are represented by the incoming wave, the particles going from one another are described by the outgoing wave. This is expressed by the asymptotics formula

$$e^{i(k,x)} \sim \frac{2\pi i}{|k|} \left(\delta(-\hat{x}, \hat{k}) \frac{e^{-i|k||x|}}{|x|} - \delta(\hat{x}, \hat{k}) \frac{e^{i|k||x|}}{|x|} \right) \tag{4.5}$$

where the δ-function $\delta(\hat{x}, \hat{k})$ is determined by

$$\int d\hat{k}\delta(\hat{x}, \hat{k})f(\hat{k}) = f(\hat{x}).$$

Above, the first term describes incoming and the second outgoing waves. To prove this relation, we consider an integral over the unit sphere

$$I(x) = \int d\hat{k}e^{i(k,x)}f(\hat{k}).$$

Using the angular coordinates $\xi = (\hat{k}, \hat{x})$, φ, by integration by parts in ξ, we get the asymptotics formula

$$I(x) \sim \frac{2\pi i}{|k|}\left(\frac{e^{-i|k||x|}}{|x|}f(-\hat{x}) - \frac{e^{i|k||x|}}{|x|}f(\hat{x})\right) \tag{4.6}$$

from whose the relation (4.5) follows immediately.

The second term (4.2) contains all information about collisions. The function $\psi^{(+)}$ represents the outgoing spherical wave so that the amplitude $f^{(+)}$ determines the probability density of scattering of a particle in $\hat{x}$ direction . Indeed, the density of the flux of particles through the area dS on a sphere with radius R in the direction $\hat{x}$ is given by

$$j = \frac{1}{(2\pi)^3}\left(\varphi^*\frac{\partial\varphi}{\partial R} - \varphi\frac{\partial\varphi^*}{\partial R}\right)R^2 d\hat{x}$$

or, when the asymptotics formula (4.4) is taken into account

$$j \sim \frac{1}{(2\pi)^3}|f(\hat{x}, k)|^2 d\hat{x}.$$

Let us recall the analogous expression (1.42) which we encountered in Chapter 1. All considerations are quite similar for the incoming waves given by the functions $\varphi^{(-)}(x, k)$.

We will call the functions $\varphi^{(\pm)}(x, k)$ the scattered waves because they describe the result of a collision. Having in mind the meaning of the first term in (4.5), we will call the functions $\exp\{i(k, x)\}$ the incident waves.

We see that solutions of the Schrödinger equation representing wave functions must have the form (4.2) with the functions $\varphi^{(\pm)}(x, k)$ having the form of the spherical wave (4.4). This means that the function describing the scattered wave must solve the inhomogeneous equation with the solution of the asymptotic form (4.4) i.e., it must satisfy the radiation condition. Instead, one can consider a non-homogeneous equation for scattered waves

$$(-\Delta + v(x) - k^2)\varphi(x,k) = -v(x)e^{i(k,x)}. \tag{4.7}$$

A solution of this equation must have the asymptotic form (4.4) i.e. it must satisfy the radiation condition

$$\lim_{R\to\infty} R\left(\frac{\partial\varphi^{(\pm)}}{\partial R} \mp ik\varphi^{(\pm)}\right) = 0.$$

Next, we must prove that the considered boundary value problem has a unique solution. This can be done most conveniently with the aid of the Fredholm integral equations in the configuration space. Let us show now how to formulate the boundary value problem (4.7) in terms of these equations.

We start with the Green formula

$$\int_\Omega (u_1\Delta u_2 - u_2\Delta u_1)d\Omega = \int \pi\Omega\left(u_1\frac{\partial u_2}{\partial n} - u_2\frac{\partial u_1}{\partial n}\right) dS; \tag{4.8}$$

here $\frac{\partial}{\partial n}$ is a derivative in the direction normal to the boundary $\partial\Omega$ of the region Ω. Let $r_0(x,x',z)$ denotes the Green function of the kinetic energy operator:

$$r_0(x,x',z) = \frac{1}{4\pi}\frac{\exp\{i\sqrt{z}|x-x'|\}}{|x-x'|}. \tag{4.9}$$

We put $u_1 = \varphi^{(\pm)}(x',k)$, $u_2 = r_0(x,x',k^2 \pm i0)$ in (4.8) and integrate over a sphere V_R of the radius R. In the limit $R\to\infty$, in the surface integral we can insert the right hand side of the asymptotics forms of the functions $\varphi^{(\pm)}(x',k)$ and $r_0(x,x',k^2\pm i0)$. Let us note that for x fixed and $x'\to\infty$, the relation

$$|x-x'| \sim |x'| - (\hat{x}',x) + O(|x|\,|x'|^{-2}) \tag{4.10}$$

is satisfied, so that we have the asymptotics representation

$$r_0(x,x',k^2\pm i0) \sim \frac{1}{4\pi}e^{\mp i|k|(\hat{x}',x)}\frac{\exp\{\pm i|k|\,|x'|\}}{|x'|} \tag{4.11}$$

and, consequently, the Green function $r_0(x,x',k^2\pm i0)$ asymptotically takes a form of a spherical wave. The function $\varphi^{(\pm)}(x',k)$ looks the same. In the integrand of (4.8), the antisymmetric combination of these functions appears.

Consequently, the corresponding terms of order $O(R^{-1})$ cancel and the rest tends to zero for $R \to \infty$. With the aid of the Schrödinger equation, the integrand on the right hand side of (4.8) transforms to

$$\varphi^{(\pm)}(x',k)\Delta_{x'}r_0(x,x',k^2 \pm i0) - r_0(x,x',k^2 \pm i0)\Delta_{x'}\varphi^{(\pm)}(x',k) =$$

$$= \varphi^{(\pm)}(x',k)\delta(x-x') - r_0(x,x',k^2 \pm i0)v(x')\varphi^{(\pm)}(x',k) -$$

$$-r_0(x,x',k^2 \pm i0)v(x')e^{i(k,x')}.$$

Denoting $\psi^{(\pm)}(x,k) = e^{i(k,x)} + \varphi^{(\pm)}(x,k)$, we get the following integral equation:

$$\psi^{(\pm)}(x,k) =$$

$$= e^{i(k,x)} - \frac{1}{4\pi}\int \frac{\exp\{\pm i|k|\,|x-y|\}}{|x-y|} v(y)\psi^{(\pm)}(y,k)\,dy. \tag{4.12}$$

Thus the problem was reduced to the discussion of the integral operator with the kernel

$$a(x,y,z) = -\frac{1}{4\pi}\frac{\exp\{i\sqrt{z}|x-y|\}}{|x-y|}v(y).$$

It is natural to consider this operator on the set of continuous decreasing functions. Let us assume that these functions satisfy the condition

$$|f(x)| \leq C(1+|x|)^{-\nu}, \quad 0 < \nu < 1.$$

Next we will always consider potentials $v(x)$ decreasing as $|x|^{-3-\epsilon}$, $\epsilon > 0$ for $x \to \infty$. In such case, the asymptotics of the integral on the right hand side of (4.12) is determined by the values of y from some neighbourhood of the origin of coordinates. (The radius of this neighbourhood depends on R). Because of (4.11) this integral has a form of a spherical wave i.e., it decreases as $(1+|x|)^{-1}$. Let us note that the smoothness of this integral regarded as a function of x is completely determined by the kernel $a(x,y,z)$, the properties of $f(y)$ being irrelevant. To be more rigorous, we can assert that the integral operator transforms the set of continuous bounded functions into the set of uniformly bounded and equicontinouous functions. For $\nu < 1$, this makes these functions decrease faster. Consequently, this integral operator is completely continuous.

Thus the Fredholm alternative can be used for equations (4.12). Similarly to the case of equation (3.4), we find that the singular points of this equation coincide with the discrete eigenvalues of the operator h. Moreover, there are no singularities on the positive real axis. Thus equation (4.12) has a unique solution. The same is true for the considered boundary values problems. This concludes our description of the differential formulation of the scattering problem.

4.1.2 Green function

Let us investigate what are the properties of the resolvent kernel in coordinate space. This kernel which is called the Green function satisfies the Schrödinger equation with the δ-like singularity:

$$-D_x + v(x) - z)r(x, x', z) = \delta(x - x'). \tag{4.13}$$

Because of the symmetry

$$r(x, x', z) = \overline{r(x', x, z)} \tag{4.13'}$$

the same equation is satisfied in the variable x'. According to equation (3.2) of perturbation theory the function r satisfies also the integral equation of second kind

$$r(x, x', z) = r_0(x, x', z) - \\ -\frac{1}{4\pi}\int \frac{\exp\{i\sqrt{z}|x-y|}{|x-y|} v(y) r(y, x', z)\, dy, \tag{4.14}$$

Consequently, investigation of the Green function leads to investigation of an integral equation similar to (4.12). The only difference is that the parameter k^2 in (4.12) can now take complex values. Everything what has been said about (4.12) is also valid for (4.14). In particular, the Fredholm alternative can be used for this equation. It follows that for $x \neq x'$, for z from Π_0 and $z \neq -\varkappa_i^2$ the Green function is a smooth bounded function. For $x = x'$, this function has a singularity $(4\pi)^{-1}|x - x'|^{-1}$ concentrated in the free term in (4.14).

Above, it has been shown that wave operators can be determined as the residua of the resolvent kernel in momentum representation $r(p, p', z)$

computed at the poles $(p'^2 - z)^{-1}$. Let us investigate how this relation can be treated in terms of the Green function.

Let us consider equation (4.14) for real positive $z = E \pm i0$, $E > 0$. Taking (4.11) into account, we take the limit $|x| \to \infty$. We obtain the asymptotics relation

$$r(x, x', E \pm i0) \; |x| \overset{\sim}{\to} \infty \; \frac{1}{4\pi} \tilde{\psi}^{(\pm)}(x', k') \frac{\exp\{\pm i\sqrt{E}|x|\}}{|x|} \tag{4.15}$$

where $k' = \mp\sqrt{E}\hat{x}$ and the function $\tilde{\psi}^{(\pm)}$ is given by

$$\tilde{\psi}^{(\pm)}(x', k') =$$

$$= e^{i(k',x')} - \exp\{i(y.k')\}v(y)r(y, x', z)\, dy. \tag{4.16}$$

On the other hand, we can also take the limit $|x'| \to \infty$ in (4.14). With the aid of (4.15) we get a similar asymptotics relation

$$r(x, x', E \pm i0) \; |x'| \overset{\sim}{\to} \infty \; \frac{1}{4\pi} \psi^{(\pm)}(x, k) \frac{\exp\{\pm i\sqrt{E}|x'|\}}{|x'|} \tag{4.17}$$

where $k = \mp\sqrt{E}\hat{x}'$ and the amplitude of this spherical wave satisfies the integral equation (4.12) and, consequently, it is identical with the wave function (4.2). Therefore, the wave functions can be constructed as follows: one calculates asymptotics of the Green function for $x' \to \infty$ and then removes the factor depending on x' which represents a spherical wave. It is clear that this spherical wave results from the presence of the pole $(p'^2 - z)^{-1}$ in the resolvent kernel in momentum representation.

Because of the self-adjointness of the resolvent (4.13) the amplitudes (4.15) and (4.17) satisfy

$$\tilde{\psi}^{(\pm)}(x, k) = \psi^{(\pm)*}(x, -k).$$

Let us also note that because of (4.13), the kernel of T-matrix is subject to the condition

$$t(k, k', z) = \overline{t(k', k, z)}.$$

Comparing the amplitudes of the of scattered waves (4.4) we find that the asymptotics $\psi^{(-)}(x, k)$ can be expressed by means of the asymptotics $\psi^{(-)}(x, k)$ as

$$\psi^{(-)}(x, k) \sim \psi^{(+)*}(x, -k) \tag{4.16'}$$

In the case of spherically symmetric potentials, this relation becomes the equality.

Relations (4.11) and (4.12) give provide an integral representation for the scattering amplitude

$$f(\hat{x},k) = -\frac{1}{4\pi}\int \exp\{-i|k|(\hat{x},y)\}v(y)\psi(y,k)\,dy \tag{4.18}$$

which reduces to (2.37) for $N = 2$.

Note that the asymptotics formulae given above give a simple prescription for evaluation of the kernel of scattering operator. We consider an integral over the unit sphere

$$I(x) = \int d\hat{k}\psi(x,k)f(k) \tag{4.19}$$

where the function $f(k)$ becomes zero in some neighbourhood of the direct ion $\hat{k} = -\hat{x}$,

$$f(k) = 0, \quad |\hat{x} + \hat{k}| \le \epsilon \tag{4.20}$$

From (4.4) and (4.6) we get the asymptotics representation

$$I(x) \sim \frac{2\pi i}{|k|}A(\hat{x},|k|)\frac{\exp\{i|k|\,|x|\}}{|x|} \tag{4.21}$$

where the amplitude of the spherical wave is explicitly given in terms of the scattering operator:

$$A(\hat{x},|k|) = \int s(k',k)f(k)\,dk. \tag{4.22}$$

We finish this section with short description of the asymptotics of the Green function for the energy operator of three-body system with only two particles forming the pair α are interacting $H_\alpha = H_0 + V_\alpha$. We will use these results in the next section. The Green function $R_\alpha(X,X',z)$ is explicitly given in terms of the two-particle Green function:

$$R_\alpha(X,X',z) =$$

$$= \frac{1}{2\pi i}\oint r_\alpha(x_\alpha,x'_\alpha,\xi)\frac{1}{4\pi}\frac{\exp\{i\sqrt{z-\xi}|y_\alpha - y'_\alpha|\}}{|y_\alpha - y'_\alpha|}. \tag{4.23'}$$

Using the identity

$$UI = \sum_A P_A + \left(I - \sum_A P_A\right) \tag{4.23}$$

we can express this function as the sum

$$R_\alpha(X,X',z) = \sum_A \psi_A(x_\alpha)\psi_A^*(x'_\alpha)\frac{\exp\{i\sqrt{E+\varkappa_A^2}|y_\alpha - y'_\alpha|\}}{|y_\alpha - y'_\alpha|} + \tilde{R}_\alpha(X,X',z) \tag{4.24}$$

where the first group of terms corresponds to the operators $P_A R_\alpha$ and the kernel $\tilde{R}_\alpha$ determines the invariant part of the operator $R_\alpha(z)$ in the subspace $((1-\sum_A P_A)\mathfrak{H}$. The asymptotics of $\tilde{R}_\alpha$ can be found by the saddle point method. Let us assume that for example X is fixed and $|x'| \to \infty$. We take a real $z = E + i0$. Then in the critical point ζ_0 determined by

$$\frac{\partial}{\partial \xi}(\sqrt{E-\zeta}|y_\alpha| + \sqrt{\zeta}|x_\alpha|) = 0, \quad \zeta_0 = \frac{|x_\alpha|}{|X|}\sqrt{E}$$

the exponent of the exponential function is $\sqrt{E}|X|$. As the result we get the formula

$$\tilde{R}_\alpha(X,X',E+i0) \sim C_E \psi_\alpha(x_\alpha,k_\alpha)e^{i(y_\alpha,p_\alpha)}\frac{\exp\{i\sqrt{E}|X|\}}{|X|^{5/2}} \tag{4.24'}$$

$$P = \{k_\alpha, p_\alpha\}, \quad P = -\sqrt{E}\hat{X}', \quad C_E = -\frac{1}{2}(2\pi)^{-3/2}E^{3/4}\exp\{i\pi/4\}.$$

Thus, for $|X'| \to \infty$, the function R_α represents a six-dimensional spherical wave. We denote here by C_E the normalizing factor in the asymptotics of the resolvent. The same notation will be used below.

4.2 Coordinate Asymptotics of Wave Functions of Three-body System

In this section, we study the three-body problem in configuration space. We will formulate the boundary-value problems for the wave functions based on the Schrödinger equation and the differential equations for the components. We will proceed following the same scheme as in the case of the two-body problem. First we consider the asymptotics of the Fourier transforms which determine the wave functions. Next, knowing these asymptotics, we formulate the boundary value problems which determine wavefunctions. These questions we discuss in Sections 2 and 3. In Section 4, we describe the properties of the three-body Green function. This section is important as it serves a justification of the differential formulation of the scattering problem.

Below, we will discuss in details only the functions with the label $(+)$; in order to make the formulae more clearer, we will omit this label whatsoever. The functions with label $(-)$ can be studied quite similarly. As in the two-body case, these two classes of function can be related to each other by merely changing the sign of the momentum variable and complex conjugating. Asymptotically, the relation similar to (4.16')

$$\Psi_A^{(-)}(X, p_A) \sim \overline{\Psi_A^{(+)}(X, -p_A)} \tag{4.16''}$$

holds.

4.2.1 Incident and scattered waves

First we introduce convenient notation for the asymptotics wave functions. Let $\rho_\alpha(X, E)$ $(\alpha = 1,2,3)$ and $\sigma_A(y_\alpha, E)$, $A = \{\alpha, i\}$, $\alpha = 1,2,3$, $i = 1,2,\ldots,N_\alpha$ be smooth functions, asymptotically equal to the six-dimensional and three-dimensional spherical waves:

$$\rho_\alpha(X, E) \sim \frac{\exp\{i\sqrt{E}|X|\}}{|X|^{5/2}}(F_0(\hat{X}, E) + O(|X|^{-\nu})), \tag{4.25}$$

$$\sigma_A(y_\alpha, E) \sim \frac{\exp\{i\sqrt{E+\varkappa_A^2}|y_\alpha|\}}{|y_\alpha|}(F_0(\hat{y}_\alpha, E) + O(|y_\alpha|^{-\nu})) \tag{4.26}$$

with $\nu > 0$. Let us recall that we denote by $-\varkappa_A^2$ the eigenvalues of the two-particle energy operators h_α.

We will denote by Φ_α a class of the vector functions $\hat{\Phi}(X, E)$ with components ρ_α and σ_A:

$$\hat{\Phi}(X, E) = \{\rho_\alpha(X, E), \sigma_B(y_\alpha, E)\}$$

$$\alpha = 1, 2, 3; \quad B = (j, \beta), \quad \beta = 1, 2, 3, \; j = 1, 2, \ldots, N_\beta. \tag{4.27}$$

We associate with these components the functions given by the sum

$$f_\alpha(X, E) = \rho_\alpha(X, E) + \sum_j \psi_B(x_\beta)\sigma_B(y_\alpha, E). \tag{4.28}$$

We will call the terms $\psi_B(x_\beta)\sigma_B(y_\alpha, E)$ the cluster spherical waves. We denote by B_E the class of the vector functions $\hat{F}(X, E)$ with components f_α:

$$\hat{F}(X, E) = \{f_1(X, E), f_2(X, E), f_3(X, E)\}.$$

These components has the asymptotics determined by (4.25)-(4.27). By S_E we denote the set of function which asymptotically take the form

$$f(X, E) \sim \sum_\alpha f_\alpha(X, E).. \tag{4.29}$$

We will not specify the class of smoothness of the function, unless it is necessary.

Below, we will consider the classes Φ_E, B_E, and S_E with the elements depending on some additional parameters. These parameters will be indicated in the symbols of these classes.

Now, we turn to description of asymptotics of the wave functions. Let us first consider the function $\Psi_A(X, p_A)$ for $A \neq 0$. We take this function in the following normalization

$$\Psi_A(X, p_A) = \frac{1}{(2\pi)^{3/2}} \int \exp\{i(X, P')\} U_A(P', p_A)\, dP'. \tag{4.30}$$

Let us show that this function can be represented as

$$\Psi_A(X, p_A) \sim \chi_A(X, p_A) + \Phi_A(X, p_A), \tag{4.31}$$

where the function $\Phi_A(X, p_A)$ belongs to the class $S_E(\hat{p}_A)$ with $E = p_A^2 - \varkappa_A^2$. We will also show that the amplitudes of the spherical waves in R^3 and R^6 can be expressed in terms of the T-matrix on energy shell as

$$F_{AB}(\hat{y}_B, p_A) =$$

$$-2\pi^2 H_{AB}\left(\sqrt{E + \varkappa_B^2}\hat{y}_B, p_A, E_A(p_A) + i0\right), \tag{4.32}$$

$$F_{AB}(\hat{X}, p_A) =$$

$$-e^{i\pi/4}(\pi/2)^{1/2}E_A^{3/2}T_{0A}\left(\sqrt{E_A}\hat{X}, p_A, E_A(p_A) + i0\right), \tag{4.32'}$$

Before we proceed to the proof of these formulae, we discuss the physical meaning of terms in the representations (4.31) and (4.28). The first term in (4.31) will be called the incident wave describes an initial state of the system. The second term describes the scattered waves. These terms are used in the same sense as in the two-body case. In (4.28), there is a particular term corresponding to any possible physical process. The cluster spherical waves $\psi_A(x_\alpha)\sigma_A(x_\alpha)$ and $\psi_B(x_\alpha)\sigma_B(x_\alpha)$, $A \neq B$ correspond to the processes of elastic scattering of the bound pair α on the third particle and to recombination of the bounded pairs ($B \to A$), respectively. Three free particles in the final state are described by the six-dimensional spherical wave. If any channel $\mathfrak{H}_A$ is closed, i.e., $E < E_A$, then the corresponding term in (4.29) decreases exponentially as $\exp\{-\sqrt{E - E_A}|y_\alpha|\}$. In particular, the six-dimensional spherical wave for $E < 0$ reduces to $|X|^{-5/2}\exp\{-\sqrt{E}|X|\}$.

The role the different components play depends on the position of the point X in configuration space: they oscillate in some directions and decrease in others. For example, if the particles of a pair α are weakly separated so that the condition

$$|x_\alpha| < a(1 + |y_\alpha|)^\nu, \quad \nu < \frac{1}{2}, \tag{4.33}$$

is satisfied, the cluster spherical waves $\psi_A(x_\alpha)\sigma_A(y_\alpha)$ play the role of leading terms. The corresponding part of the configurations space will be denoted by $\Omega_\alpha(a, \nu)$. The correction term corresponding to six-dimensional spherical wave is square integrable in $\Omega_\alpha(a, \nu)$ and therefore its explicit form is unimportant. We will use this property when formulating the boundary value problems for wave functions.

In the region $\Omega_0(a,\nu)$, where all particles are strongly separated and for all $\alpha = 1,2,3$, conditions opposite to (4.33) are fulfilled, the eigenfunction $\psi_A(x_\alpha)$ are rapidly decreasing and destroy contribution from the cluster spherical waves. The asymptotics $\Psi_A(X,p_A)$ is then determined by a six-dimensional spherical wave. Figure 11 shows the division of the configuration space into these two regions.

We prove now the asymptotics formulae. Inserting equation (3.63) into the integral (4.30), we get representation (4.31) where the function Φ_A has a form of the sum of integrals

$$\Phi_{\beta A}(X,p_A) = -\frac{1}{2\pi)^{3/2}} \int dP' e^{i(X,P')} \frac{T_{\beta A}(P',p_A,E_A(p_A)+i0)}{P'^2 - E_A(p_A) - i0}. \tag{4.34}$$

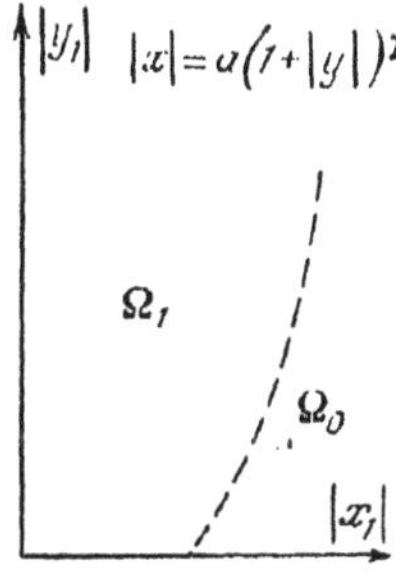

Fig. 11

According to (3.62), the kernels $T_{\beta A}$ are equal to the sum of smooth and singular terms. The latter we transform by means of the identity

$$\varphi_A(k'_A)(P'^2 - E_A(p_A) - i0)^{-1}(E_B(p'_B) - E_A(p_A) - i0)^{-1} =$$

$$-\psi_A(k'_A)\left((P'^2 - E_A(p_A) - i0)^{-1} - (E_B(p'_B) - E_A(p_A) - i0)^{-1}\right)$$

where, on the right hand side, the three-particle and two-particle singularities are separated. As a result, we express the function $\Phi_{\beta A}(X,p_A)$ as a sum

$$\Phi_{\beta A}(X,p_A) = \rho_{\beta A}(X,p_A) + \sum_j \psi_B(x_\beta)\sigma_{BA}(y_\beta,p_A) \tag{4.35}$$

Let us note now that the asymptotics of integrals $\sigma_{BA}(y_\beta, p_A)$ is determined by the two-particle poles $(E_B(p'_B) - E_A(p_A) - i0)^{-1}$ and can be found by means of the formula analogous to (4.4). In this way, we get a spherical wave with the amplitude given by the elements of T-matrix on energy shell which are determined by equation (4.32).

The asymptotics of the six-dimensional integral $\Phi_{\beta A}$ is determined by the pole $(P'^2 - E_A(p_A) - i0)^{-1}$. It has a form of the six-dimensional spherical wave with bounded amplitude (4.32'). Collecting all contributions, we get the desired expression.

4.2.2 The wave functions $\Psi_0(X, P)$

Now we turn to the study of asymptotics of the wave function $\Psi_0(X, P)$. We define this function by the integral (4.30) where we replace the factor $(2\pi)^{-3/2}$ by one. By virtue of equation (3.63), the function $\Psi_0(X, P)$ can be expressed as the sum

$$\Psi_0(X,P) = e^{i(X,P)} + \Phi_0(X,P),$$

$$\Phi_0(X,P) = \sum_\alpha (\Phi_{\alpha 0}(X,P) + W_{\alpha 0}(X,P)) \tag{4.36}$$

where the functions $\Phi_\alpha(X, P)$ and $\Phi_{\alpha 0}(X, P)$ are given by the integrals (4.34) in which the kernels $T_{\alpha A}$ should be replaced with the respective kernels $T_\alpha(P', P, P^2 + i0)$ and $\sum_\alpha W_{\alpha\beta}(P', P, P^2 + i0)$ and insert $E_A = P^2$.

Because of equation (4.3), the functions $\Phi_\alpha(X, P)$ can be expressed in terms of two-particle wave functions $\psi_k(x, k)$ as

$$\Phi_\alpha(X,P) = e^{i(y_\alpha, p_\alpha)} \varphi_\alpha(x_\alpha, k_\alpha). \tag{4.37}$$

If the coordinate x_α is bounded, this function does not decrease. For $|x_\alpha| \to \infty$, the scattered waves $\varphi_\alpha(x_\alpha, k_\alpha)$ can be replaced by the asymptotics (4.4), so that Φ_α decreases like $|x_\alpha|^{-1}$.

Further, let us consider the terms $W_{\alpha 0}(X, P)$. Contrary to the case of the wave functions $\Psi_A(X, P_A)$, the integrand $W_{\alpha 0}(X, P)$ has besides the pole $(P'^2 - P^2 - i0)^{-1}$ the three-particle singularities corresponding to the first iteration of the integral equations (3.28) $Q^{(0)}_{\alpha\beta}$. We separate these terms and express them by means of two-particle wave functions. To this end, to the kernel $Q^{(0)}_{\alpha\beta}(P', P, P^2 + i0)$ we add and subtract the expressions

$$T^{(0)}_{\alpha\beta}(P', P) = t_\alpha \left(k'_\alpha, \sqrt{P^2 - p'^2_\alpha}\hat{k}_\alpha(p'_\alpha, p_\beta), P^2 - {p'_\alpha}^2 + i0\right) \times$$

$$\times \frac{t_\beta(k_\beta(p'_\alpha, p_\beta), k^2_\beta, k^2_\beta + i0)}{k_\beta(p'_\alpha, p_\beta) - k^2_\beta - i0} \frac{1}{|s_{\alpha\beta}|^2}$$

and note that the kernel

$$\Delta Q^{(0)}_{\alpha\beta}(P', P) = Q^{(0)}_{\alpha\beta}(P', P, P^2 + i0) - T^{(0)}_{\alpha\beta}(P', P)$$

does not have three-particle singularities anymore, because the numerator and denominator of $Q^{(0)}_{\alpha\beta}$ vanish simultaneously. If $p'^2_\alpha > P^2$, then the kernel $Q^{(0)}_{\alpha\beta}$ is non-singular as well. It follows that all three-particle singularities are contained in the kernel $T^{(0)}_{\alpha\beta}$ for $p'^2_\alpha \leq P^2$.

We insert the kernel $T^{(0)}_{\alpha\beta}$ into the integral (4.34) and we take as the independent integration variables the pair $\{k'_\alpha, p'_\alpha\}$. Taking the relations (4.3) in account, we obtain the following expression for the corresponding term in the wave function

$$\varphi_{\alpha\beta}(X, P) =$$

$$= \frac{1}{|s_{\alpha\beta}|^2} \int_{p'^{-1}_\alpha \leq E} dp'_\alpha e^{i(y_\alpha, p'_\alpha)} \varphi_\alpha \left(x_\alpha, \hat{k}_\alpha(p'_\alpha, p_\beta)\sqrt{E - {p'_\alpha}^2}\right) \times$$

$$\times \hat{\varphi}_\alpha(k_\beta(p'_\alpha, p_\beta), k_\beta), \quad E = P^{1/2} \tag{4.38}$$

where $\hat{\varphi}_\alpha$ is the Fourier transform of the function $\varphi_\alpha(x, k)$:

$$\hat{\varphi}_\alpha(k, k') = -\frac{t(k, k', k'^2 + i0)}{k^2 - k'^2 - i0}.$$

As a result, we get the wave function $\Psi_0(X, P)$ expressed as the sum

$$\Psi_0(X, P) = \exp\{i(X, P)\} + \sum_\alpha (\tilde{\Phi}^{(s)}_\alpha(X, P) + \tilde{W}_\alpha(X, P)) \tag{4.39}$$

where the term $\tilde{\Phi}^{(s)}_\alpha$ is given in terms of the two-particle wave functions as

$$\tilde{\Phi}^{(s)}_\alpha(X, P) = \Phi_\alpha(X, P) + \sum_{\beta \neq \alpha} \Phi_{\alpha\beta}(X, P).$$

The function $\tilde{W}_\alpha(X, P)$ is given by the integral (4.34) where the integrand has only two types of poles:$(P'^2 - P^2 - i0)^{-1}$ and $(E_A(p'_A) - P^2 - i0)^{-1}$.

Thus the asymptotics of $\tilde{W}_\alpha(X,P)$ can be found by the same scheme as the asymptotics of the integral (4.34) in the case of the functions $\Psi_A(X,p_A)$, $A \neq 0$. We present now the final result.

In almost all directions of the configuration space the term $\tilde{W}_\alpha(X,P)$ has the asymptotics of cluster and six-dimensional waves of the form (4.35). The amplitudes of these waves $F_{A0}^{(\alpha)}(\hat{y}_\alpha, P)$ and $F_{00}^{(\alpha)}(\hat{X}, P)$ are related to the elements of T-matrix on energy shell by the relations

$$F_{A0}^{(\alpha)}(\hat{p}'_\alpha, P) = -2\pi \sum_\beta J_{A\beta}(p'_A, P, E + i0),$$

$$p'^{\,2}_A = E + \varkappa_A^2, \tag{4.40}$$

and

$$F_{00}^{(\alpha)}(P,P') = C_0(E)\left(\sum_\beta M_{\alpha\beta}(P,P',P'^2 + i0) - \right.$$

$$\left. - T_\alpha(P,P',P'^2 + i0) - \sum_{\beta\neq\alpha} \tilde{T}_{\alpha\beta}(\hat{P},p')\right), \quad P^2 = p'^2 = E. \tag{4.41}$$

Here $C_0(E) = -e^{i\pi/4}E^{3/2}\pi(2\pi)^{5/2}$ and the kernel $\tilde{T}_{\alpha\beta}$ is given by

$$\tilde{T}_{\alpha\beta}(\hat{P},P') =$$

$$= -\frac{1}{|s_{\alpha\beta}|^3}\frac{t_\alpha(k'_\alpha, |k'_\alpha|\hat{k}'^2_{\alpha\beta}, k'_{\alpha\beta} + i0)t_\beta(|k_\beta|\hat{k}_{\beta\alpha}, k_\beta, k_\beta^2 + i0)}{k_{\alpha\beta}^2 - k_\beta^2 - i0} \tag{4.42}$$

where $k'_{\alpha\beta} = k_\alpha(p'_\alpha, p_\beta)$, $k_{\beta\alpha} = k_\beta(p'_\alpha, p_\beta)$, $p'_\alpha = |x|^{-1}E^{1/2}y_\beta$.

In correspondence with the interpretation of the spherical waves given above, the cluster spherical waves can be associated with processes of particle capture $(3 \to 2)$, and the six-dimensional waves with "true" three-particle elastic scattering $(3 \to 3)$. The term $\tilde{\Phi}^{(s)} = \sum_\alpha \tilde{\Phi}_\alpha^{(s)}$ generated by the first iterations of equations (3.28) correspond to the processes of single and double pair collisions.

Let us note, however, that the six-dimensional spherical waves do not describe correctly the asymptotics $\tilde{W}_{\alpha 0}(X,P)$ in all directions of the configuration space. We have seen in the last chapter that the connected part of T-matrix on energy shell $\sum_{\alpha\beta} W_{\alpha\beta}(P', P'p^2 + i0)$ has besides poles also weaker singularities which correspond to iterations $Q^{(n)}$ up to the fourth order. Thus the amplitude $F_{00}^{(\alpha)}$ may become infinite if the vector $\hat{X}$ takes some particular directions. The integrals corresponding to these iterations can be transformed by means of a suitable substitution into the one-dimensional rapidly oscillating integrals with power singularities. The asymptotics of these integrals will be constructed in Section 5. However, we will neither describe the tedious substitutions nor analyse the asymptotics in detail, because we do not need the explicit form of these integrals for the problems discussed in this book.

To avoid specifying singularities in asymptotics formulae in every single case, we consider wider classes of functions $\Phi_{B,\epsilon}$, $B_{B,\epsilon}$, and $S_{B,\epsilon}$. In these cases, the six-dimensional waves may have weak singularities in some directions of configuration space and they may be absolutely integrable on the unit sphere $|p| = 1$ with the power ϵ.

Thus, we see that the asymptotic behaviour of the three-body wave functions substantially depends on the type of initial state. Let us suppose that the asymptotics of functions with two clusters in the initial state is given by a simple superposition of spherical and cluster waves corresponding to the final state. However, to describe wave functions with three free particles, we must consider also some more detailed processes of the pair rescattering. The physical reason why these processes take place is that the region in configuration space where the pair collisions may occur is not bounded. On the other hand, the region where the bounded pair collides with a third particle lies just in the neighbourhood of the origin, its size being determined by the range of forces. This fact defines the privileged, from the point of view of asymptotics, behaviour of the two-cluster wave functions in the initial states of the three-body system as well as in the initial states of systems composed by an arbitrary number of particles.

The coordinate asymptotics of the terms describing the processes of pair rescattering will be studied in detail in the next section. Now we formulate the boundary value problem for the wave functions.

4.2.3 Boundary values problems

Knowing asymptotics of the wave functions, we can formulate the boundary value problem based on the Schrödinger equation. As in the case of two-body system, by separating the incoming waves

$$\Psi_A(X, P_A) = \chi_A(X, p_A) + \Phi_A(X, p_A)$$

we get the inhomogeneous Schrödinger equation for scattered waves

$$\left(-\Delta + \sum_\alpha v_\alpha(x_\alpha) - E_A\right)\Phi_A(X, P_A) = -\bar{V}_A\chi_A(X, p_A). \qquad (4.43)$$

As in chapter 2, we denote by $\bar{V}_A$ the part of the interaction missing in the channel A, i.e.

$$\bar{V}_A = \sum_{\beta\neq\alpha} V_\beta \quad \text{for } A = \{\alpha, i\}, \quad \bar{V}_A = \sum_\beta V_\beta \quad \text{for } A = 0.$$

If there are only two clusters in the initial channel, the asymptotic conditions for the corresponding wave functions are determined by giving the sum of the cluster and six-dimensional waves. In the other words, for $A \neq 0$, the solution $\Phi_A(X, p_A)$ will belong to the class $S_{E_A}(\hat{p}_A)$.

For the functions $\Psi_0(X, P)$, the asymptotics conditions are more complicated. In this case, besides the plane wave $\exp\{i(X, P)\}$ we must also separate the slowly decreasing terms corresponding to the single and double collisions. The latter are known explicitly, so that only the "true" scattered three-body waves must be found. These waves must be looked for in the class $S_{E,\epsilon}(\hat{P})$, $\epsilon < 2$. We do not write here rather complicated equations for these functions. All relations necessary for their derivation were given above.

Next, we derive the differential equations fulfilled by components of the wave functions. We write the system (3.59) in coordinate representation and then apply the operator $H_\alpha - z$ to both sides. We get the following system of equations for components of the Green function:

$$(-\Delta_X + v_\alpha(x_\alpha) - z)R_{\alpha\beta}(X, X', z) = -v_\alpha(x_\alpha)R_0(X, X', z)\delta_{\alpha\beta} -$$

$$-v_\alpha(x_\alpha)\sum_{\gamma\neq\alpha} R_{\gamma\beta}(X, X', z). \qquad (4.44)$$

Using definitions (3.60), (3.61) determining components of wave functions we obtain the desired equations

$$(-\Delta + v_\beta(x_\beta) - E_A)\Phi_{\beta A}(X) =$$

$$= -v_\beta(x_\beta)\left(\chi_A^{(\beta)}(X) + \sum_{\gamma\neq\alpha}\Phi_{\gamma\beta}(X)\right). \tag{4.45}$$

For anyγ, the components $\chi_0^{(\beta)}$ are given by

$$\chi_0^{(\beta)} e^{i(X,P)}\delta_{\beta\gamma}$$

and the components $\chi_A^{(\beta)}(X)$ for $A \neq 0$ are

$$\chi_A^{(\beta)}(X, p_A) = \delta_{\alpha\beta}\chi_A(X, p_A).$$

The wave functions are given by the sum

$$\Psi_A(X, p_A) = \chi_A(X, p_A) + \sum_\beta(\Phi_{\beta A}(X, p_A)$$

and they satisfy the Schrödinger equation

$$\left(-\Delta + \sum_\alpha v_\alpha(x_\alpha) - E_A\right)\Psi_A(X, P_A) = 0. \tag{4.43'}$$

The functions $\sum_\beta \Phi_{\beta A}$ satisfies the inhomogeneous equation (4.43).

Now, we formulate the initial value problem based on equations (4.45) for components.

First we consider the processes with two clusters in the initial state. We have seen that the components $\Phi_{\beta A}(X, p_A)$ are asymptotically equal to the sum (4.35) of the cluster spherical waves corresponding to the operator h_α and the six-dimensional spherical waves. Consequently, we must look for the corresponding solutions of the system of equations (4.45) in the class of vector functions $B_{E_A}(\hat{p}_A)$. Let us recall that the element $\hat{\Phi}_A$ of this class is a set of three components $\hat{\Phi}_A - = \{\Phi_{1A}, \Phi_{2A}, \Phi_{3A}\}$ with fixed asymptotics (4.28) at the infinity.

The asymptotic boundary conditions for components of the wave functions $\Psi_0(X, P)$ are analogous to the conditions for two-particle collisions. We must only separate slowly decreasing components describing the rescattering. The remaining part of the components $\tilde{W}_{\alpha 0}$ given by (4.39) is to be looked for in the class $B_{E,\epsilon}(\hat{P})$, $0 < \epsilon < 2$.

We outline the proof of our boundary value problems. We have e.g., the boundary value problem for components of the wave functions Ψ_A for

$A \neq 0$. Let $R_\alpha(X, X', z)$ be the Green functions for the operator H_α (4.23). We use the Green formula (4.8) with $\Omega = V_R$, $u_1 = \Phi_{\beta A}(X', p_A)$, and $u_2 = R_\beta(X, X', E_A + i0)$. We know that the components $\Phi_{\beta A}(X', p_A)$ have the asymptotics of the form (4.35). According to (4.24) and (4.24'), the Green function $R(X, X', E_A + i0)$ has for $|X'| \to \infty$ the same asymptotics. Thus, we can evaluate the asymptotics of the integrals on the right hand side of (4.8) as in the case of two bodies. In the limit $R \to \infty$ these integrals vanish. The integral on the left hand side can be computed with the aid of relations (4.45). As a result, we get for the wave functions the following system of the integral equations:

$$\Phi_{\beta A}(X', p_A) = -\int dX' R_\beta(X, X', E_A + i0) v_\beta(x'_\beta) \times$$

$$\times \left(\sum_{\gamma \neq \beta} \Phi_{\gamma A}(X', p_A) + \chi_A^{(\beta)}(X', p_A) \right).$$

This system corresponds to formal inversion of the operator $H_\beta - E$ in (4.45). We will see in Section 4 that for this system considered in the class of functions with asymptotics (4.28), the Fredholm alternative can be used. It follows that our boundary value problem is correct and has a unique solution. This is again quite analogous to the two-body case.

Thus we have formulated the boundary value problems for the wave functions and their components which form a foundation for an alternative way of determining the wave operators. We mentioned before that the differential formulation is very convenient for practical calculations. Now we consider the boundary value problems from this point of view.

The Schrödinger equations is the simplest differential relation satisfied by wave functions. However, the asymptotics conditions are rather involved in this case. In all the domains Ω_α ($\alpha = 1, 2, 3$) there are both the two-particle potentials different from zero and the slowly decreasing cluster spherical waves. All three pairs of the Jacobi coordinates are necessary for their description and this is very troublesome for the numerical calculations.

These difficulties disappear when the wave functions are decomposed into their components. There is only one potential corresponding to a given pair β in each equation (4.45). The asymptotics of components $\Phi_{\beta A}$ contains only one term generated by the eigenfunction of the energy operator of the two-particle subsystem (and not the sum over all pairs as in the case of the Schrödinger equation). However, the equations for these components possess all advantages of the differential equations. In particular, the input data are the potentials and eigenfunctions and not the T-matrix as it is the case in integral equations. Thus the differential equations for components combine the rigorousness of the compact integral equations which correctly deal with the structure of wave functions with simplicity of form which is the virtue of the Schrödinger equation. We will show in Chapter 7 how these merits of differential equations for components can be used to develop an effective numerical methods of the scattering theory.

4.3 Contribution of Elementary Two-particle Collisions

In this section, we describe the coordinate asymptotics of the functions Φ_α and $\Phi_{\alpha\beta}$ which correspond to single and double two-particle collisions. We will show that these terms have an intermediate position between plane and spherical waves. The term Φ_α is of order $O(|x_\alpha|^{-1})$ and order of the term $\Phi_{\alpha\beta}$ varies from $O(|X|^{-2})$ to $O(|X|^{-5/2})$.

In heuristic considerations we will use the language of the diffraction theory. It should be noted that the geometry as well as the mathematical apparatus of the quantum theory of scattering resembles has a lot in common with the theory of diffraction on semi-transparent wedge. We compare these two problems at the end of this section.

4.3.1 Single and double collisions of classical particles

We will show below that the exponents of rapidly oscillating exponential functions which determine the coordinate asymptotics of the functions Φ_α and $\Phi_{\alpha\beta}$ can be described in terms of action. This is why it is useful to consider first the single and double collisions of classical particles and to express the asymptotic action in terms of kinematical variables.

A single collision of the pair α is pictured schematically on Figure 12. where k_α and p_α denote the relative momenta of particles before the collision and k'_α and p'_α are the relative momenta after collision. Asymptotically, outside the range of forces, the relations $p'_\alpha = p_\alpha$ and $k'_\alpha = |k_\alpha|\hat{x}_\alpha$ are satisfied. The action $\sqrt{E}Z_\alpha$ corresponding to the asymptotic motion of particles can be found by means of these relations to be

$$\sqrt{E}Z_\alpha = |k_\alpha|\,|x_\alpha| + (p_\alpha, y_\alpha).$$

Next, we consider the process of double collision of particles. Let us assume that the particles of the pair β collide first and followed by the collision of particles belonging to the pair α. This process is schematically showed on Figure 13. We denote by $k_{\alpha\beta}$ and $p_{\alpha\beta}$ the relative momenta of particles after collision. Then the corresponding action is

$$\sqrt{E}Z_{\alpha\beta} = |k_{\alpha\beta}|\,|x_\alpha| + (p_{\alpha\beta}, y_\alpha). \tag{4.46}$$

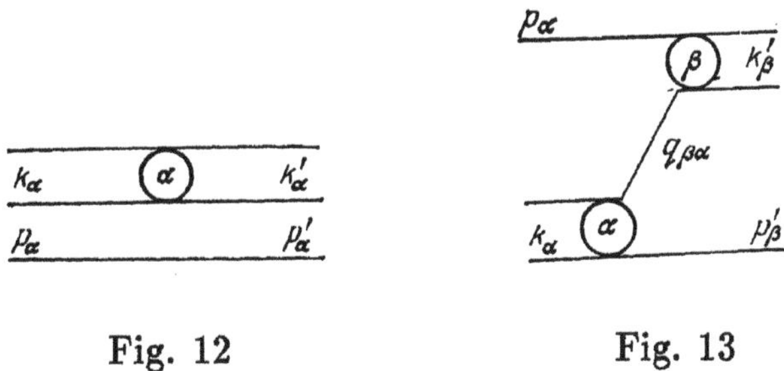

Fig. 12 Fig. 13

The momenta $k_{\alpha\beta}$ and $p_{\alpha\beta}$ can be found from the least action principle and energy and momentum conservation law applied to both successive two-particle collisions. They are given by

$$k_{\alpha\beta} = c_{\alpha\beta} q_{\beta\alpha} + s_{\alpha\beta} p_\beta,$$

$$p_{\alpha\beta} = -s_{\alpha\beta} q_{\beta\alpha} + c_{\alpha\beta} p_\beta,$$

$$k^2_{\alpha\beta} + p^2_{\alpha\beta} = E \tag{4.47}$$

where $E = k^2_\alpha + p^2_\alpha$ is the kinetic energy, and k_α and p_α are the relative momenta before collision. The intermediate relative momentum $q_{\beta\alpha}$ of the pair β is determined by the relations

$$q_{\beta\alpha} = |q|(\cos\theta_{\alpha\beta}\hat{p}_\beta + \cos\theta_{\alpha\beta}\sin\varphi_{\alpha\beta}e^{(2)}_\beta +$$

$$+ \sin\theta_{\alpha\beta}\cos\varphi_{\alpha\beta}e^{(3)}_\beta), \quad q^2 = k^2_\beta \tag{4.47'}$$

where the vectors $\hat{p}_\beta$, $e^{(2)}_\beta$, $e^{(3)}_\beta$ form an orthogonal basis in R^3:

$$e^{(2)}_\beta = -\cot\tilde{\theta}_\beta\hat{p}_\beta + \operatorname{cosec}\tilde{\theta}_\beta\hat{p}_\beta, \quad \cos\tilde{\theta}_\beta = (\hat{k}_\beta, \hat{p}_\beta),$$

$$e_\beta^{(3)} = [\hat{p}_\beta, e_\beta^{(2)}].$$

The angles $\theta_{\alpha\beta}$ and $\varphi_{\alpha\beta}$ are defined by the equations

$$\frac{\partial}{\partial\theta_{\alpha\beta}} Z_{\alpha\beta} = 0, \quad \frac{\partial}{\partial\varphi_{\alpha\beta}} Z_{\alpha\beta} = 0 \tag{4.48}$$

which follow from the least action principle; the function $Z_{\alpha\beta}$ is determined in terms of $\theta_{\alpha\beta}$ and $\varphi_{\alpha\beta}$ by the formulae (4.46) - (4.48).

It follows from these equations that the point $q_{\beta\alpha}$ is situated on the plane spanned by the vectors $\hat{p}_\beta$ and y_α, and $Z_{\alpha\beta}$ is given by

$$\sqrt{E}Z = |x_\alpha|(E_{\alpha\beta} + 2s_{\alpha\beta}c_{\alpha\beta}|k_\beta|\,|p_\beta|\cos\theta_{\alpha\beta})^{1/2} -$$

$$-s_{\alpha\beta}|y_\alpha|\,|k_\beta|\cos(\theta_{\alpha\beta} - \tilde{\theta}_{\alpha\beta}) + c_{\alpha\beta}(p_\beta, y_\alpha)$$

where $E_{\alpha\beta}^2 = c_{\alpha\beta}^2 k_\beta^2 + s_{\alpha\beta}^2 p_\beta^2$ and $\cos\tilde{\theta}_{\alpha\beta} = (\hat{y}_\alpha, \hat{p}_\beta)$. The angle $\theta_{\alpha\beta}$ is defined by equation (4.48), the explicit solution of whose can be found only in few cases. For example, if $\tilde{\theta}_{\alpha\beta} = 0$, then for $\theta_{\alpha\beta}$ we find two solutions: $\theta_{\alpha\beta}^{(1,2)} = 0, \pi$.

It follows from definition that the functions Z_α and $Z_{\alpha\beta}$ satisfy the eikonal equation

$$|\nabla L(X)|^2 = 1 \tag{4.49}$$

We will call the function Z_α single and the function $Z_{\alpha\beta}$ double eikonals. Note that the phases of plane and spherical waves $(X, \hat{P})$ and $|X|$ fulfil the eikonal equation as well. We will call these functions plane and spherical eikonals.

4.3.2 Asymptotics of the functions Φ_α and $\Phi_{\alpha\beta}$

We will now discuss the coordinate asymptotics of the terms Φ_α and $\Phi_{\alpha\beta}$. We mentioned it already in the preceding section that for $|x_\alpha| \to \infty$, the two-particle functions $\varphi_\alpha(x, k)$ in (4.37) can be replaced by their asymptotics forms (4.4). The leading term of the resulting expression is

$$\Phi_\alpha(X, P) \sim \frac{\exp\{i\sqrt{E}Z_\alpha\}}{|x_\alpha|}(f_\alpha(\hat{x}_\alpha, k_\alpha) + O(|x_\alpha|^{-\nu})) \tag{4.50}$$

where $f_\alpha(\hat{x}_\alpha, k_\alpha)$ is the scattering amplitude of the wave functions $\_\alpha(x_\alpha, k_\alpha)$. If the potentials $v_\alpha(x_\alpha)$ decrease faster than any power $|x_\alpha|^{-N}$, then the lower order terms $|x_\alpha|^{-j}$, $j = 2, 3, \ldots$ can be added to the asymptotics decomposition of the two-particle wave functions such that the asymptotics solution Φ_α satisfies the Schrödinger equation up to the arbitrary power Z_α^{-N}.

It is more complicated to find asymptotics of the functions $\Phi_{\alpha\beta}$. We consider only the most typical case when the particles are widely separated before and after collision. More precisely, we will assume that the conditions

$$|x_\alpha| > a(1 + |y_\alpha|)^\nu \quad \nu > half,$$

$$k_\beta^2 > |X|^{-1-\nu'}, \quad \nu' > \frac{1}{2} \tag{4.51}$$

are fulfilled.

Using the asymptotics form of the functions $\varphi_\alpha(x, k)$ we write the integral $\Phi_{\alpha\beta}$ as

$$\Phi_{\alpha\beta}(X, P) \sim$$

$$\sim \int_{p'^2_\alpha \le E} dp'_\alpha e^{i\sqrt{E} Z_\alpha(p'_\alpha, X)} \frac{\tilde{T}_{\alpha\beta}(\hat{x}_\alpha, p'_\alpha, P)}{k_\beta^2(p'_\alpha, p_\beta) - k_\beta^2 - i0}. \tag{4.52}$$

Here

$$Z_\alpha(p'_\alpha, X) = \frac{1}{\sqrt{E}} \left((p'_\alpha, y_\alpha) + \sqrt{E - p'^2_\alpha} |x_\alpha| \right) \tag{4.53}$$

and

$$\tilde{T}_{\alpha\beta}(\hat{x}_\alpha, p'_\alpha, P) =$$

$$\frac{2\pi^2}{|s_{\alpha\beta}|^3} t_\alpha \left(\sqrt{E - p'^2_\alpha} \hat{x}_\alpha, \sqrt{E - p'^2_\alpha} \hat{k}'_{\alpha\beta}, E - p'^2_\alpha + i0 \right) \times$$

$$\times t_\beta(k_{\alpha\beta}, k_\beta, k_\beta^2 + i0). \tag{4.54}$$

The principal value of this integral is generated for $|X| \to \infty$ by the two sorts of characteristic points. The first set of characteristic points is represented by a neighbourhood of the single point of the stationary phase $p_\alpha^{(0)}$, $p_\alpha^{(0)} = E^{1/2}|X|^{-1} y_\alpha$ which can be found from the equation

$$\nabla_{p'_\alpha} Z_\alpha(p'_\alpha, X) = 0.$$

The second set is the set of zeros of the singular denominator $\Omega^{(s)}$, where

$$k_\beta^2(p'_\alpha, p_\beta) - k_\beta^2 = 0.$$

If the point $p_\alpha^{(0)}$ is sufficiently far from the manifold $\Omega^{(s)}$ so that

$$\left|k_\beta^2(p_\alpha^{(0)}, p_\beta) - k_\beta^2\right| \geq C > 0, \tag{4.55}$$

then the contributions of the characteristic points to the asymptotics $\Phi_{\alpha\beta}$ can be investigated separately.

In the critical point $p_\alpha^{(0)}$, we have the equation

$$Z_\alpha(p_\alpha^{(0)}, X) = |X|$$

and a neighbourhood of this point generates a spherical wave in R^3 with the amplitude equal $C_0(E)T_{\alpha\beta}^{(0)}(|P|\hat{X}, P)$. This wave can be united with the "true" three-particle term $\sum_\alpha \tilde{W}_{\alpha 0}$ in (4.39). As a result, we get a spherical wave with the amplitude $C_0(E)T_c(\hat{X}, P)$:

$$T_c(\hat{X}, P) = T(\sqrt{E}\hat{X}, P, P^2 + i0) -$$
$$- \sum_\alpha T_\alpha(\sqrt{E}\hat{X}, P, P^2 + i0). \tag{4.56}$$

Let us note that if the critical point $p_\alpha^{(0)}$ is close to $\Omega^{(s)}$, then the described form of asymptotics lose its meaning. In this case the amplitude of the spherical wave (4.56) diverges because of the three-particle poles of the kernel $T_{\alpha\beta}^{(0)}$. We will show later that in this case the asymptotics $\Phi_{\alpha\beta}$ is given by Fresnel integrals.

We will turn now to investigation of the contribution of the pole singularity (4.52). Substitution $q = k_\beta(p'_\alpha, p_\beta)$ in the integral (4.52) leads to the expression

$$\Phi_{\alpha\beta} \sim \int_{V_E} dq \frac{\exp\{i\sqrt{E}|X|\tilde{Z}_\alpha(q, \hat{X})\}}{q^2 - k_\beta^2 - i0} \tilde{T}_{\alpha\beta}(q, \hat{x}_\alpha, P). \tag{4.57}$$

Here $\tilde{T}_{\alpha\beta}$ is the function (4.54) expressed in terms of q. The function $|X|\tilde{Z}$ is given by (4.53) where

$$p'_\alpha = -s_{\alpha\beta} q + c_{\alpha\beta} p_\beta.$$

The integration is performed over the sphere V_E of radius $E^{1/2}|s_{\alpha\beta}|^{-1}$ with the centre at the point $q_0 = c_{\alpha\beta}(s_{\alpha\beta})^{-1} p_0$. We introduce the spherical coordinates $|q|$, θ, φ with respect to the vector p_β, so that

$$q = |q|(\cos\theta \hat{p}_\beta + \sin\theta\cos\varphi e_\alpha^{(2)} + \sin\theta\sin\varphi e_\alpha^{(3)}).$$

We perform the integration with respect to the angular variables θ and φ. By virtue of condition (4.50), the manifold $\Omega^{(s)}$ and the point $p_\alpha^{(0)}$ are separated from the boundary of the integration region and the vector $\hat{q}$ takes all values on unit sphere. Since the integrand depends on θ and φ smoothly, the coordinate asymptotics of the integral can be found by means of the method of stationary phase. Observing that the critical point $\hat{q}_0 = \{\theta_0, \varphi_0\}$ is determined by equations (4.48), we get asymptotics formula

$$\Phi_{\alpha\beta} \sim \frac{A_{\alpha\beta}(\hat{X}, P)}{|x_\alpha|\,|y_\alpha|} \Phi_{\alpha\beta}^{(1)}(X, P) \tag{4.58}$$

where

$$A_{\alpha\beta}(\hat{X}, P) = \left(\frac{|k_\beta|\,|y_\alpha| \sin\theta_{\alpha\beta}}{\sin\tilde{q}_{\alpha\beta}|s_{\alpha\beta}|} \right)^{1/2} \left| \frac{\partial^2 z_{\alpha\beta}^{(1)}}{\partial\theta_{\alpha\beta}^2} \right|^{-1/2} \tag{4.59}$$

and the function $\Phi_{\alpha\beta}^{(1)}$ is given by the integral

$$\Phi_{\alpha\beta}^{(1)} = -\frac{2\pi i}{|k_\beta|^{1/2}} \int_0^\infty dt\, t^{3/2} \frac{\exp\{iZ_{\alpha\beta}(t)\}}{t^2 - k_\beta^2 - i0} \tilde{\jmath}(\hat{x}_\alpha, t\hat{q}_0, P).$$

The function $Z_{\alpha\beta}(t)$ is determined by the formulae (4.46)-(4.48) where one sets $|q| = t$ and the differentiation of $Z_{\alpha\beta}(t)$ with respect to $\theta_{\alpha\beta}$ is performed for fixed $\varphi_{\alpha\beta}$, x_α, and y_α.

Since the contribution of the critical point was discussed above, we confine ourselves to description of the second term only. According to (4.121) of Section 6 below, the form of the asymptotics of this integral depends on the relative position of the stationary point $t = |k_{\alpha\beta}|$ and of the pole $t = |k_\beta|$. The configuration space is divided into two parts depending on the sign $\hat{\omega}_{\alpha\beta} = \operatorname{sign}(k_{\alpha\beta}^2 - k_{\alpha\beta}^2)$. The set of points where $\hat{\omega}_{\alpha\beta} = -1$ will be denoted by $\Omega_{\alpha\beta}^{(-)}$ and will be called the domain of light; its complement $\Omega_{\alpha\beta}^{(-)}$ will be called the domain of shadow. The reason for this terminology will be apparent after we compare the problem of scattering with the problem of the diffraction of the plane waves on a wedge.

In the domain of light, the asymptotics $\Phi_{\alpha\beta}^{(s)}$ is determined by the residua in the pole and has the order $O(|x_\alpha|^{-1}|y_\alpha|^{-1})$; in the domain of shadow, this term is exponentially small. Thus we have the asymptotic formula

$$\Phi_{\alpha\beta}^{(s)}(X, P) \sim \frac{A_{\alpha\beta}(\hat{X}, P)}{|x_\alpha|\,|y_\alpha|} T_{\alpha\beta}^{(s)}(\hat{X}, P) \exp\{i\sqrt{E} Z_{\alpha\beta}\} E(\hat{\omega}_{\alpha\beta}) \tag{4.60}$$

where $E(t)$ is the unit step function: $E(t) = 1$, $t > 0$; $E(t) = 0$, $t < 0$. The amplitude $T^{(s)}_{\alpha\beta}$ can be expressed in terms of the two-particle T-matrices on energy shell:

$$T^{(s)}_{\alpha\beta}(\hat{X}, P) =$$

$$= \frac{4\pi^2}{|s_{\alpha\beta}|^3} t_\alpha(\hat{x}_\alpha |k_{\alpha\beta}|, k_{\alpha\beta}, k^2_{\alpha\beta} + i0) t_\beta(|k_\beta| \hat{q}_{\alpha\beta}, k_\beta, k^2_\beta + i0)$$

where the momenta $k_{\alpha\beta}$ and $q_{\alpha\beta}$ are given by (4.47) and (4.47').

If the critical point belongs to the set $\Omega^{(s)}$, then, as it was mentioned above, the formulae (4.56) and (4.60) lose their meaning. In this case, the new asymptotics representations are valid which make use of the Fresnel integrals $\Phi(t) = \int_t^\infty e^{i\tau^2} d\tau$:

$$\Phi_{\alpha\beta} \sim \frac{e^{i\pi/4} A_{\alpha\beta}(\hat{X}, P)}{\sqrt{\pi}|x_\alpha|\,|y_\alpha|} T^{(s)}_{\alpha\beta}(\hat{X}, P) \Phi(E^{1/4} \xi^{1/2}_{\alpha\beta} \hat{\omega}_{\alpha\beta}). \tag{4.61}$$

Here be $\xi_{\alpha\beta}$ we denote the variable $\xi_{\alpha\beta} = |X| - Z_{\alpha\beta}$. This variable vanishes on the boundary $\Omega^{(0)}_{\alpha\beta}$ between $\Omega^{(+)}_{\alpha\beta}$ and $\Omega^{(-)}_{\alpha\beta}$. If $\xi_{\alpha\beta} \to \infty$ and $\hat{\omega} = 1$, i.e., if the point lies in the shadow, then we can transform $\Phi_{\alpha\beta}$ into the spherical wave by replacing the Fresnel integral by its asymptotics $\Phi(t) \sim \frac{i}{2t} e^{it^2}$. The amplitude of this wave is $C_0(E) T^{(0)}_{\alpha\beta}(|P|\hat{X}, P)$. In other extreme case, when the point X falls into the domain of light, i.e., when $\hat{\omega} = 1$, the argument of the Fresnel integral is negative and the integral can be expressed as a sum of a constant and of the term tending to zero:

$$\Phi(-t) \sim \sqrt{\pi} e^{-i\pi/4} - \frac{i}{2t} e^{it^2}, \quad t > 0.$$

As a result, we get the sum of the spherical wave and of the term $\Phi^{(s)}_{\alpha\beta}$. This ensures a smooth matching of the transient asymptotics with the standard regime (4.60).

Let us note that we can regard the Fresnel integral $\Phi(E^{1/4} \xi^{1/2}_{\alpha\beta} \hat{\omega}_{\alpha\beta})$ as a function of $E^{1/4} \xi^{1/2}_{\alpha\beta}$ only, if we consider this variable as a point on the complex plane with the cut Π_0 (Figure 14). The domain $\Omega^{(+)}_{\alpha\beta}$ corresponds to the real values of $\xi_{\alpha\beta}$ on the upper side of the cut, $\arg \xi_{\alpha\beta} = 0$. The transposition of the variable X in the domain of shadow corresponds to the displacement of $\xi_{\alpha\beta}$ around the origin to the lower side of the cut where $\arg \xi_{\alpha\beta} = 2\pi$,

$$\Phi(E^{1/4} \xi^{1/2}_{\alpha\beta} \hat{\omega}_{\alpha\beta}) \to \Phi(E^{1/4} \xi^{1/2}_{\alpha\beta}), \quad \xi_{\alpha\beta} \in \Pi_0.$$

These results lead to the conclusion that the wave function $\Psi_0(X,P)$ can be written as the sum:

$$\Psi_0(X,P) = \\ = e^{i(X,P)} + \sum_{\alpha} \Phi_\alpha(X,P) + \sum_{\substack{\alpha,\beta \\ \alpha\neq\beta}} \Phi_{\alpha\beta}(X,P) + \tilde{\Phi}(X,P) \qquad (4.62)$$

where the component $\tilde{\Phi}(X,P)$ belongs to the class $\Phi_{E,\epsilon}(\hat{P})$ and the amplitude of the six-dimensional spherical wave $\tilde{F}_{00}(\hat{X},P)$ is expressed in terms of the scattering matrix as

$$\tilde{F}_{00}(\hat{X},P) = C_0(E)\left(T_c(\hat{X},P) - \sum_{\substack{\alpha,\beta \\ \alpha\neq\beta}} T_{\alpha\beta}(\hat{X}|P|,P)\right) \qquad (4.63)$$

where the kernel $T_{\alpha\beta}$ is given by (3.79). The amplitudes of spherical waves in R^3 which describe the processes of the particle capture are determined in terms of elements of T-matrix by equation (4.49) and the asymptotics of the function $\Phi_{\alpha\beta}$ is described by the formula (4.61).

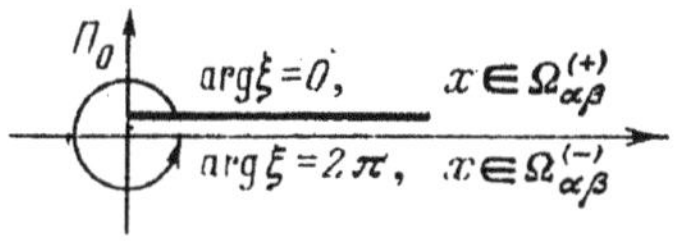

Fig. 14

Let us note that if the potentials $v_\alpha(s)$ decrease faster than any power $|x|^{-r}$, the asymptotics terms of higher orders can be given by means of the recurrent formulae. In this case the Schrödinger equation will be satisfied up to the arbitrary power $|x|^{-r}$. These formulae recurrent will be given in Section 1 of the next chapter.

Thus we see that the coordinate asymptotics of the wave functions $\Psi_0(X, P)$ cannot be described in terms of elementary functions. In many directions in coordinate space, the asymptotics is given in terms of a special function like the Fresnel integrals.

In order to understand the physical reason why the Fresnel integrals appear, we compare the asymptotics of the term $\Phi_{\alpha\beta}(X, P)$ with the asymptotics of field in the problem of diffraction of the plane wave on a semi-transparent wedge. For the sake of simplicity we will assume that the particles move in the same direction before and after collision. We will also assume that all particles have the same masses $m_1 = m_2 = m_3 = 1$.

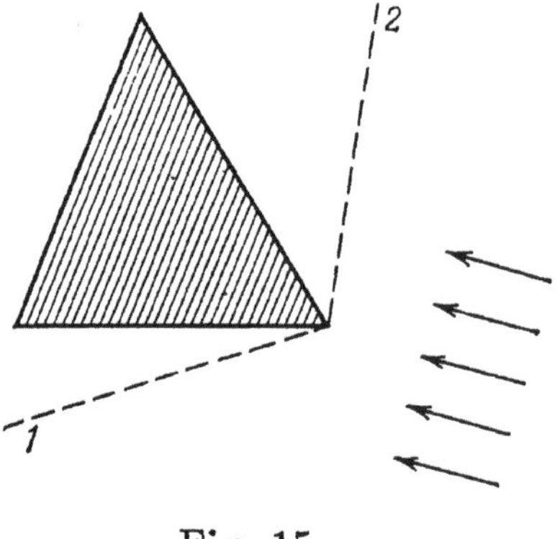

Fig. 15

We will first describe the asymptotics of the wave field in the case of the Fraunhoffer diffraction on an opaque wedge. Suppose we have a plane wave moving perpendicularly to the edge of a wedge (see Figure 15). The asymptotics of the field of the reflected waves far from the borders 1 and 2 is given by a sum of the plane waves (incident and reflected or only incident) and of the diffracted cylindric wave. However, if we come close to either boundary 1 or 2, the amplitude of the cylindric wave is infinitely growing so that the asymptotics we have described loses its meaning. The reason is that in the neighbourhood of the boundaries 1 and 2 of the half-shadow domain, the scattered field is not given only by the cylindrical wave bus has a more complex form and must be described by the Fresnel integral.

In the case of three identical particles moving along a line, the motion of particles takes place in the subspace R^2 of configuration space R^3. This subspace is shown on Figure 16. Here, the hatching marks the regions in which the interaction potentials between the particles 2, 3 and 3, 1 differ substantially from zero. These regions border the wedges formed by the lines $x_1 = 0$ and $x_2 = 0$. The interaction between the particles 1 and 2 is not represented on Figure 16. Let the motion of particles before collision be described by the plane wave with momentum P. The boundary $\Omega_{12}^{0)}$ is then formed by the following four rays. The ray A corresponds to the plane wave travelling in the original direction P. It represents the boundary of the wave which passed without reflection (the "diffracting body" is the wedge (102') or (201')). The rays B and C have the direction coinciding with the plane waves, reflected by the boundaries (11') and (22'). These rays are the boundaries of the waves reflected by the sides of the wedges (102') and (201'). Finally, the ray D coincides with the direction of the double reflected wave, first from the side (02) and then from the side (01) of the wedge (201). This ray represents the boundary of such waves. Let us stress that the wedges are to be considered semi-transparent.

Far from $\Omega_{12}^{(0)}$, the term Φ_{12} plays a role of the cylindric scattered wave corresponding to the case of diffraction on wedge. In vicinity of $\Omega_{12}^{(0)}$, this term is given by the Fresnel integral quite similarly to the diffraction on wedge.

The term $\Phi_{\alpha\beta}^{(s)}$ differs from zero only in the sections AOP and COD. Thus the sections COB and COA are the domains of shadow. In the domain AOP, the phase factor $\Phi_{\alpha\beta}^{(s)}$ has the form $\exp\{i(|x_1|\,|k_1| + (y_1, p_1)\}$ and corresponds to the plane wave travelling in the direction P for $x_1 > 0$ or in the direction OB for $x_1 < 0$. In the region COD, the phase is $|x_1|\left|\frac{\sqrt{3}}{2}p_2 - \frac{1}{2}k_1\right| - |y_1|\left|\frac{\sqrt{3}}{2}k_2 - \frac{1}{2}p_2\right|$ which corresponds to the plane wave moving in direction of the ray C for $x_1 > 0$ and in direction of D for $x_1 < 0$. The amplitudes of such waves decrease as $|x_1|^{-1}|y_1|^{-1}$.

Let us note that the directions OA, OB, OC, and OD can be also described in terms of the classical particle mechanics. The laws of energy and momentum conservation permit the trajectories situated in the domains of light only. The rays OA, OB, OC, and OD represent their boundaries. The directions of motion of classical particles coincide with these rays.

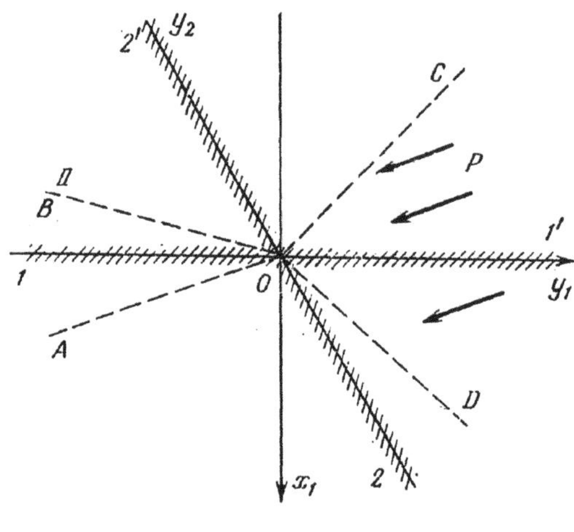

Fig. 16

4.3.3 Exponential decreasing of eigenfunctions

In many problems, in addition to asymptotics of wave functions, the coordinate asymptotics of eigenfunctions of the energy operator is used. The corresponding formulae as well as the methods of their justification share many common features with the theory of wave functions described above. We introduce here these formulae without proof.

The eigenfunctions $\psi_i(x)$ for a two-particle system decrease in all directions of configuration space as $|x|^{-1}\exp\{-\varkappa_i|x|\}$. For three-particle system, the eigenfunctions decrease exponentially as well, but their distance dependence is different in different domains of the configuration space.

In order to describe asymptotics of the three-body eigenfunctions, we introduce some conventions. Let $\hat{Q}(X)$ be a function which we obtain from a six-dimensional spherical wave as the analytic continuation to the point $-E$, $E > 0$:

$$\hat{Q}_0 = \frac{\exp\{-\sqrt{E}|X|\}}{|X|^{5/2}} F_0(\hat{X}).$$

Further, let

$$\hat{Q}_\alpha = \frac{\exp\{-\varkappa_\alpha|x_\alpha|\sqrt{E-\varkappa_\alpha^2}|y_\alpha|\}}{|x_\alpha|\,|y_\alpha|} F_\alpha(\hat{x}_\alpha, \hat{y}_\alpha)$$

be a result of the analytic continuation of the cluster spherical wave corresponding to the minimal two-particle eigenvalue $-\varkappa_\alpha^2$. These functions determine the asymptotics of eigenfunctions in different parts of configuration space.

In analogy to the coordinate $\xi_{\alpha\beta}$ introduced above, let us by ξ_α the difference of exponents of $\hat{Q}_0$ and $\hat{Q}_\alpha$:

$$\xi_\alpha = \sqrt{E}|X| - \left(\varkappa_\alpha|x_\alpha| + \sqrt{E-\varkappa_\alpha^2}|y_\alpha|\right).$$

We note that

$$\xi_\alpha = \omega_\alpha\left(\sqrt{E}|X| + \varkappa_\alpha|x_\alpha| + \sqrt{E-\varkappa_\alpha^2}|y_\alpha|\right)$$

where $\omega_\alpha = \sqrt{E-\varkappa_\alpha^2}|x_\alpha| - \varkappa_\alpha|y_\alpha|$. This last relation implies that the variable ξ_α is positive everywhere in configuration space with exception of the directions $\partial\Omega_\alpha^{(0)}$, which are determined by the condition $\omega_\alpha = 0$. In these directions $\xi_\alpha = 0$. We denote by $\Omega_\alpha^{(+)}$ ($\Omega_\alpha^{(-)}$) the domains in configuration space where the quantity ω_α is positive, $\omega_\alpha > 0$ (negative, $\omega_\alpha < 0$) and by $\Omega_\alpha^{(0)}$ the neighbourhood of the direction $\partial\Omega_\alpha^{(0)}$ where $\omega_\alpha| \leq |X|^{-\nu}$, $0 < \nu < 1/2$.

The eigenfunction $\Psi(X)$ is equal to the sum of the components Φ_α which satisfy the homogeneous equations (4.45):

$$\Psi(X) = \sum_\alpha \Phi_\alpha(X). \tag{4.64}$$

These components have the following asymptotics form. In the domain $\Omega_\alpha^{(+)}$, the function Φ_α has the form of the spherical wave $\Phi_\alpha(X) \sim \hat{Q}_0(X)$, and in the domain $\Omega_\alpha^{(-)}$, they are described by the cluster functions $\Phi_\alpha(X) \sim \hat{Q}_\alpha(X)$. In the transient regions $\Omega_\alpha^{(0)}$, the coordinate asymptotics $\Phi_\alpha(X)$ is given in terms of the function $\mathrm{erfc}(t) = \frac{1}{\sqrt{\pi}}\int_t^\infty e^{-\tau^2}d\tau$:

$$\Phi_\alpha(X) \sim \hat{Q}_\alpha(X)\mathrm{erfc}(\hat{\omega}_\alpha\sqrt{\xi_\alpha}) + \hat{Q}_0(X) \tag{4.65}$$

$$\hat{\omega}_\alpha = \operatorname{sign} \omega_\alpha.$$

Let us discuss in more in details the character of the asymptotics $\Psi(X)$ when the point X moves on the plane $x_\alpha = t_1 c$, $y_\alpha = t_2 c$, where c is some fixed vector, $c \in R^3$ The corresponding two-dimensional subspace of configuration space is shown on Figure 17.

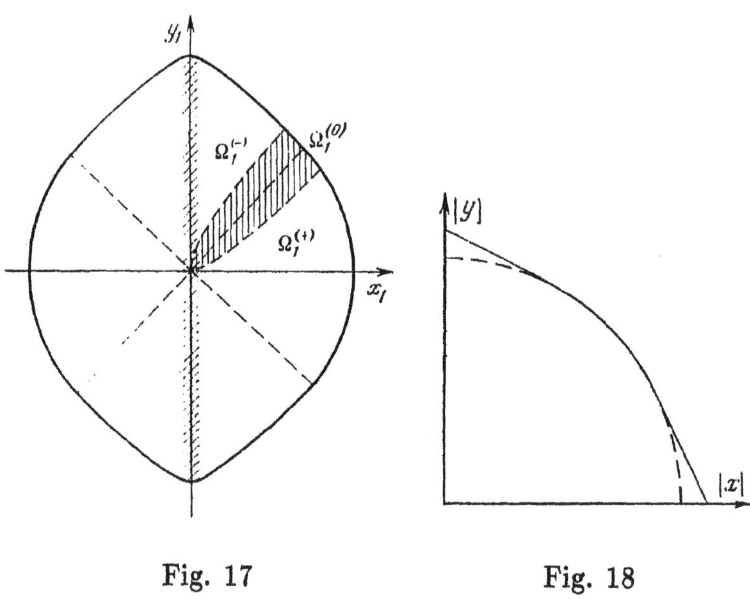

Fig. 17 Fig. 18

The geometric locus of points where the exponent of the exponentials factors are constant will be called the asymptotic front. On Figure 16 the asymptotics front is marked by the solid line. In the domain $\Omega_\alpha^{(+)}$, the function Φ_α has a spherical front. If $|x_\alpha| \to \infty$ and $X \in \Omega_\alpha^{(-)}$, then the front is flat. The amplitude coefficient depending on $\hat{x}_\alpha$ are equal to the corresponding two-particle amplitudes $f_\alpha(\hat{x}_\alpha)$ of the eigenvalues of the ground state $\psi_\alpha(x_\alpha)$. While crossing the direction $x_\alpha = 0$, the plane front smoothly changes the orientation and the coefficient depending on x_α is equal to the eigenfunction $\psi_\alpha(x_\alpha)$. In the domain $\Omega_\alpha^{(0)}$, the plane front becomes spherical and the change of the shape of the front is described by the function erfc(t). The asymptotics of the eigenfunction $\Psi(X)$ is determined by the leading terms in the sum (4.64). Its front is shown on Figure 18.

4.4 The Green Function

In the section, we describe the properties of the three-particle Green function. This function can be determined as a solution of the differential equation

$$\left(-\Delta_X + \sum_\alpha v_\alpha(x_\alpha) - z\right) R(X, X', z) = \delta(X - X')$$

or of the analogous equation with respect to the variable X'.

On the other hand, the Green function can be separated into components which satisfy the compact equations (3.59). These equations are most convenient for the study of properties of the Green function.

4.4.1 Singularities and asymptotics

The integral equations in configuration space (3.59) can be investigated along the same lines as the system of compact equations for the T-matrix (3.28). The only difference is that in momentum representation, the kernels of iteration had a form of integrals with poles, whereas after transition to coordinate representation, the corresponding singular behaviour is expressed by the slow decrease and oscillations of kernels at infinity. Consequently, instead of the theorems concerning singular integrals, the main technical tool is now the method of stationary phase. In order to get a heuristic insight into the problem, we compare both types of singularities in some examples we have considered before.

The simplest example of the relation between the asymptotics and the singularities is given by the formula (4.4). According to this formula, the poles $(p^2 - E - i0)^{-1}$ generate an approximate solution of the Schrödinger equation of the form of spherical waves.

The strongest singularity in momentum representation is contained in the kernel of the free resolvent $\delta(P - P')(P'^2 - z)^{-1}$. The corresponding Green function explicitly is known:

$$R_0(X, X', z) = \frac{iz}{(4\pi)^2} \frac{H_2^{(1)}(\sqrt{z}|X - X'|)}{|X - X'|^2}. \tag{4.66}$$

Its asymptotics is determined by the eikonal $|X - X'|$ corresponding the process of spreading of a spherical wave from the source an the point X' to the point X:

$$R_0(X, X', z) \sim C_z \frac{i\sqrt{z}|X - X'|}{|X - X'|^{5/2}},$$

$$C_z = -\frac{e^{i\pi/4} z^{3/4}}{2(2\pi)^{5/2}}. \qquad (4.67)$$

The next strongest singularity is generated by the process of single scattering. It has a form of the product $\delta(p_\alpha - p'_\alpha)(k'^2_\alpha - z)^{-1}$. In configuration space, these singularities generate slowly decreasing terms. These terms have been described in detail in the case of the wave functions Ψ_0. The asymptotics of the corresponding terms in the Green function is

$$(R_\alpha - R_0)(X, X', E + i0) \sim$$

$$\sim C_z \frac{|x_\alpha| + |x'_\alpha|}{|x_\alpha|\,|x'_\alpha|} f_\alpha \left(\hat{x}_\alpha, \frac{|x_\alpha| + |x'_\alpha|}{Z_\alpha(X, X')} \sqrt{E} \hat{x}'_\alpha \right) \times$$

$$\times \frac{\exp\{\pm i\sqrt{E} Z_\alpha(X, X')\}}{Z_\alpha^{3/2}}. \qquad (4.68)$$

Here Z_α is the eikonal describing propagation of a ray from the point X' to the point X with the refraction point on the hypersurface $x_\alpha = 0$:

$$Z_\alpha(X, X') = \left((|x_\alpha| + |x'_\alpha|)^2 + |y_\alpha - y'_\alpha|^2\right)^{1/2}.$$

This relation can be obtained from (4.23) by the saddle-point method.

Besides the singularities mentioned before, the resolvent kernel possesses also secondary singularities, which depend not only on the magnitudes but also on the directions of momentum variables. We have seen in Section 3 that the strongest of them, the poles, generate asymptotic solutions of the Schrödinger equation, which correspond to the double eikonals $Z_{\alpha\beta}$. Analogous terms are encountered in asymptotics of the Green function. These terms are described in terms of the eikonals $Z^{(n)}_{\alpha\alpha_1\ldots\alpha_n\beta}(X, X')$, $n = 0, 1, \ldots$ which correspond to the propagation of the spherical wave from the point X' to the point X by the multiple reflection f rom the hypersurfaces $x_\alpha = 0$, $x_{\alpha_1} = 0$, ..., $x_\beta = 0$, $\alpha \neq \alpha_1$, $\alpha_1 \neq \alpha_2$, ..., $\alpha_n \neq \beta$. The explicit expression for $Z^{(n)}_{\alpha\alpha_1\ldots\alpha_n\beta}(X,X')$ can be found by means of minimization of the optical ray

$$Z^{(n-2)}_{\alpha\alpha_1\ldots\alpha_n\beta}(X,X^{(1)},X^{(2)},\ldots,X^{(n)},X') =$$
$$|X-X^{(1)}|+|X^{(1)}-X^{(2)}|+\cdots+|X^{(n)}-X'|$$

with respect to the intermediate refraction points $X^{(i)}$ $(i=1,2,\ldots,n)$ situated on the hypersurfaces $x^{(n)}_{\alpha_n}=0$, i.e., by solving the equations

$$\nabla_{X^{(n)}}\tilde{Z}^{(n-2)}(X,X^{(1)},X^{(2)},\ldots,X^{(n)},X')=0, \quad x^{(n)}_{\alpha_n}=0,$$
$$X^{(k)}=\{x^{(k)}_{\alpha_k},y^{(k)}_{\alpha_k}\} \quad k=1,2,\ldots,n \tag{4.69}$$

It should be mentioned that the number of reflections from the sides of the wedges is bounded by some constant which depends only on the angles between the hyperplanes $x^{(k)}_{\alpha_k}=0$ i.e., only on the relations between masses of particles. For example, in the problem of the plane propagation of the ray from the point X to the point X' with n reflections, the maximal number of reflections is $N_{\max}=E(\pi/\varphi)$ where φ is the angle between the sides of the wedge and $E(t)$ is the integer part of t. If the process of the multiple reflections is forbidden, equation (4.69) means that the rays go through the vertex $X^{(k)}=0$ $(k=1,2,\ldots,n)$. In this case the solution of (4.69) is a sum of spherical eikonals $Z_s(X,X')=|X|+|X'|$. This follows from the fact that the kernels of iterations $R^{(n)}_{\alpha\alpha_1\ldots\alpha_n\beta}$ in momentum representation become smooth starting from $n=4$ so that the asymptotics of the corresponding terms of the wave functions have a form of spherical waves.

4.4.2 Iterations of $R^{(n)}_{\alpha\beta}$

Let us turn to the investigation of solutions of the system of equations for components of the resolvents (3.59). These equations read:

$$R_{\alpha\beta}(X,X',z)=(R_\alpha-R_0)(X,X',z)\delta_{\alpha\beta}-$$
$$-\int R_\alpha(X,X'',z)v_\alpha(x'_\alpha)\sum_{\gamma\neq\alpha}R_{\gamma\beta}(X'',X',z)dX''. \tag{4.70}$$

First we consider the iterations

$$R^{(n)}_{\alpha\beta}(z)=\sum_{\gamma_1,\gamma_2,\ldots,\gamma_n}{}' R^{(n)}_{\alpha\gamma_1\gamma_2\ldots\gamma_n\beta}(z), \quad n=0,1,2,\ldots; \tag{4.70'}$$

where the operators $R^{(n)}_{\alpha\gamma_1\gamma_2\ldots\gamma_n\beta}$ are

$$R^{(n)}_{\alpha\gamma_1\gamma_2\ldots\gamma_n\beta}(z) =$$

$$= (-1)^n R_\alpha(z) v_\alpha R_{\gamma_1}(z) v_{\gamma_1} \cdots R_{\gamma_n}(z) v_{\gamma_n} (R_\beta(z) - R_0(z)). \tag{4.71'}$$

The kernels of these operators have a form of multiple integrals

$$R^{(n)}_{\alpha\gamma_1\gamma_2\ldots\gamma_n\beta}(X, X'z) =$$

$$= (-1)^{n-1} \int dX^{(1)} dX^{(2)} \ldots dX^{(n+1)} \times$$

$$\times R_\alpha(X, X^{(1)}, z) v_\alpha(x^{(1)}_\alpha) R_{\gamma_1}(X^{(1)}, X^{(2)}, z) v_{\gamma_1}(x^{(2)}_{\gamma_1}) \cdots$$

$$\cdots v_{\gamma_n}(x^{(n+1)}_{\gamma_n})(R_\beta - R_0)(X^{(n+1)}, X', z). \tag{4.71}$$

These kernels can be expressed with the help of (4.29) as the sum

$$Q(X, X', z) = F(X, X', z) + \sum_j G_B(X, y'_\beta, z) \psi^*_B(x'_\beta) +$$

$$\sum_j \psi_A(x_\alpha) J_A(y_\alpha, X', z) + \sum_i \sum_j \psi_A(x_A) H_{AB}(y_A, y'_B, z) \psi^*_B(x'_B). \tag{4.72}$$

Here we omitted the indices $\alpha\gamma_1\gamma_2\ldots\gamma_n\beta$ which enumerate the kernels, i.e., $R^{(n)}_{\alpha\gamma_1\gamma_2\ldots\gamma_n\beta} = Q$. The meaning of this expression is analogous to representation (3.36) for components of T-matrix. Similarly to momentum representation, we will call the kernels which can be expressed in this form the kernels of type $\mathcal{D}_{\alpha\beta}$ and the functions F, G, J, and H the components of these kernels. Let us investigate properties of these components.

First we consider the components F, which, from the technical point of view, is the most difficult case. We note that the asymptotics of integrals which corresponds to these terms is determined by these parts of domains of integration where the potentials $v_{\gamma_i}(x_{\gamma_i})$ are concentrated. In this case the first $X^{(\gamma_i - 1)}$ and the second $X^{(\gamma_i)}$ arguments of the kernels are situated in the different parts of the configuration space. Thus it is sufficient to consider in detail only the asymptotics with the special values of arguments. It simplifies the study of the kernels of the iterations $R^{(n)}_{\alpha\beta}$.

We start with $n = 2$. In accordance with the previous considerations, we assume that the point X lies in the domain where the particles are strongly separated. By virtue of equation (4.68), the kernel $R^{(0)}_{\alpha\beta}(X, X', z)$ can be written in the form

$$R^{(0)}_{\alpha\beta}(X, X', z) =$$
$$= \int dX^{(1)} \frac{\exp\{i\sqrt{z}(|X - X^{(1)}| + Z_\beta(X^{(1)}, X'))\}}{|X - X^{(1)}|^{5/2} z_\beta^{3/2}(X^{(1)}, X')} \times$$
$$\times v_\alpha(x^{(1)}_\alpha) \frac{|x^{(1)}_\alpha| + |x'_\beta|}{|x^{(1)}_\alpha|\,|x'_\beta|} \tilde{f}_{\alpha\beta}(X^{(1)}, X, X', z) \tag{4.73}$$

where $\tilde{f}_{\alpha\beta}$ is a smooth function. It is clear that the integration variable $X^{(1)}$ is concentrated in the region Ω_α, where the potential $v_\alpha(x_\alpha)$ substantially differs from zero. In order to determine the asymptotics of the integral (4.73), three types of points are to be considered. First of all, we have the critical point X_0, $X_0 = \{0, y^{(0)}_\alpha\}$ given by the equation

$$\nabla_{y^{(0)}_\alpha}(|X - X^{(1)}| + Z_\beta(X^{(1)}, X')) = 0.$$

Next, there are singular points X and X' of the integrand and, finally, points in the neighbourhood of the origin of the coordinate system. As the latter, we take the sphere of the radius R, $R = (|X| + |X'|)^\nu$, $\nu < 1/2$ with the centre at the origin.

In the integral I_n over the sphereV_R, the term $|X^{(1)}|(|X| + |X'|)^{-1}$ is a small parameter. Consequently, the functions $|X - X^{(1)}|$ and $Z_\beta(X^{(1)}, X')$ can be expanded in the series with respect to this parameter. In this way, we find that the asymptotics is a product of the functions

$$I_R \sim \frac{\exp\{i\sqrt{z}(|X| + |X'|)\}}{|X|^{5/2}|X'|^{5/2}} \tilde{A}(\hat{X}, \hat{X}', z). \tag{4.74}$$

which exponentially decrease for $|X|, |X'| \to \infty$. If z is real, the asymptotics I_R is the product of the spherical waves in variables X and X'.

To estimate integrals generated by neighbourhoods of the points X and X', we integrate by parts with respect to the variable $\tau_1 = |X - X^{(1)}|$ or $\tau_2 = |X^{(1)} - X'|$. It can be shown that these integrals are rapidly decreasing and do not influence the leading terms in the asymptotics of the kernel $R^{(0)}_{\alpha\beta}(X, X', z)$..

Finally, we consider the integral I_0 over a neighbourhood of the critical point $X^{(0)}$. We will study this asymptotics for real $z = E + i0$. Making the change of integration variable $y = \lambda y_\alpha^{(1)}$ where $\lambda = |X| + |X'|$, we get the function I_0 as the integral

$$I_0 = \int dx_\alpha^{(1)} \int dy e^{i\sqrt{E}\varphi_{\alpha\beta}(y)} v_\alpha(x_\alpha^{(1)}) \tilde{f}_{\alpha\beta}(y, x_\alpha^{(1)}, X, X', E).$$

Consequently, the asymptotics can be found with the help of the method of stationary phase. Note that the equation

$$\nabla_y \varphi_{\alpha\beta}(y) = 0 \tag{4.75}$$

defining the critical point, has a solution $X^{(0)}$ only if the points X and X' can be connected by a polygon with two breaking points $y_\alpha^{(0)}$ and $y_\beta^{(1)}$ on the hypersurfaces x_α and x_β, respectively. Then the exponent of the exponential function $\lambda\varphi_{\alpha\beta}(y)$ equals the eikonal $Z_{\alpha\beta}(X, X')$.

The explicit form of the asymptotics can be found with the help of formula (4.115) of Sec give the final result which can be deduced by means of the identical transformation of this formula:

$$I_0 = C_E \frac{A_{\alpha\beta}(X, X')}{|x_\alpha|\,|x'_\beta|} f_\alpha(\hat{x}_\alpha, k_\alpha) f_\beta(\hat{k}_\beta, |k_\beta| \hat{x}'_\beta) \frac{\exp\{i\sqrt{E} Z_{\alpha\beta}\}}{Z_{\alpha\beta}^{5/2}}. \tag{4.76}$$

Here

$$C_E = -e^{i\pi/4}(2\pi)^{-5/2}\frac{1}{2}E^{3/4},$$

$$k_\alpha = -s_{\alpha\beta}\frac{|x_\alpha|}{Z_\alpha(X, X^{(0)})} y_\alpha^{(0)}, \quad k_\beta = -s_{\beta\alpha}\frac{|x'_\beta|}{Z_\beta(X^{(1)}, X')} y_\beta^{(1)},$$

$$X^{(1)} = \{0, y_\beta^{(1)}\}.$$

The function $A_{\alpha\beta}$ is smooth and bounded:

$$A_{\alpha\beta}(X, X') = \frac{(1+\sigma_\alpha)(1+\sigma_\beta)(1 - c_{\alpha\beta}^2 \sigma_\alpha \sigma_\beta)^{1/2}}{\left((1 - c_{\alpha\beta}^2\sigma_\alpha\sigma_\beta)^2 + c_{\alpha\beta}^2 s_{\alpha\beta}^4 \sigma_\alpha^3 \sigma_\beta^3 \frac{[y_\alpha, y'_\beta]}{x_\alpha^2 x'^2_\beta}\right)^{1/2}},$$

$$\sigma_\alpha = \frac{Z_\alpha(X, X_0^{(\alpha)})}{Z_\beta(X_0^{(\alpha)}, X')}, \quad \sigma_\beta = \frac{Z_\beta(X', X^{(1)})}{Z_\alpha(X^{(1)}, X)}$$

We will denote by $\Omega_{\alpha\beta}^{(-)}$ the set of points $\{X, X'\}$ in which formula (4.76) is satisfied. In these domains of configuration space, the function I_0 decreases more slowly than the product of spherical waves (4.74).

If equation (4.75) has a solution only for $y = 0$ so that the process with two breaking points is impossible, the asymptotics of the integral I_0 assumes the factorized form (4.74). The set of points $\{X, X'\}$ will be denoted by $\Omega^{(-)}_{\alpha\beta}$.

As in the case of wave functions, in the transient region the asymptotics of the integral I_0 is given in terms of the Fresnel integral. In this case, the right hand side of equation (4.76) has to be multiplied by the function

$$\pi^{-1/2}e^{i\pi/4}\Phi(\hat{\omega}^{(\pm)}_{\alpha}E^{1/4}(Z_s(X,X') - Z_{\alpha\beta}(X,X'))^{1/2})$$

where the parameter $\hat{\omega}^{(\pm)}_{\alpha}$ takes the value 1 if the points $\{X, X'\}$ lay in the domain $\Omega^{(+)}_{\alpha\beta}$ and -1 if they belong to $\Omega^{(-)}_{\alpha\beta}$.

Similarly we can estimate the integral I_0 in the case of complex z. In this case, the oscillating exponents are replaced by the exponentially decreasing ones.

We have studied the component F and shown that for z complex its kernel decreases as an arbitrary power of $|Z_s(X, X')|^{-N}$. If the variable z is on the real axis, it decreases slowly and its leading terms are given by equations (4.74) and (4.76).

The integrals corresponding to the components G, J and H can be examined similarly. In this case the terms $\psi_\alpha(x^{(1)}_\alpha)\psi_\beta(x^{(1)}_\beta)$ or $v_\alpha(x^{(1)}_\alpha)\psi_\beta(x^{(1)}_\beta)$ appear in the integrands of the integrals similar to (4.73). These terms decrease rapidly in all direction of configuration space. Thus the asymptotics of these integrals is determined by the neighbourhood of the origin of the coordinate system only.

Consequently, it can be shown that for real z the components $G(X, y'_B, z)$ and $J(y_A, X', z)$ asymptotically assume a form of the product of cluster and the six-dimensional spherical waves which depend on the corresponding variables, e.g.

$$G(X, y'_B, E + i0) \sim |X|^{-5/2}|y'_B|^{-1}\exp\{i\sqrt{E}|X| + i\sqrt{E + \varkappa^2_B}|y'_\beta|\}\tilde{g}.$$

The asymptotics of the components $H(y_A, y'_B, E + i0)$ can be expressed in terms of the product of cluster spherical waves

$$H(y_A, y'_B, E + i0) \sim$$

$$\sim \exp\{i\sqrt{E + \varkappa^2_A}|y_\alpha| + i\sqrt{E + \varkappa^2_B}|y'_\beta|\}\times$$

$$\times |y_A|^{-1}|y'_B|^{-1}\tilde{h}(\hat{y}_A, \hat{y}'_B, E).$$

For complex z, all these components decrease rapidly. We give a short description of the behaviour of the kernels $R^{(n)}_{\alpha\gamma_1\ldots\gamma_n\beta}(z)$ for $n \geq 1$ We consider the kernels $R^{(1)}_{\alpha\gamma_1\beta}(z)$ which correspond to the second iterations. They are given in terms of the kernels $R^{(0)}_{\alpha\beta}(z)$ by

$$R^{(1)}_{\alpha\gamma_1\beta}(X, X', z) =$$

$$= \int R_\alpha(X, X'', z) v_\alpha(x''_\alpha) R^{(0)}_{\gamma_1\beta}(X'', X', z)\, dX''.$$

It is not hard to check that these kernels are of the type $\mathcal{D}_{\alpha\beta}$. Knowing the properties of the kernels R_α and $R^{(0)}_{\gamma_1\beta}(z)$, we can study the asymptotics of the individual components in the way similar to the case of the kernels $R^{(0)}_{\alpha\beta}(z)$.

It can be shown that the asymptotics of the component G, J, and H has the same forms as the asymptotics of the analogous components of the kernel $R^{(0)}_{\alpha\beta}(z)$. The asymptotics of the components $F(X, X', z)$ of these kernels is generated by the same singular points as in the case of the kernels $R^{(0)}_{\alpha\beta}(X, X', z)$. The most dangerous terms are those which result from the critical point of the exponential function; an this point the equation

$$\nabla_{y''_\alpha}(|X - X'| + Z_{\gamma_1\beta}(X'', X'))|_{x''_\alpha=0} = 0$$

is satisfied. These terms decrease more slowly than the product of the spherical waves (4.74) in many singular directions of configuration space.

The higher-order kernels $R^{(n)}_{\alpha\gamma_1\ldots\gamma_n\beta}(z)$ can be studied analogously. It can be shown that these kernels are also of the type $\mathcal{D}_{\alpha\beta}$ and, moreover, starting with $n = 4$ the asymptotics of the components similar to F stabilizes; it does not contain the terms decreasing more slowly than the product of the spherical waves for z real any more. For $z = E + i0$ real, and β fixed, the kernels $R^{(n)}_{\alpha\beta}$ are functions depending on X which belong to the class B_E. The asymptotic representations (4.25) and (4.29) are uniform with respect to X'. The kernels which possess these properties will be called kernels of the type $\tilde{\mathcal{D}}_{\alpha\beta}$.

4.4.3 Compact equations

After investigating behaviour of the kernels of the iterations $R^{(n)}_{\alpha\beta}$, we can proceed to proving compactness of the integral equations (4.70).

The formal procedure which reduce these equations to compact equations in the suitable functional space is quite similar to the procedure we have used in the case of the integral equations for T-matrix (3.28). Namely, the structure of the kernels of iterations makes it necessary to consider the set of continuous vector-functions (4.27) which decrease at infinity as follows:

$$\rho_\alpha(X) \sim (1+|X|)^{-5/2+\nu}, \quad \sigma_A(y_A) \sim (1+|y_\alpha|)^{-1+\nu}, \quad \nu > 0. \qquad (4.77')$$

The set of components $\omega_\beta(X, X', z) = \{F_{\alpha\beta}(X, X', z), J_{\alpha\beta}(y_a, X', z)\}$ and $\omega_B(X, y'_B, z) = \{G_{\alpha\beta}(X, y'_B, z), J_{\alpha\beta}(y_a, y'_B, z)\}$ can be regarded as elements of this class. To obtain compact integral equations of the second kind from (4.70), we divide the kernels $R_{\alpha\beta}$ and R_α into components (4.70) and (4.29) and equal the corresponding components on the right and left hand sides. The operator $\omega' = A\omega$ is associated with these equation; this means that

$$\sigma'_A(y_A) = -\frac{1}{4\pi}\int dX' \frac{\exp\{i\sqrt{z+\varkappa_A^2}|y_A - y'_A|\}}{|y_A - y'_A|} \times$$

$$\times \psi^*_A(x'_A) v_A(x'_A) \sum_{\beta\neq\alpha} \Phi_\beta(X'), \qquad (4.77)$$

$$\rho'_\alpha(X) = -\int dX' \tilde{R}_\alpha(X, X', z) v_\alpha(x'_\alpha) \sum_{\beta\neq\alpha} \Phi_\beta(X').$$

Here $\tilde{R}_\alpha(X, X', z)$ is the invariant part of the Green function (4.23) acting in the subspace $\mathfrak{H}^{(c)}_\alpha = \left(I - \sum_j P_A\right)\mathfrak{H}$. The functions $\Phi_\alpha(X)$, $\alpha = 1, 2, 3$ are expressed in terms of the components ρ_α and σ_A as

$$\Phi_\alpha(X) = \rho_\alpha(X) + \sum_j \psi_A(x_\alpha)\sigma_A(y_\alpha).$$

It follows from the estimates given above that the operator $A(z)$ transforms the set of functions with the asymptotics (4.77') into the class $\Phi_E(X')$.

By making use of these results, it can be shown that a sufficiently high power of the operator determined by these equations is a completely continuous operator. The singular points where the homogeneous equation (4.77) has a nontrivial solution coincide with the points of discrete spectrum of the operator H. We will not repeat the considerations necessary for proving this assertion; they are quite similar to those presented in Chapter 3.

Now we list the properties of the Green function which follow from the results given above.

The Green function is symmetric with respect to interchanging the arguments $R(X, X', z) = \overline{R(X', X, z)}$.

The following relation holds

$$R(z) = R_0(z) = \sum_{\alpha}(R_\alpha(z) - R_0(z)) +$$

$$+ \sum_{n=0}^{4} \sum_{\alpha,\beta} R_{\alpha\beta}^{(n)}(z) + \sum_{\alpha,\beta} \tilde{R}_{\alpha\beta}^{(4)}(z) \tag{4.78}$$

where the kernels of the operators $\tilde{R}_{\alpha\beta}^{(n)}(z)$ in configuration space are of the class $\tilde{\mathcal{D}}_{\alpha\beta}$. The asymptotic behaviour of other terms have been described above.

For $X \neq X'$, the Green function $R(X, X', z)$ is a smooth function of X and X' and there are finite intervals where z takes real values which differ from the singular points of the integral equations.

Finally we state an asymptotics formula

$$R(X, X', E \pm i0) \; |X'| \overset{\sim}{\to} \infty \; C_B \Psi_0(X, P) \frac{\exp\{i\sqrt{E}|X|\}}{|X|^{5/2}} +$$

$$+ \sum_{A \neq 0} \frac{1}{4\pi} \Psi_A(X, p_A) \frac{\exp\{i\sqrt{E + \varkappa_A^2}|y_A|\}}{|y_A|} \psi_A(x_A) \tag{4.79}$$

which provides the explicit relation between the Green function and the wave functions. In analogy with (4.21), this formula can be used for an alternative definition of the wave functions independent of the boundary value problems under study. The analogous relation exists for components of the Green function $R_{\alpha\beta}$ and components of the wave functions $\Phi_{\alpha\beta}$.

4.5 Differential Equations for Components of N-body Wave Functions

In this section, we will describe the differential equations for the components of Green functions and N-body wave functions. These equations are equivalent to the compact integral equations and they can be used for an alternative formulation of the scattering problem. We derive in detail the differential equations for the four-body system. In the general case, we only list the principal results and outline the scheme of the proofs.

4.5.1 Formal derivations of differential equations for three- and four-particle systems

We start with the formal derivation of the differential equations for the three-body system. We divide the wave function $\Psi(X)$ into three components which correspond to the possible asymptotics states of the system:

$$\Phi_\alpha = -R_0(E+i0)V_\alpha\Psi. \tag{4.80}$$

Applying the operator $H_0 - E$ to both parts of this equation, we get its differential counterpart:

$$(H_0 - ER)\Phi_\alpha = -V_\alpha\Psi.$$

Next, we express the wave function in terms of the components $\Psi = \sum_\alpha \Phi_\alpha$. Moving the diagonal terms with $\alpha = \beta$ onto the left hand side of the equation, we get desired equality

$$(H_0 + V_\alpha - E)\Phi_\alpha = -V_\alpha \sum_{\alpha\neq\beta} \Phi_\alpha. \tag{4.81}$$

Thus we have passed directly from the Schrödinger equation to the differential equations for the components, avoiding the analysis of the cumbersome integral equations. Of course, in order to give a rigorous justification of these equations and to prove that their solutions coincide with the solutions of the compact equations the lengthy considerations of Section 2 must be repeated. However, we see that the algebraic part of the problem is simple and is based on the definition of the components and their connection with the wave functions.

The similar procedure can be applied in the case of the differential equations for a few-body system. In particular, we consider the a_3-components of the wave functions

$$\Phi_{a_3} = -R_0(E+i0)V_{a_3}\Psi. \tag{4.82}$$

As in the three-body system, the wave function equals their sum:

$$\Psi = \sum_{a_3} \Phi_{a_3}$$

and the components satisfy the equations

$$(H_0 + V_{a_3} - E)\Phi_{a_3} = -V_{a_3} \sum_{b_3 \neq a_3} \Phi_{b_3}. \tag{4.83}$$

With the help of these components, we find the A_2-components according to the formulae

$$\Phi_{a_3 a_2} = -R_{a_3}(E+i0)V_{a_3} \sum_{(b_3 \neq a_3) \subset a_2} \Phi_{b_3}. \tag{4.84}$$

The inverse relation has the form

$$\Phi_{a_3} = \sum_{a_2 \supset a_3} \Phi_{a_3 a_2}.$$

Multiplying both sides of equation (4.84) by the operator $H_{a_3} - E$, we get the relation

$$(H_0 + V_{a_3} - E)\Phi_{a_3 a_2} = -V_{a_3} \sum_{(b_3 \neq a_3) \subset a_2} \Phi_{b_3}.$$

Next, we express the b_3-components on the right hand side in terms of the B_2-component $\Phi_{b_3 b_2}$ and move the term with $b_2 = a_2$ to the left hand side. We get the following system of differential equations for A_2-component:

$$(H_0 + V_{a_3} - E)\Phi_{a_3 a_2} + V_{a_3} \sum_{(b_3 \neq a_3) \subset a_2} \Phi_{b_3 a_2} =$$

$$= -V_{a_3} \sum_{b_2 \neq a_2} \sum_{(b_3 \neq a_3) \subset a_2} \Phi_{b_3 b_2}. \tag{4.85}$$

This is the desired generalization of the equations (4.83) for the case of four-body system. To determine components of the wave functions uniquely, we must specify the asymptotic boundary conditions. These conditions will be described on the end of this section.

We write the equations (4.85) in details in the case of four identical particles. We denote by a_2 the three-particle partition (123)(4) and by b_2 the two-particle partition (12)(34). We denote by a_3 and b_3 the partitions into three clusters following a_2 and b_2, respectively and corresponding to the pair (12): $a_3 = (12)(3)(4)$, $b_3 = = (12)(3)(4)$. Since all particles are identical, the system of 16 equations is equivalent to two linked equations for the components $\Phi_{a_2 a_3}$ and $\Phi_{b_2 b_3}$ which, however, depend on the Jacobi coordinates of different kinds:

$$
\begin{aligned}
&(-\Delta_X + v(x_{12}) - E)\Phi_{A_2}(x_{12}, y_3^{(12)}, y_4) + \\
&+v(x_{12})(\Phi_{A_2}(x_{23}, y_1^{(23)}, y_4) + \Phi_{A_2}(x_{31}, y_2^{(31)}, y_4)) = \\
&= -v(x_{12})(\Phi_{A_2}(x_{23}, y_4^{(23)}, y_1) + \Phi_{A_2}(x_{13}, y_4^{(13)}, y_2) + \\
&+\Phi_{B_2}(x_{23}, y_{14}, y_{14}) + \Phi_{B_2}(x_{13}, y_{24}, y_{13})); \qquad (4.86) \\
&(-\Delta + v(x_{12}) - E)\Phi_{B_2}(x_{12}, y_{34}, y_{12}) + \\
&+V(x_{12})\Phi_{B_2}(x_{34}, y_{12}, y_{12}) = \\
&= -v(x_{12})(\Phi_{A_2}(x_{34}, y_1^{(34)}, y_2) + \Phi_{A_2}(x_{34}, y_2^{(34)}, y_1))
\end{aligned}
$$

The Jacobi coordinates which are used in individual terms on the right hand sides of these equations correspond to chains of partitions over which the summation in (4.85) runs. Here

$$
y_4 = y_{123,4}, \quad y_3^{(12)} = y_{12,3}, \quad y_{12} = y_{12,34}
$$

and the similar shortened notation is used for the other indices.

The next problem we are going to consider is the relation between these equations and the compact integral equations for the wave functions. To solve it, we consider the differential equations for components of the resolvent.

4.5.2 Differential equations for components of the resolvent for the four-body system

Let $R(z)$ be the resolvent of the energy operator of the four-body system. In analogy with (3.57), we determine the a_3-components of the resolvent by

$$R_{a_3b_3}(z) = -R_0(z)M_{a_3b_3}(z)R_0(z) \tag{4.87}$$

where the T-matrix components $M_{a_3b_3}(z)$ are given in terms of $R(z)$ by equation (3.83) for $N = 4$. The resolvent can be expressed as

$$R(z) = R_0(z) + \sum_{a_3b_3} R_{a_3b_3}(z).$$

Next we introduce the A_2-components $R_{A_2B_2}$ which correspond to the chains of partitions:

$$R_{A_2B_2} = -R_{a_3}V_{a_3}R_0\delta_{a_3b_3}\delta_{a_2b_2} - R_{a_3}V_{a_3} \sum_{(d_3\neq a_3)\subset a_2} R_{d_3b_2}. \tag{4.88}$$

These components are connected with the components of T-matrix $M_{A_2B_2}$ above by the equations

$$R_{A_2B_2} = -R_0M_{A_2B_2}R_0.$$

Note that from the sum rule (3.101) follows the relation inverse to (4.88)

$$R_{a_3b_3} = \sum_{a_2} R_{a_3a_2,b_3b_2}$$

for any b_2. This relation makes it possible to express $R(z)$ as

$$R(z) = R_0(z) + \sum_{a_2}\sum_{a_3,b_3} R_{a_3a_2,b_3b_2}(z).$$

where, when summing over b_2, we choose partitions b_2 satisfying the condition $b_2 \supset b_3$.

Now we deduce the differential equations for the components $R_{A_2B_2}$. To this end, as in the case of wave functions, we replace a_2-components of $R_{d_3b_3}$ on the right hand side of (4.88) by the corresponding expressions in terms of A_2-components $R_{A_2B_2}$. Then we multiply from left by the operator $H_0 + V_{a_3} - z$ and transfer to the left hand side the diagonal terms with $d_2 = a_2$. As a result, we obtain the system of differential equations similar to (4.85) for kernels of the operators $R_{A_2B_2}$ in configuration space called A_2-components of the Green functions

$$\begin{aligned}
&(-\Delta_X + v_{a_3}(x_{a_3}) - z)R_{A_2B_2}(X,X',z)+\\
&+v_{a_3}(x_{a_3}) \sum_{(c_3\neq a_3)\subset a_2} R_{c_3a_2,B_2}(X,X',z) =\\
&= -v_{a_3}(x_{a_3})R_0(X,X',z)\delta_{a_3b_3}\delta_{a_2b_2}-\\
&-v_{a_3}(x_{a_3}) \sum_{d_2\neq a_2} \sum_{(d_3\neq a_3)\subset a_2} R_{d_3d_2,B_2}(X,X',z).
\end{aligned} \tag{4.89}$$

Now we show that this system is formally equivalent to the system of compact integral equations (3.103).

Let $R_{A_2B_2}(X,X',z)$ be a solution of equation (4.89) which is smooth for $X \neq X'$. Multiplying (4.89) by the operator $R_{a_3}(z)$ and using the relation $R_{a_3}V_{a_3} = R_0T_{a_3}$, we rewrite this system in the form

$$\begin{aligned}
&R_{A_2B_2} + R_0T_{a_3} \sum_{(c_3\neq a_3)\subset a_2} R_{c_3a_2,B_2} =\\
&= -R_0T_{a_3}R_0\delta_{a_3b_3}\delta_{a_2b_2} - R_0T_{a_3} \sum_{d_2\neq a_2} \sum_{(d_3\neq a_3)\subset a_2} R_{d_3d_2,B_2}
\end{aligned} \tag{4.90}$$

Further, let us note that the matrix integral operator generated by the left hand side of this system has a block structure which corresponds to the two-cluster partitions a_2. We encountered a similar operator already when deducing the integral equations. In order to find the inverse of this operator, we consider once more the degenerate four-body problems corresponding to the partitions a_2. The corresponding components of T-matrix determined by relation (3.97) for $k = 2$ and $N = 4$ fulfil the system of compact equations

$$M^{(a_2)}_{a_3c_3} + T_{a_3}R_0 \sum_{(d_3\neq a_3)\subset a_2} M^{(a_2)}_{d_3c_3} = T_{a_3}\delta_{a_3c_3}. \tag{4.91}$$

Comparing the left hand sides of (4.91) and (4.90), we observe that the matrix integral operator corresponding to a certain block in (4.90) equals, up to the factor R_0, the operator on the left hand side of (4.91) which corresponds to the same partition a_2. Thus one can invert the left hand side of (4.90) in terms of the operator $M^{(a_2)}_{a_3c_3}$; we encountered a similar situation when we deduced the integral equations (3.103). As a result, we get the system of integral equations for the components $R_{A_2B_2}$:

$$R_{A_2B_2} = R^{(a_2)}_{a_3b_3} - R_0 \sum_{\substack{d_2 \neq a_2 \\ (d_3 \neq c_3) \subset a_2}} M^{(a_2)}_{a_3c_3} R_{d_3d_2,B_2} \tag{4.92}$$

where $R^{(a_2)}_{a_3b_3} = -R_0 M^{(a_2)}_{a_3b_3} R_0$. Thus we have shown that the solution of the differential equations (4.89) satisfies also the integral equations (4.92), which are equivalent to the system (3.103).

The inverse is also true. Let $R_{A_2B_2}$ be a solution of the system of integral equations (4.92). Since the inverse to the transformations from (4.90) to (4.92) exists, the components $R_{A_2B_2}$ satisfy equations (4.(90). Differentiating these equations, we find that solutions of the compact integral equations satisfy also the differential equations (4.89).

The differential equations for the A_2-components of the wave function can be obtained in a usual manner from the differential equations for components of the resolvent (4.90). Recall that the wave function $\Psi^{(\pm)}_A$ corresponding to the channel A is defined by relation (2.3). In accordance with this equation, we introduce the components of the wave function according the formulae

$$\Phi^{(\pm)}_{B_2A} = \lim_{\epsilon \downarrow 0} \mp i\epsilon \sum_{c_3} R_{B_2C_2}(E_A \pm i\epsilon) L_A(E_A). \tag{4.93}$$

The wave function $\Psi^{(\pm)}_A$ can be expressed as

$$\Psi^{(\pm)}_A(X, p_A) = e^{i(P,X)} \delta_{k4} + \sum_{B_2} \Phi^{(\pm)}_{B_2A}(X, p_A).$$

For $k = 4$, the plane wave $e^{i(P,X)}$ appears on the right hand side; this case corresponds to four free particles in the initial state. Taking in account this definition, we get from (4.89) the compact equations in differential form for the components Φ_{B_2A}:

$$(-\Delta_X + v_{b_3}(x_{b_3}) - E_A)\Phi_{AB_2}(X, p_A) +$$

$$+ v_{b_3}(x_{b_3}) \sum_{(c_3 \neq b_3) \subset b_2} \Phi_{c_3b_2, B_2}(X, p_A) =$$

$$= -v_{b_3}(x_{b_3}) e^{i(P,X)} \delta_{k4} \delta_{b_2 b_2'} -$$

$$-v_{b_3}(x_{b_3}) \sum_{\substack{d_2 \neq b_2 \\ (d_3 \neq b_3) \subset b_2}} \Phi_{D_2 A}(X, p_A). \tag{4.94}$$

Here $E_A = p_A^2 - \varkappa_A^2$ and b_2' is an arbitrary partition satisfying the condition $b_2' \supset b_2$. Note that summation of this equation over all B_2 leads to the Schrödinger equation for the functions $\Phi_A(X, p_A) = \Psi_A(X, p_A) - e^{i(P,X)}\delta_{k4}$, to wit

$$(-\Delta_X + V(X) - E)\Phi_A(X, p_A) = -V(X)e^{i(P,X)}\delta_{k4}.$$

These equations are non-homogeneous for $k = 4$. We will denote by $\Phi_{B_2}^{(0)}$ the components of wave function in this case. In order to get the homogeneous differential equations, we introduce new components by the relations

$$\Phi_{B_2,0}(X, P) = \Phi_{B_2}^{(0)} - e^{i(P,X)} \left(\sum_{(c_3 \neq b_3) \subset b_2} \delta_{c_3 f_3} - \delta_{b_2 b_2'} \right)$$

where f_3 is an arbitrary partition into three subsystems and the partition b_2 is determined as in (4.94). Inserting this expression into equation (4.94), we find that the components $\Phi_{B_2,0}$ satisfy the system of homogeneous equations (4.85). The complete wave function Ψ_0 is obtained from the components $\Phi_{B_2,0}$ by the summation over all chains:

$$\Psi_0(X, P) = \sum_{B_2} \Phi_{B_2,0}(X, P).$$

It can be shown that these components are equal to the components defined by equation (4.84).

Now we turn to deducing the differential equations for N-body system.

4.5.3 Differential equations for components for the N-body system

Let $R_{a_{N-1} b_{n-1}}$ be components of the resolvent of the N-body problem, constructed with the help of the formulae (4.87) in which the symbols a_3 and b_3 should be replaced by a_{N-1} and b_{N-1}, respectively. We define A_N-components of the resolvent $R_{A_k B_k}$, $2 \leq k < N-1$ by means of the recurrent formulae

$$R_{A_k B_k} = R^{(a_{k+1})}_{A_{k+2} B_{k+2}} \delta_{a_{k+1} b_{k+1}} \delta_{a_k b_k} -$$

$$-\sum_{\substack{d_{k+1}\neq a_{k+1}\\ d_{k+1}\subset a_k}} \sum_{(D_{k+2}\neq C_{k+2})\subset a_{k+1}} R_0 M^{(a_{k+1})}_{A_{k+2}C_{k+2}} R_{D_{k+1}B_{k+1}} \tag{4.95}$$

where $R^{(a_{k+1})}_{A_{k+2}B_{k+2}} = -R_0 M^{(a_{k+1})}_{A_{k+2}B_{k+2}} R_0$. The operators $M^{(a_{k+1})}_{A_{k+2}B_{k+2}}$ are determined by relations (3.98). Using the sum rule (3.101), we get the inverse relation

$$R_{A_{k+1}B_{k+1}} = \sum_{a_k} R_{A_k B_k} \quad \text{for all } b_k \ . \tag{4.96}$$

According to (3.103), the operators $R_{A_k B_k}$ satisfy the system of integral equations

$$R_{A_k B_k} = R^{(a_k)}_{A_{k+1}B_{k+1}} \delta_{a_k b_k} -$$

$$-\sum_{d_k \neq a_k} \sum'_{(D_{k+1}\neq C_{k+1})\subset a_k} R_0 M^{(a_k)}_{A_{k+1}C_{k+1}} R_{d_k D_{k+1} B_k} .$$

Let us show that the components $R_{A_k B_k}$ satisfy also the differential equations. First we prove the following theorem.

For any $k, 2 \leq k \leq N-1$, the kernels of operators $R_{A_k B_k}$ satisfy the following system of differential equations

$$(-\Delta_X + v_{a_{N-1}}(x_{a_{N-1}}) - z) R_{A_k B_k}(X, X', z) +$$

$$+ v_{a_{N-1}}(x_{a_{N-1}}) \sum''_{(C_{k+1}\neq A_{k+1})\subset a_k} R_{a_k C_{k+1}, B_k}(X, X', z) =$$

$$= -v_{a_{N-1}}(x_{a_{N-1}}) R_0(X, X', z) \delta_{a_k b_k} \ldots \delta_{a_{N-1} b_{N-1}} -$$

$$-v_{a_{N-1}}(x_{a_{N-1}}) \sum'_{\substack{d_k \neq a_k\\ (A_{k+1}\neq D_{k+1})\subset a_k}} R_{d_k D_{k+1} B_k}(X, X', z), \tag{4.97}$$

where

$$\sum''_{(C_{k+1}\neq A_{k+1})\subset a_k} R_{a_k C_{k+1}, B_k} = \sum_{i=k+1}^{N-1} \sum_{(C_i \neq A_i)\subset a_{i+1}} R_{a_k a_{k+1} \ldots a_{i-1} C_i, B_k} .$$

First we note that the assertion is true for $k = N-1$ and $k = N-2$, because in these cases, the system (4.97) is identical with the system of differential equations for a_{N-1}-component (4.83) and the system (4.89), respectively.

In the next step we show that it is valid also for $k = N-3$. Let us note that, as a consequence of (4.95) and (4.96), the components $R_{A_{N-3}B_{N-3}}$ satisfy the equations

$$R_{A_{N-3}B_{N-3}} = R^{(a_{N-2})}_{a_{N-1}b_{N-1}}\delta_{a_{N-3}b_{N-3}}\delta_{a_{N-2}b_{N-2}} -$$

$$- \sum_{(d_{N-2}\neq a_{N-2})\subset a_{N-3}} \sum_{(d_{N-1}\neq c_{N-1})\subset a_{N-2}} \sum_{d_{N-2}} R_0 R^{(a_{N-2})}_{a_{N-1}b_{N-1}} \times$$

$$\times R_{d_{N-3}D_{N-2},B_{N-3}}. \tag{4.98}$$

To get the differential equations (4.97), we use the fact that the components $R^{(a_{N-2})}_{a_{N-1}b_{N-1}}$ satisfy also the system of differential equations

$$(-\Delta + v_{a_{N-1}} - z)R^{(a_{N-2})}_{a_{N-1}b_{N-1}} +$$

$$+ v_{a_{N-1}} \sum_{c_{N-1}\neq a_{N-1}} R^{(a_{N-2})}_{a_{N-1}b_{N-1}} =$$

$$= -v_{a_{N-1}} R_0 \delta_{a_{N-1}b_{N-1}}, \tag{4.99}$$

where, for simplicity, we omit the arguments of the functions $v_{a_{N-1}}(x_{a_{N-1}})$ and $R^{(a_{N-2})}_{a_{N-1}b_{N-1}}(X, X', z)$. Clearly, that the kernels of the operators $R_0(z)M^{(a_{N-2})}_{a_{N-1}c_{N-1}}(z)$ satisfy the system (4.99) with the free term replaced by $V_{a_{N-1}}\delta_{a_{N-1}b_{N-1}}$. Further, we see that the matrix integral operator on the right hand side of (4.98) has a block structure corresponding to the partitions a_{N-1} and fixing a_{N-2}, we multiply this equation by the differential operator generated by the left hand side of (4.99) for the corresponding a_{N-2}. Using (4.99) on the right hand side of the system (4.98), we obtain the following system of differential equations for $R_{A_{N-3}B_{N-3}}$:

$$(-\Delta_X + v_{a_{N-1}} - z)R_{A_{N-3}B_{N-3}} +$$

$$+ v_{a_{N-1}} \sum''_{(c_{N-1}\neq A_{N-1})\subset a_{N-2}} R_{a_{N-3}a_{N-2}c_{N-1},B_{N-3}} =$$

$$= -v_{a_{N-1}} R_0 \delta_{a_{N-3}b_{N-3}}\delta_{a_{N-2}b_{N-2}}\delta_{a_{N-1}b_{N-1}} -$$

$$- v_{a_{N-1}} \sum_{d_3} \sum'_{(D_{N-2}\neq A_{N-2})\subset a_k} R_{d_{N-3}D_{N-2}B_{N-3}}.$$

Moving the term with $d_{N-3} = a_{N-3}$ to the left hand side, we obtain the differential equations (4.97) for $R_{A_{N-3}B_{N-3}}$.

Repeating this procedure, we prove the validity of (4.97) for all k.

Thus, we have described the family of the differential equations satisfied by A_k-component of resolvents of different kinds. For $k = 2$, the system of equations (4.97) represents a generalization of the compact equations in the differential form to the case of N bodies.

We will describe shortly the differential equations for components of wave functions. These equations can be obtained in a usual way from the differential equations for components of the resolvents. We define components of the wave function Ψ_A corresponding to the l-cluster channel by the relations

$$\Phi^{(\pm)}_{B_k A} = \lim_{\epsilon \downarrow 0} \mp i\epsilon \sum_{c_{N-1}} R_{B_k C_k}(E_A \pm i\epsilon) L_A. \tag{4.100}$$

The wave function Ψ_A is given in the terms of these components by

$$\Psi^{(\pm)}_{B_k A}(X, p_A) = \exp\{i(P, X)\}\delta_{lN} + \sum_{B_k} \Phi^{(\pm)}_{B_k A}(X, p_A). \tag{4.100'}$$

Using equations (4.97) and the definition of components, we obtain the following system of equations

$$\begin{aligned}
&(-\Delta_X + v_{b_{N-1}}(x_{b_{N-1}}) - E_A)\Phi_{B_k A}(X) + \\
&+ v_{b_{N-1}}(x_{b_{N-1}}) \sum_{(C_{k+1} \neq A_{k+1}) \subset a_k}^{\prime\prime} \Phi_{b_k C_{k+1} A}(X) = \\
&= e^{i(P,X)} - v_{b_{N-1}}(x_{b_{N-1}})\delta_{lN}\delta_{b_k c_k} \ldots \delta_{b_{N-1} c_{N-1}} - \\
&- v_{b_{N-1}}(x_{b_{N-1}}) \sum_{\substack{d_k \neq b_k \\ (B_{k+1} \neq D_{k+1}) \subset b_k}}^{\prime} \Phi_{D_k A}(X),
\end{aligned} \tag{4.101}$$

where the choice of c_{N-2} is restricted by the condition $c_{N-2} \subset b_{N-1}$. For $k = 2$, we get the desired generalization of the compact equations in the differential form which are fulfilled by components of the wave function in the case of a N-body system. As in the case of four bodies, the summation in (4.100) over all B_k leads to the Schrödinger equation for $\Phi^{(\pm)}_A = \Phi^{(\pm)}_A - \delta_{lN} e^{i(P,X)}$.

4.5.4 Asymptotic boundary conditions

In order to determine the solution representing components of the wave function uniquely, we must add the asymptotics boundary conditions to the differential equation for the components. As in the case of the three-body system the boundary conditions can be determined starting from the coordinate asymptotics of the Fourier transform of the wave operator kernels. However, in contrast to the three-body problem, a complete description of singularities of these kernels (including the poles of order two which correspond to rescattering of clusters) has not be performed up to now. Therefore, we restrict ourselves to the description of the asymptotics of wave functions for the processes with two clusters in the initial state in which case the singularities of the order two are disappear.

Thus we will consider the wave functions $\Psi_A^{(\pm)}(X, p_A)$ where the symbol A corresponds to two-cluster partition. We perform the procedure in the case of functions with $(+)$ sign. The asymptotics of the functions with $(-)$ sign is obtained by means of relation (4.16"). First of all, we perform partition of the wave function and of the function of initial state to their components. Let $\psi_A^{(B_j)}$ be components of eigenfunctions of the internal Hamiltonian determined by the equation

$$\psi_A^{(B_j)} = r_0^{(int)}(-\varkappa_A^2)\Phi_A^{(B_j)} \tag{4.102}$$

where $\Phi_A^{(B_j)}$ are components of the form-factor Φ_A. We define the components of the initial state by the formulae

$$\chi_A^{(B_j)} = \psi_A^{(B_j)}(x_A)\exp\{p_A, y_A)\}.$$

The components of the wave function $\Phi_{B_2 A}(X, p_A)$ are given by the relations

$$\Phi_{B_2 A}(X, p_A) =$$

$$= (2\pi)^{\frac{3}{2}(l_A - N)} \int dP' \exp\{i(X, P')\} U_{B_2 A}(P', p_A)$$

where l_A is the number of clusters of the detailed partition A and $U_{B_2 A}$ are components of wave operators kernels (3.98'). Moreover, the wave function is equal to the sum (4.100'). Let us note that the kernels $U_{B_2 A}(P', p_A)$ have singularities generated by the poles extracted in representations (2.23) and (3.108). These components can be written in a form of the sum

$$U_{B_2A}(P',p_A) = L_A^{(B_3)}(P',p_A)\delta_{a_2b_2} + \frac{T_{B_2A}(P',p_A)}{P'^2 - E_A(p_A) - i0} \tag{4.103}$$

where $L_A^{(B_3)}(P',p_A)$ are the Fourier transforms of the functions $\chi_A^{(B_j)}(P',p_A)$. The kernels $T_{B_2A}(P',p_A)$ possess cluster singularities which are reflected in the relation

$$T_{B_2A}(P',p_A) = G_{B_2A}(P',p_A) + \\ + \sum_{l=2}^{N-1} \frac{\Phi_A^{(B_{l+1})}(k'_B)}{E_B(p'_B) - E_A(p_A) - i0} H_{BA}^{(b_2\cdots b_{l-1}}(p_B,p_A) \tag{4.104}$$

where the kernels $H_{BA}^{(b_2\cdots b_{l-1}}$ and G_{B_2A} are smooth bounded functions. The index B corresponds to the detailed partition I_{B_l}

Let us recall that the denominators in the formulae (4.103) and (4.104) vanish in different parts of the space R^6. It allows for division the region of integration into parts in such a way that the function U_{B_2A} has only one pole in each of them. The asymptotics of integrals generated in this way can be computed by means of the formulae (4.119) which will be presented in Section 6. Each pole $(E_B(p'_B) - E_A(p_A) - i0)^{-1}$ generates a spherical wave in $(3(l-1)$-dimensional space

$$Q_B(y_B,E) = \frac{\exp\{i\sqrt{E + \varkappa_B^2}|y_B|\}}{|y_B|^{(3l-4)/2}}$$

where l is a number of clusters of detailed partition B. The case $l = N$ $(B = 0,\ \varkappa_B^2 = 0)$ corresponds to the case of N free particles. As a result we obtain the following representation:

$$\Phi_{B_2A}(X,p_A) = \chi_A^{(B_3)}(X,p_A)\delta_{a_2b_a} + \\ + \sum_{l=3}^{N}\sum_{B} \psi_B^{(B_l)}(x_B)\sigma_{BA}^{(b_2\cdots b_{l-1}}(y_B,p_A) + \rho_{B_2A}(X,p_A). \tag{4.105}$$

The functions $\sigma_{BA}^{(b_2\cdots b_{l-1}}$ turn asymptotically into the spherical waves in a $3(l-1)$-dimensional space

$$\sigma_{BA}^{(b_2\cdots b_{l-1}} \sim \\ \sim (2\pi)^{\frac{3}{2}(l_A - l_B)} C_l H_{BA}^{(b_2\cdots b_{l-1}}(|p_A|\hat{y}_B,p_A)Q_B(y_B,E_A). \tag{4.106}$$

Here $p_B^2 = E_A(p_A) + \varkappa_B^2$, $C_i = i\pi(2\pi)^{(3l-4)/2}\frac{E^{(3l-4)/4}}{l^{i\pi(3l-4)/4}}$. These functions correspond to the processes $(A \to B)$ of formation of l clusters in the final state $l \equiv l_B$. The term ρ_{B_2A} represents decay of the system into a system of N free particles. This term is asymptotically equal to the spherical wave in the space $R^{3(N-1)}$:

$$\rho_{B_2A}(X, P_A) \sim C_N T_{B_2A}(\sqrt{E_A}\hat{X}, p_A) Q_0(X, E_A). \tag{4.107}$$

From (4.105) we see that the asymptotics of the component Φ_{B_2A} for the partition B_2 contains only those spherical waves which correspond to the partition following B_2.

The wave functions $\Psi_A(X, p_A)$ are the sum of components $\Phi_{B_2A}(X, p_A)$. Thus, their asymptotics will contain spherical waves corresponding to all opened channels:

$$\Psi_A(X, p_A) = \chi_A(X, p_A) + \\ + \sum_{\substack{B \\ -\varkappa_B^2 \leq E_B}} F_{BA}(\hat{y}_A, p_A) Q_B(y_B, E_A). \tag{4.108}$$

The amplitudes of these waves are determined in terms of the components by

$$F_{BA}(\hat{y}_A, p_A) = \\ = \sum_{b_2 \ldots b_{l-1}} (2\pi)^{\frac{3}{2}(l_A - l_B)} C_{l_B} H_{BA}^{(b_2 \cdots b_{l-1}}(|p_B|\hat{y}_B, p_A), \quad B \equiv I_M,$$

$$F_{0A}(\hat{y}_A, p_A) = \sum_{B_2} C_N T_{B_2A}(\sqrt{E_A}\hat{X}, p_A).$$

Finally, let us note that if three or more clusters are present in the initial state, the asymptotics of the wave functions is not given by the superposition of cluster plane waves and scattered spherical waves only; the slowly decreasing components corresponding to the rescattering of clusters also appear.

After we have described the coordinate asymptotics of wave functions and of their components, we can formulate the boundary value problems for them using the Schrödinger equation and differential equations for the components.

The differential equations for components (4.101) possess a unique solution in the class of smooth functions with the asymptotics (4.105) – (4.107). At the same time, the wave function is given by the sum over all partitions

(4.100'). The boundary conditions for the Schrödinger equation are determined by equations (4.108). The solution of this equation is also determined uniquely by these conditions.

Differential formulation of the scattering problem can be given with the help of considerations similar to those we were using in the case of the three-body problem. In particular, the asymptotics representation of the Green function is used, which is given by the solution of the equation

$$\left(-\Delta_X + \sum_{a_{N-1}} v_{a_{N-1}}(x_{a_{N-1}}) - z\right) R(X, X', z) = \delta(X - X'). \tag{4.109}$$

We summarize shortly the properties of these functions. For $X \neq X'$, the Green function $R(X, X', z)$ is a smooth bounded function; for $X = X'$ it has a singularity of the form

$$\frac{\Gamma\left(\frac{3N-5}{4}\right)}{4}(2\pi)^{\frac{-3N+3}{2}}|X - X'|^{-3N+5}.$$

If $\operatorname{Im} z \neq 0$, the function rapidly decreases for $|X - X'| \to \infty$. Its asymptotics for real $z = E \pm i0$ and $|X'| \to \infty$ is given by the sum of the spherical waves

$$R(X, X', E \pm i0) \underset{|X'| \to \infty}{\sim} \sum_{\substack{B \\ -\varkappa_B^2 \leq E_B}} A_B^{(\pm)}(X, p_B) \times$$

$$\times \psi_B(x'_B) Q_B^{(\pm)}(y'_B, E), \tag{4.110}$$

where the summation runs over all opened channels. The momentum variables p_B are given by

$$p_B = \mp \frac{\sqrt{E}}{|X|} y'_B; \quad Q_B^{(-)} = Q^{(-)*}_B, \quad Q_B^{(+)} \equiv Q_B.$$

The amplitudes of spherical waves $A_B^{(\pm)}$ can be explicitly expressed in terms of wave functions corresponding to the detailed partition B:

$$A_B^{(\pm)}(X, p_B) = \frac{1}{4\pi}\left(\frac{|p_B|}{2\pi}\right)^{(3l_B-6)/2} e^{\mp\frac{\pi}{4}(3l_B-6)} \Psi_B^{(\pm)}(X, p_B). \tag{4.111}$$

Finally, the asymptotics for $|X| \to \infty$ can be obtained from these formulae with the aid of the symmetry conditions

$$R(X, X', z) = \overline{R(X, X', \bar{z})}.$$

In many cases, it is necessary to know the asymptotics of the Green function of the Laplace operators in the $3(N-1)$-dimensional space $R_0^{(N)}(X, X', z)$. These function are known in the closed form

$$R_0^{(N)}(X, X', z) = \frac{1}{4}\left(\frac{\sqrt{z}}{2\pi}\right)^{(3N-5)/2} \frac{H^{(1)}_{(3N-5)/2}(\sqrt{z}|X - X'|)}{|X - X'|^{(3N-5)/2}} \tag{4.112}$$

so that their asymptotics is determined by the Hankel function. For example, if $|X'|rar\infty$ and the variable X is fixed, we get

$$R_0^{(N)}(X, X', E + i0) \sim C_0^{(N)} e^{i(P,X)} \frac{\exp\{i\sqrt{E}|X|\}}{|X|^{(3N-4)/2}}$$

where $P = -\sqrt{E}\hat{X}'$ and

$$C_0^{(N)} = \frac{1}{4\pi}\left(\frac{\sqrt{z}}{2\pi}\right)^{(3N-5)/2} e^{-i\frac{\pi}{4}(3N-6)}.$$

Let us note that these functions can be determined with the help of the Fourier transforms

$$R_0^{(N)}(X, X', z) = (2\pi)^{-3(N-1)/2} \int dP \frac{\exp\{i(P, X - X')\}}{P^2 - z}. \tag{4.113}$$

At this point we finish our discussion of the problem of N neutral particles.

4.6 Rapidly Oscillating Integrals

This section represents a mathematical appendix in which we describe asymptotics of integrals of the form

$$I(\lambda, q) = \int e^{i\lambda\varphi(q,q')} f(q, q', \lambda)\, dq' \tag{4.114}$$

for $\lambda \to \infty$

The main part of the results contained here is well known and we present them only for completeness.

To avoid complicated estimates of the remaining terms in the integral (4.114), we assume that the function f is finite.

We consider first the simplest case of functions φ and f depending on the integration variable q' only.

If the function $\varphi(q')$ has no critical points, the asymptotics $I(\lambda)$ is determined by the behaviour of the function f on the boundary of the integration region. For a finite function $f(q')$ the integral $I(\lambda)$ decreases rapidly, as arbitrary power of λ^{-N}, $N \gg 1$ If there exists a critical point, the principal value is generated by its neighbourhood. In order to describe the asymptotics of $I(\lambda)$ in this case, we make several assumptions concerning the behaviour of the function $\varphi(q')$.

Let $\varphi(q)$ has only one non-degenerate critical point q_0, so that

$$\nabla_q \varphi(q)|_{q=q_0} = 0, \quad \det \frac{\partial^2 \varphi}{\partial q^2}\bigg|_{q=q_0} \neq 0$$

and the matrix $\frac{\partial^2\varphi}{\partial q^2}$ has ν_+ positive and ν_- of negative eigenvalues. We denote by sign $\frac{\partial^2\varphi}{\partial q^2}$ the signature of the quadratic form determined by the matrix $\frac{\partial^2\varphi}{\partial q^2}$:

$$\operatorname{sign} \frac{\partial^2 \varphi}{\partial q^2} = \nu_= - \nu_-, \quad \left(\frac{\partial^2 \varphi}{\partial q^2}\right)_{ij} = \frac{\partial^2 \varphi}{\partial q_i \partial q_j},$$

$$q = (q_1, q_2, \ldots, q_n).$$

The following asymptotics formula holds:

$$I(\lambda) \sim \left(\frac{2\pi}{\lambda}\right)^{n/2} \left|\det \frac{\partial^2 \varphi(q_0)}{\partial q^2}\right|^{-1/2} f(q_0) \times$$

$$\times \exp\left\{i\lambda\varphi(q_0) + i\frac{\pi}{4}\operatorname{sign}\frac{\partial^2\varphi}{\partial q^2}\right\}\left(1 + O\left(\frac{1}{\lambda}\right)\right). \tag{4.115}$$

We make several comments concerning the proof of this formula. The main contribution to the integral comes from the points inside the sphere of the radius $\lambda^{-1/2+\epsilon}$ with the centre at the point q_0. Inside this sphere the function $\varphi(q)$ is given to the higher order by the equation

$$\varphi(q) = \varphi(q_0) + \frac{1}{2}\left(\frac{\partial^2\varphi}{\partial q^2}(q - q_0), q - q_0\right) + O(|q - q_0|^3).$$

Introducing the basis which reduces the matrix $\frac{\partial^2\varphi}{\partial q^2}$ to the diagonal form, the integral $I(\lambda)$ can be expressed to the higher order as the product of n standard integrals

$$\int_{-\infty}^{\infty} e^{i\lambda\nu_i t^2}\, dt = \sqrt{\frac{2\pi}{\lambda|\nu_i|}}\, e^{i\frac{\pi}{4}\operatorname{sign}\nu_i} \tag{4.116}$$

and of the factor $f(q_0)\exp\{i\lambda\varphi(q_0)\}$. The resulting expression takes the form (4.115).

In most cases considered above, the function f depend on some additional variables and, moreover, on the large parameter λ. The function φ usually also depends on additional parameters. In order to justify the usage of the formula (4.115) based on the method of stationary phase, we must make some assumptions concerning the character of oscillations of the function $f(q, q', \lambda)$.

We will say that a smooth function $f(q, q', \lambda)$ is a slowly oscillating function of a large parameter λ if it satisfies together with its derivatives the relations

$$\left|D_q^{\varkappa} D_{q'}^{\varkappa'} \frac{\partial^n}{\partial\lambda^n} f(q, q', \lambda)\right| \le C\lambda^{m-n}, \quad D_q^{\varkappa} = \frac{\partial^{|\varkappa|}}{\partial q_1^{\varkappa_1}\ldots\partial q_n^{\varkappa_n}},$$

$$|\varkappa| = \varkappa_1 + \varkappa_2 + \cdots + \varkappa_n \tag{4.117}$$

where m is a fixed number. This estimation guarantees that the ordinary asymptotic formulae can be used.

Indeed, it can be shown that if the function $\varphi(q, q')$ has a unique non-degenerate critical point for all q from the domain of definiteness and the function $f(q, q', \lambda)$ is a slowly oscillating function, then the asymptotics of the integral (4.114) is also given by the formula (4.115). The large parameter λ can take its values also in the complex plane. The corresponding asymptotics formulae can be found with the help of the saddle-point method. We are interested in the only case which has a practical significance, i.e., the case of the parameter in the form of product $\lambda\sqrt{z}$, where z belongs to the complex plane with the cut Π_0. It can be shown that if all eigenvalues of the matrix $\frac{\partial^2\varphi}{\partial q^2}$ are negative, the contribution of the critical point is described by the formula (4.115), where the branch of the function $(\sqrt{z})^{-n/2}$ is chosen on the basis of the conditions $\arg(\sqrt{z})^{-n/2} = 0$ for $z = 0$.

If the function f possesses singularities, the main contribution to the asymptotics is given both by the critical points and by the singular points. The most complicated problem is to describe the asymptotics in the case of critical points approaching the singular points. We consider only some special examples of the integrals of this type.

In this section we have been often encountering the integrals of the form

$$I(q,z) = \int dq \frac{e^{i(q,x)}g(q)}{q^2 - E \mp i0} \tag{4.118}$$

where g is a smooth function. The following propositions describe the asymptotics of these integrals for $|x| \to \infty$.

Proposition A. *The asymptotics formula*

$$I(x,E) \sim i\left(\frac{2\pi}{|x|}\right)^{\frac{n-1}{2}} \pi E^{\frac{n-3}{4}} e^{\pm i\sqrt{E}|x| \mp i\frac{(n-1)\pi}{4}} \times$$

$$\times (f(\pm\sqrt{E}\hat{x}) + O(|\sqrt{E}x|^{-1})). \tag{4.119}$$

is satisfied for $|\sqrt{E}x| \to \infty$.

We perform the proof of this formula in two steps. We introduce the spherical coordinates $|q|$ and $\hat{q}$ and we first integrate over the $(n-1)$-dimensional sphere S^{n-1}. The asymptotics of this integral can be found with the help of the saddle-point method. The principal contribution from the critical points $\hat{q} = \pm\hat{x}$ can be represented in the form of the sum

$$I(x,E) \sim \left(\frac{2\pi}{|x|}\right)^{\frac{n-1}{2}} (I^{(+)}(x,E) - I^{(-)}(x,E))$$

where $I^{(\pm)}$ are integrals over the radial variable:

$$I^{(\pm)}(x,E) =$$

$$= \int_0^\infty d\zeta \frac{\exp\left\{\pm i\zeta\sqrt{E}|z| \mp i\frac{(n-1)\pi}{4}\right\}}{\zeta^2 - E \mp i0} \zeta^{\frac{n-1}{2}} f(\pm\zeta\hat{x}).$$

We next note that the principal contributions to the integrals $I^{(\pm)}$ are generated by the poles. They are easily evaluated with the aid of the residue theorem if the path of integration is completed to the closed contour in the neighbourhood of poles. For the integral $I^{(+)}$ the path has to be closed in the upper half-plane and in the case of the integral $I^{(+)}$ in the lower half-plane. If the path of integration is closed in this way, the integrals over the added curves are of lower order. As a result we get the expression (4.119).

The situation is more complicated in the case of nonlinear function $\varphi(q)$. We will not examine this type of integrals in the general case. We consider only the asymptotics of one-dimensional integrals of the form

$$F(\lambda,E) = \int_a^b dt \frac{e^{i\lambda\varphi(t)}}{t - E - i0} f(t) \tag{4.120}$$

where φ and f are smooth functions.

Let us assume first that the function $\varphi(t)$ has no critical points in the interval of integration. In this case, the principal contribution generated by the pole can be evaluated by the same method as in the case of φ linear. The asymptotics formula

$$F(\lambda,E) \sim 2\pi i e^{i\lambda\varphi(E)} f(E)\mathcal{E}(\operatorname{sign}\varphi'(E)) \tag{4.121}$$

is satisfied; here $\mathcal{E}(t)$ is the unit step function

$$\mathcal{E}(t) = \begin{cases} 1, & t > 0 \\ 0, & t < 0. \end{cases}$$

Further, let us assume that the function $\varphi(t)$ has a single critical point t_0 and $\varphi''(t_0) < 0$. The following proposition describes the asymptotics of the integral (4.120) in this case.

Proposition B. *Let $t_0 \neq E$. Then for $\lambda \to \infty$*

$$F(\lambda,E) \sim F_0(\lambda,E) + F_\epsilon(\lambda,E), \tag{4.122}$$

where the function F_ϵ is given by (4.121) and

$$F_0(\lambda, E) = \left(\frac{2\pi}{|\varphi''(t_0)|\lambda}\right)^{1/2} \frac{f(t_0)\exp\{i\lambda\varphi(t_0) - i\frac{\pi}{4}\}}{t_0 - E}.$$

If the critical point t_0 is located in the neighbourhood of a pole so that

$$|t_0 - E| = o(\lambda^{-\nu}), \quad \nu > 0,$$

then the asymptotics of $I(\lambda)$ is given by

$$F(\lambda, E) \sim \tilde{F}_0(\lambda, E) + F_s(\lambda, E), \tag{4.123}$$

where

$$\tilde{F}_0(\lambda_0, E) = \frac{f(t_0) - f(E)}{t_0 - E}\sqrt{\frac{2\pi}{\lambda|\varphi''(t_0)|}}e^{-i\frac{\pi}{4}+i\lambda\varphi(t_0)}, \tag{4.124}$$

and the function F_s is expressed by means of the Fresnel integral $\Phi(t) = \int_t^\infty e^{i\tau^2}d\tau$:

$$F_s(\lambda, E) =$$
$$= 2\sqrt{\pi}e^{3i\frac{\pi}{4}}f(E)\Phi(-\lambda^{1/2}\,\mathrm{sign}\,(t_0 - E)|\varphi(E) - \varphi(t_0)|^{1/2})$$

We will outline the proof of this assertion. We split the integral $F(\lambda, E)$ into two terms according to the formula

$$f(t) = f(E) + (f(t) - f(E)).$$

The integrand of the integral $\tilde{F}_0$ containing the difference $f(t) - f(E)$ is non-singular so that the asymptotics of (4.124) can be found with the help of the method of stationary phase.

The second term F_s can be expressed in the form of a product:

$$f_s(\lambda, E) = e^{i\lambda\varphi(E)}f(E)\tilde{I}(\lambda, E).$$

Differentiating the integral $\tilde{I}$ with respect to the parameter λ, we get

$$\frac{\partial\tilde{I}}{\partial\lambda} = i\int_a^b dt\frac{\varphi(t) - \varphi(E)}{t - E}e^{i\lambda(\varphi(t)-\varphi(E))}.$$

The integrand of the integral $\frac{\partial\tilde{I}}{\partial\lambda}$ is non-singular and its asymptotics is given by

$$\frac{\partial\tilde{I}}{\partial\lambda} \sim i\left(\frac{2\pi}{\lambda(\varphi''(t_0))}\right)^{1/2}\frac{\varphi(t_0) - \varphi(E)}{t_0 - E}\times$$

$$\times \exp\left\{ i\lambda(\varphi(t_0) - \varphi(E) - i\frac{\pi}{4}\right\}.$$

We integrate this equation with respect to λ. Let us note that for $t_0 \neq E$ the relation (4.122) holds; in this way the boundary conditions are determined so that the constant of integration is determined. As a result, we get the asymptotics formula

$$\tilde{I}(\lambda, E) \sim \left(\frac{2\pi}{|\varphi''(t_0)|}\right)^{1/2} \frac{|\varphi(t_0) - \varphi(E)|^{1/2}}{t_0 - E} \times$$

$$\times e^{-i\frac{\pi}{4}} \Phi(\lambda^{1/2}|\varphi(E) - \varphi(t_0)|^{1/2}) + 2\pi i \mathcal{E}(t_0 - E). \tag{4.125}$$

If the critical point t_0 is sufficiently far from the pole, $|t_0 - E| \geq \delta > 0$, the Fresnel integral in (4.125) can be replaced by its asymptotics. In this way we get the expression of the form (4.122) where the contributions from the pole and from the critical point are separated. In the case when the critical point and the pole merge, $|t_0 - E| = o(\lambda^{-\nu})$, in leading order we have the relation

$$\frac{|\varphi(t_0) - \varphi(E)|^{1/2}}{t_0 - E} \sim \sqrt{\frac{|\varphi''(t_0)}{2}} \operatorname{sign}(t_0 - E).$$

Equality (4.123) is a consequence of this relation.

We are particularly interested in the integrals (4.114) in which the functions f has weak singularities. In order to understand the nature of asymptotics of such expressions, we will consider as an example the one-dimensional integral

$$I(\lambda) = \int_{-a}^{a} \frac{f(t)}{t^{1-r}} e^{i\lambda\varphi(t)}\, dt \tag{4.126}$$

where f and φ are holomorphic functions. We will assume that the branch of the function t^{r-1} defined on the plane with the cut Π_0 is determined by the condition $\arg t^{r-1} = 0$ for $\arg t = 0$.

If in the interval of integration $\varphi'(t) \neq 0$ then the leading order terms of the asymptotics are generated by the singular point $t = 0$; they are described by the formula analogous to (4.121):

$$I(\lambda) \sim (i\lambda)^{-r}\Gamma(r) f(0) e^{i\lambda\varphi(0)}. \tag{4.127}$$

If the critical point t_0 exists but $t_0 \neq 0$, then its contributions adds to the contribution from the singular point. At the same time, the representation of the form (4.122) is valid. If the critical point approaches the singular point, then the asymptotics $I(\lambda)$ is described by a special function. In order to find it we expand the function $\varphi(t)$ into the power series in t and retain the terms up to the second order. For $t_0 = O(\lambda^{-1/2+\nu})$ the representation

$$I(\lambda) \sim f(0)\frac{e^{i\lambda\varphi(0)}}{\lambda^{r/2}}B_r(-\Delta) + f'(0)\frac{e^{i\lambda\varphi(0)}}{\lambda^{r/2+1/2}}B_{r+1}(-\Delta) \qquad (4.128)$$

is valid; here $B_r(\xi)$ is the standard integral

$$B_r(\xi) = \int_0^{\infty} t^{1+r}e^{-i\left(\frac{t^2}{2}+\xi t\right)}\,dt; \quad \Delta_\lambda = \lambda^{1/2}(\varphi(t_0) - \varphi(0))^{1/2}.$$

Let us note that $B_r(\xi)$ can be expressed in terms of the parabolic cylindric function $D_\nu(\xi)$:

$$B_r(\xi) = \Gamma(r)\exp\left\{-ir\frac{\pi}{4} + i\frac{\xi^2}{4}\right\}D_{-r}(-e^{i\frac{\pi}{4}}\xi).$$

If the critical point t_0 moves away from the singular point and $\lambda|\varphi(t_0) - \varphi(0)|^{1/2} \to \infty$, then the function $D_r(\xi)$ can be replaced by its asymptotics for $|\xi| \to +\infty$, $\arg\xi = \frac{3\pi}{4}$:

$$D_{-r}(\xi) \sim \frac{\sqrt{2\pi}}{\Gamma(r)}e^{-r\pi i}\xi^{r-1}e^{\frac{\xi^2}{4}}\left(1 + O\left(\frac{1}{\xi}\right)\right) +$$
$$+\xi^{-r}e^{-\frac{\xi^2}{4}}\left(1 + O\left(\frac{1}{\xi}\right)\right).$$

In this case, the right hand side of (4.128) is the sum of two terms which correspond to the contribution from critical and singular points.

CHAPTER 5

Charged Particles in Configuration Space

In this chapter, we will study the wave functions and the Green functions for systems of charged particles. The two-body potential will be assumed to be equal to the sum of the Coulomb and short-range parts

$$v_\alpha(x) = \frac{n_\alpha}{|x|} + v_\alpha^{(s)}(x)$$

where the coefficient n_α is given in terms of charges of particles and their masses by $n_\alpha = \gamma q_i q_j (2\mu_{ij})^{-1/2}$, $\alpha = i, j$. As in the case of neutral particles, in this chapter, we will present the asymptotic expressions for the wave functions in coordinate representation and will state their boundary problems based on the Schrödinger equation and the differential equations for their components. Moreover, we will obtain compact integral equations in configuration space which can be used for the study of wave operators and the scattering matrix for arbitrary energies, including the case $E > 0$, when the channel for disintegration into N free particles is opened. Basically, we will carry out the calculations for the case of three bodies ; for N particles only the final results will be described.

5.1 Two Charged Particles

Let us start with the study of a system of two charged particles to show some peculiarities of the scattering problem in the configuration space.

5.1.1 Wave functions

First of all, let us note that the wave function of a system of two particles interacting according to Coulomb law can be found explicitly, because the Schrödinger equation allows separation of variables. In fact, if we use parabolic coordinates ξ and ζ:

$$\xi = |x| - (\hat{k}, x), \quad \zeta = |x| + (\hat{k}, x)$$

where $\hat{k}$ is the direction of the z-axis, and denote by φ the angle around this axis, the Schrödinger equation reads

$$\left(-\frac{4}{\xi+\zeta}\left(\frac{\partial}{\partial \xi}\xi\frac{\partial}{\partial \xi} + \frac{\partial}{\partial \xi}\zeta\frac{\partial}{\partial \zeta}\right) - \frac{1}{\xi\zeta}\frac{\partial^2}{\partial \varphi^2} + \frac{2n}{\xi+\zeta}\right)\psi_c = k^2\psi_c$$

and it has a solution of the form of a product of a plane wave and a function, depending only on the ξ-coordinate:

$$\psi_c(x,k) = e^{i(k,x)}\varphi_c(\xi,k) \tag{5.1}$$

which satisfies the confluent hypergeometric equation

$$\xi\varphi_c''(\xi) + (1 - i|k|\xi)\varphi_c'(\xi) - \frac{1}{2}n\varphi_c(\xi) = 0.$$

To the wave function $\psi_c(x,k)$ corresponds the regular solution of this equation, which will normalize as follows

$$\varphi_c(\xi,k) = e^{-\pi\eta/2}\Gamma(1+i\eta)\Phi(-i\eta, 1, i|k|\xi), \tag{5.2}$$

where the confluent hypergeometric function $\Phi(a,b,t)$ is given by the series

$$\Phi(a,b,t) = \frac{\Gamma(b)}{\Gamma(a)}\sum_{j=0}^{\infty}\frac{\Gamma(a+j)}{\Gamma(b+j)}\frac{t^j}{j!}$$

and η is the Coulomb parameter $\eta = \dfrac{n}{2|k|}$.

For $t \to \infty$, we have the following asymptotic representation of Φ:

$$\Phi(a,b,t) = \frac{\Gamma(b)}{\Gamma(b-a)}e^{-a\log(-t)}(1 + O(t^{-1})) +$$

$$+\frac{\Gamma(b)}{\Gamma(a)}e^{t+(a-b)\log t}(1 + O(t^{-1})). \tag{5.3}$$

Therefore, the wave function $\psi_c(x,k)$ for large values of $|k|\xi$ is given by the sum

$$\psi_c(x,k) = \exp\{i(k,x) + iw(k,x)\}(1 + O((|k|\xi)^{-1}) +$$
$$+ f_c(\hat{x},k)\frac{\exp\{i|k|\,|x| + iw_0(k,x)\}}{|x|}(1 + O((|k|\xi)^{-1}) \qquad (5.4)$$

Here the first term is called the distorted plane wave and the second the distorted spherical wave. The Coulomb phases w and w_0 distorting the plane and spherical waves are

$$w(x,k) = \eta \log(|k|\,|x| - (k,x)),$$

$$w_0(x,k) = -\eta \log 2|k|\,|x|,$$

and the amplitude of the distorted spherical wave is

$$f_c(\hat{x},k) = \frac{\eta}{2|k|\sin^2\frac{\theta}{2}} \exp\left\{-i\eta \log \sin^2\frac{\theta}{2} + 2i\delta_0\right\} \qquad (5.5)$$

where $cos\theta = (\hat{x},\hat{k})$ and $\delta_0 = \arg\Gamma(1+i\eta)$.

Let us note that these asymptotic formulae are valid only for sufficiently large values of the momentum satisfying the inequality

$$|k| > x^{-2/3+\epsilon}, \quad \epsilon > 0. \qquad (5.6)$$

In other words, when $|k| \to 0$ the size of the region of validity of the asymptotic approximation (5.3) to infinity. At small momenta k, bounded by the condition

$$|k| > O(x^{-3/4+\epsilon}), \qquad (5.6')$$

the wave function $\psi_c(x,k)$ takes the semiclassical form:

$$\psi_c(x,k) \sim e^{i|k|\xi/2}\Gamma(1+i\eta)e^{-\pi\eta/2}J_0(\sqrt{2n\xi}) \qquad (5.7)$$

where J_0 denotes the Bessel function. The transient region is described by very complicated expressions which will not be given here. Let us note that in the case of attraction $n < 0$, the wave function is rapidly oscillating and increasing as $k \to 0$:

$$\psi_c(x,k) \sim |2\pi\eta|^{1/2}e^{-i|\eta|(\log|\eta|^{-1})}e^{i|k|\xi/2}J_0(\sqrt{2|n|\xi}); \qquad (5.8)$$

in the case of repulsion $n > 0$, it decreases exponentially:

$$\psi_c(x,k) \sim (2\pi\eta)^{1/2} e^{-\pi\eta + i\eta(\log\eta - 1)} e^{i|k|\xi/2} J_0(\sqrt{2n\xi}). \tag{5.9}$$

We have constructed wave functions describing outgoing scattered waves. The wave functions $\psi^{(-)}(x,k)$ corresponding to incoming spherical waves, are given by

$$\psi^{(-)}(x,k) = \psi^*(x,-k). \tag{5.10}$$

It can be shown that up to the coefficient $(2\pi)^{-3/2}$, these functions coincide with the kernels of the modified wave operators $U_c^{(\pm)}$ given in this case by

$$U_c^{(\pm)} = \lim_{t\to\mp\infty} e^{iht} L \exp\{-ik^2 t - iw_t(k)\}, \tag{5.11}$$

where

$$w_t(k) = \operatorname{sign} t\eta \log 4k^2|t|.$$

The proof of this fact will be given in Chapter 6.

Thus, we see that the wave functions $\psi_c(x,k)$ are substantially different from the analogous solutions of the Schrödinger equation for short-range potentials. As it has been shown in the previous chapter, the asymptotic behaviour of the wave function for neutral particles reduces to the sum of a plane wave describing the free motion of non-interacting particles and an outgoing spherical wave with smooth amplitude. In the case of charged particles, the plane and spherical waves are distorted by phase factors logarithmically depending on the distance between the particles. This distortion can be understood from the physical circumstances have mentioned when constructing the modified non-stationary wave operators: the asymptotic motion of particles in a Coulomb field is never free, the particles are interacting at arbitrary large distances. The second characteristic property of the Coulomb wave functions is that the scattering amplitude $\hat{f}_c(\hat{x},k)$ has a singularity in the direction of forward scattering. This singularity is so strong that the function $\hat{f}_c(\hat{x},k)$ is not integrable over the angular variables θ, φ and the integrals where this singularity appears must be understood in a distributional sense. We will not make this remark more precise here; we will return to the question of the class of generalized functions appropriate for the problem in the next chapter. Finally, in contrast to neutral particles, the asymptotic formula (5.4) does not describe behaviour of the wave function in all directions correctly. For bounded $|k|\xi$, i.e., in the direction of forward scattering, it is necessary to use the exact expressions (5.1), (5.2) in terms of the confluent hypergeometric function.

In spite of these differences, the asymptotic expressions (4.2) - (4.4) and (5.4) have one common property, namely, the rapidly oscillating exponents, generating the plane and the spherical waves, are the same in both cases. These expressions differ only in the phase factors. Note, that the additional Coulomb phases have the semiclassical origin.

Indeed, consider the eikonal approximation to the wave function $\psi_c(x,k)$ corresponding to the linear motion with the momentum k. The following representation

$$\psi_c(x,k) \sim \exp\{i(x,k) + iw(x,k)\},$$

holds where w is the phase shift in the Coulomb field along a linear trajectory starting at distant points and passing through the point of observation x:

$$w(x,k) = \frac{1}{2|k|}\int_{z_0}^{z} dt v_c(u+\hat{k}t),$$

$$z = (\hat{k},x), \quad v_c(x) = \frac{n}{|x|}. \tag{5.12}$$

Here u is the projection of the vector x on the plane perpendicular to the momentum k, $u = x-(\hat{k},x)\hat{k}$. By computing this integral we obtain the phase w distorting the plane wave (5.4) up to a constant factor. Analogously, w_0 can be defined as the phase shift of the wave function along the trajectory with the direction vector $\hat{x}$. Moreover, the vectors $\hat{k}$ or $\hat{x}$ along which the particle moves, are related to the reduced actions $(\hat{k},x)$, and $|x|$ defining the motion by the relations $\hat{k} = \nabla(\hat{k},x)$, $\hat{k} = \nabla|x|$. We will call the functions $(\hat{k},x)$ and $|x|$ the plane and spherical eikonals, bearing in mind the above mentioned analogy between the quantum scattering problem and the problems of wave propagation.

Thus, the distorted plane and spherical waves (5.4) can be interpreted as the eikonal approximations to the wave function constructed along the trajectories of motion which ae defined by the eikonals $(\hat{k},x)$ and $|x|$, respectively.

We see that the semiclassical considerations which led us to the modified definition of the non-stationary wave operators in Chapter 1, made it also possible to find the correct asymptotics of the wave functions in the stationary formulation. In the problem of two bodies this fact is of no principal importance because the Coulomb wave function is known in explicitly. But it can be expected that the semiclassical approach will be fruitful in the problem of three and more charged particles where no exact solution of Schrödinger equation exists.

5.1.2 Green function

Let us describe properties of the Green function for the Hamiltonian operator of two Coulomb particles h_c, $h_c = h_0 + v_c$.

A remarkable property of the two particle Coulomb scattering problem is that the Green function $r_c(x, x', z)$, satisfying the non-homogeneous Schrödinger equation

$$\left(-\Delta_+\frac{n}{|x|} - z\right) r_c(x, x', z) = \delta(x - x')$$

is, together with the wave functions, explicitly known:

$$r_c(x, x', z) = \frac{\sqrt{z}}{4\pi|x - x'|}\left(\frac{\partial}{\partial t} - \frac{\partial}{\partial s}\right) ts\varphi_c^{(1)}(s, z)\varphi_c^{(2)}(t, z) \tag{5.13}$$

Here, the following notation is used. The functions $\varphi_c^{(1)}$, $\varphi_c^{(2)}$ called Coulomb radial functions, can be expressed in terms of the regular $\Phi(a, b, t)$ and irregular $\Psi(a, b, t)$ solutions of the confluent hypergeometric equation as

$$\varphi_c^{(1)}(s, z) =$$

$$= e^{-\frac{n\pi}{2\sqrt{z}}}\Gamma\left((1 + i\frac{n}{2\sqrt{z}})\right) e^{is\sqrt{z}}\Phi\left(1 + i\frac{n}{2\sqrt{z}}, 2, -2is\sqrt{z}\right),$$

$$\varphi_c^{(2)}(t, z) =$$

$$= e^{-\frac{n\pi}{2\sqrt{z}}}\frac{e^{it\sqrt{z} - i\pi/2}}{\Gamma\left((1 - i\frac{n}{2\sqrt{z}})\right)}\Psi\left(1 + i\frac{n}{2\sqrt{z}}, 2, -2it\sqrt{z}\right),$$

where the parabolic coordinates s and t in this case are defined as

$$2s = |x| + |x'| - |x - x'|, \quad 2t = |x| + |x'| + |x - x'|.$$

The Green function simplifies asymptotically, when $|x - x'| \to \infty$, and

$$|x|\,|x'| + (x, x') \geq \left||x| + |x'|\right|^{1+\nu}, \quad n > 2/3,$$

so that the points x and x' do not lie on the opposite sides of the origin on a straight line passing through it. Using the asymptotic representation (5.3) for the confluent hypergeometric function and an analogous equation for the linearly independent solution $\Psi(a, b, t) \sim t^{-a}$, we have

$$r_c(x, x', z) \sim \frac{1}{4\pi} \frac{\exp\{i\sqrt{z}|x - x'| + iw_z(x, x')\}}{|x - x'|} \tag{5.14}$$

where

$$w_z(x, x') = i\frac{n}{2\sqrt{z}} \log \frac{|x|\,|x - x'| + (x, x - x')}{|x'|\,|x - x'| + (x', x - x')} \tag{5.14'}$$

By comparing this formulae with the definition of the eikonal phase (5.12), we reach the conclusion that the function $w_z(x, x')$ can be obtained with the help of this formula by replacing the momentum $\hat{k}$ by the direction of the vector $\widehat{x - x'}$ which joins two points x and x'. Hence, as in the case of the wave functions, the Coulomb Green function $r_c(x, x', z)$ differs from the free Green function $r_0(x, x', z)$ by a phase factor having semiclassical character.

Finally, let us note that by analogy with the case of neutral particles, for $|x'| \to \infty$, the following relation between the Green function and the wave functions holds

$$r_c(x, x', E \pm i0) \sim$$

$$\sim \frac{1}{4\pi} \psi_c^{(\pm)}(x, k) \frac{\exp\{\pm i|k|\,|x'| \pm iw_0(x')\}}{|x'|}. \tag{5.15}$$

Here the momentum variable is given by $k = \mp\sqrt{E}\hat{x}'$.

The existence of explicit representation of the Green function solves the problem of spectral analysis of the Hamiltonian h_c completely. In particular, in the case of attraction $n < 0$, the poles of the Green function at the points $E_N = -\dfrac{n^2}{4N^2}$, $N = 1, 2, \ldots$, where the Γ-function goes to infinity, give the eigenvalues of the operator h_c. These eigenvalues are N^2 times degenerated and the corresponding eigenfunctions are given by

$$\psi_{Nl}^m = R_{Nl}(r) Y_l^m(\hat{x}),$$

where $r = |x|$, Y_l^m are the spherical functions, and the function R_{Nl}, depending on the radial variable, is expressed by means of the confluent hypergeometric functions as

$$R_{Nl} = \frac{2}{N^{l+2}(2l+1)!}\left(\frac{(N+l)!}{(N-l-1)!}\right)^{1/2}(2rn)^l \times$$
$$\times \exp\left\{-\frac{rn}{N}\right\}\Phi\left(-N+l+1, 2l+2, \frac{2rn}{N}\right). \tag{5.16}$$

Let us note that the kernel $r_c(x, x', z)$ can be defined by the spectral decomposition

$$r_c(x, x', z) = \sum_{N,l,m}\frac{\psi_{Nl}^m(x)\overline{\psi_{Nl}^m(x')}}{z+\frac{n^2}{4N^2}}+$$
$$+\frac{1}{(2\pi)^3}\int\frac{\psi_c^{(\pm)}(x,k)\overline{\psi_c^{(\pm)}(x',k)}}{k^2-z}\,dk_1 \tag{5.17}$$

where the integral corresponds to the continuous and the sum to the discrete part of the spectrum of the operator h_c.

5.1.3 Superposition of Coulomb and short-range potentials

Let us now turn to study of the problem of scattering of charged particles, the interaction potential of whose contains, besides the Coulomb part, a short-range term $v^{(s)}(x)$. If in the identity (2.8) in terms of the operators A and B we take the pair $h_c - z$ and $h - z$, $h = h_c + v^{(s)}$, we obtain the following modified perturbation equation:

$$r(z) = r_c(z) - r_c(z)v^{(s)}r(z). \tag{5.18}$$

This equation can be written in terms of the Green function as an integral equation of second kind:

$$r(x, x', z) =$$
$$= r(x, x', z) - \int r_c(x, y, z)v^{(s)}(y)r(y, x', z)\,dy. \tag{5.19}$$

This equation can be studied in the same way as the analogous equation (4.14) for the case of neutral particles.

Let us give, without proof, some results following from this equation.

For $|x'| \to \infty$, the Green function $r(x, x', z)$ can be represented in the form (5.15), where the amplitude of the distorted spherical wave coincides with the wave function $\psi^{(\pm)}(x, k)$ of the operator h. This wave function satisfies the compact integral equation

$$\psi^{(\pm)}(x, k) = \psi_c^{(\pm)}(x, k) -$$
$$- \int r_c(x, y, k^2 \pm i0) v^{(s)}(y) \psi^{(\pm)}(y, k)\, dy \tag{5.20}$$

and can be represented as the sum

$$\psi^{(\pm)}(x, k) = \psi_c^{(\pm)}(x, k) + \psi_{cs}^{(\pm)}(x, k) \tag{5.21}$$

where the term $\psi_{cs}^{(\pm)}$ is asymptotically equal to the distorted spherical wave with the smooth amplitude $f_{cs}^{(\pm)}(\hat{x}, k)$. Since the asymptotics of the function $\psi_c^{(\pm)}(x, k)$ contains a distorted spherical wave as well, the full scattering amplitude for the function $\psi(\pm)(x, k)$ is equal to the sum of the singular purely Coulomb part $f_c^{(\pm)}(\hat{x}, k)$ and the smooth term $f_{cs}^{(\pm)}$, called the short-range part of the amplitude:

$$f^{(\pm)}(\hat{x}, k) = f_c^{(\pm)}(\hat{x}, k) + f_{cs}^{(\pm)}(\hat{x}, k). \tag{5.22}$$

From equation (5.20) follows the integral representation of $f_{cs}^{(\pm)}(\hat{x}, k)$:

$$f_{cs}^{(\pm)}(\hat{x}, k) = \frac{1}{4\pi} \int \psi_c^{(\mp)}(y, -|k|\hat{x}) v^{(s)}(y) \psi^{(\pm)}(y, k)\, dy. \tag{5.23}$$

As an alternative to (5.20), the wave functions $\psi^{(\pm)}(x, k)$ can be defined as solutions of the Schrödinger equation

$$\left(-\Delta + \frac{n}{|x|} + v^{(s)}(x) - k^2\right) \psi(x) = 0 \tag{5.24'}$$

in the class of smooth functions equal to the sum (5.21), where the asymptotics of the unknown function $\varphi_{cs}(x, k)$ has the form of distorted spherical wave with bounded amplitude:

$$\varphi_{cs}^{(\pm)}(x, k) \sim f_{cs}^{(\pm)}(x, k) \frac{\exp\{\pm i|k|\,|x| \pm i w_0(x, k)\}}{|x|}. \tag{5.24}$$

The reasoning leading to justification of the differential formulation of the scattering problem does not differ from that used in the case of neutral particles.

5.1.4 Angular singularities of the scattering amplitude

Above, we assumed that the short-range potential $v^{(s)}(x)$ decreases sufficiently fast for $|x| \to \infty$, at least as $|x|^{-3-\epsilon}$, $\epsilon > 0$. In this case, the short-range part of the scattering amplitude turns out to be a bounded function. If this potential decreases more slowly, e.g., if there are multipole parts:

$$v^{(s)}(x) \sim \frac{\mu_1(\hat{x})}{|x|^2} + \frac{\mu_2(\hat{x})}{|x|^3} + \cdots$$

then the function f_{cs} has singularities in the forward direction $\hat{x} = \hat{k}$. Let us briefly describe them.

We write the wave functions in the form of the sum:

$$\psi(x,k) = \psi_c(x,k) + \sum_{l=1}^{2}(r_c v^{(s)})^l \psi_c(x,k) + \tilde{\varphi}(x,k), \tag{5.25}$$

where the terms $(r_c v^{(s)})^l \psi_c$ correspond to the first two iterations of the integral equations (5.19). For the scattering amplitudes we find

$$f(\hat{x},k) = f_c(\hat{x},k) + \sum_{j=1}^{3} f_{j1}(\hat{x},k) + f_{12}(\hat{x},k) + \tilde{f}_{cs}(\hat{x},k) \tag{5.26}$$

where $f_c(\hat{x},k)$ is the Coulomb amplitude (5.5) and $f_{cs}(\hat{x},k)$ is a smooth function. Here, the functions f_{j1} and f_{12} contain singularities and describe the contributions from the multipole terms $\frac{\mu_1(\hat{x})}{|x|^2}$ and $\frac{\mu_2(\hat{x})}{|x|^3}$ to the first iteration $(r_c v^{(s)})^l \psi_c$:

$$f_{j1}(\hat{k}',k) =$$

$$= -\frac{1}{4\pi} \lim \epsilon \downarrow 0 \int dx \psi_c(x, -|k|\hat{k}') \frac{\mu_l(\hat{x})}{|x|^{l+1}} \psi_c(x,k) e^{-\epsilon|x|}. \tag{5.27}$$

The function f_{12} contains the contribution of the dipole potential to the second iteration

$$f_{12}(\hat{k}',k) =$$

$$= -\frac{1}{4\pi} \lim \epsilon \downarrow 0 \int dx dy e^{-\epsilon|x|-\epsilon|y|} \psi_c(x, -|k|\hat{k}') \times$$

$$\times \frac{\mu_1(\hat{x})}{|x|^2} \frac{\mu_2(\hat{y})}{|y|^2} r_c(x,y,k^2+i0)\psi_c(x,k). \tag{5.28}$$

As an example, let us consider the function f_{11} which contains the leading (after the Coulomb) singularity. It is clear that the singularities appear in those points k' and k where the integral (5.27) becomes divergent. The integral in (5.27) over a finite region is a smooth function of the vectors k and k'. Therefore, the divergence is determined by the behaviour of the integrand at large distances.

For the qualitative description of the singularity, we confine ourselves to the examination of the integral over regions which do not contain neighbourhoods of the directions $x = \hat{k}$ and $\hat{x} = -\hat{k}'$. In such a domain, the asymptotics of the Coulomb wave functions in the leading order is given by the distorted plane wave (5.4). Let us replace in (5.27) the Coulomb wave functions with their asymptotics and integrate over a new variable $y = |q|x$, where $q = k - k'$. As a result, we get

$$f_{11}(\hat{k}', k) =$$

$$= |q|^{-1-2i\eta} \lim_{\epsilon \downarrow 0} \int \exp\{i(\hat{q}, y) - \epsilon|y|\} \, |y|^{2i\eta} g(\hat{y}) d\hat{y} d|y|$$

with $g(\hat{y})$ is a smooth function. The integral over the angular variable can be estimated by the method of stationary phase:

$$\int d\hat{y} g(haty) \exp\{i(\hat{q}, y)\} \sim$$

$$\sim \frac{1}{i|y|} \left(e^{i|y|} g(\hat{y}) - e^{-i|y|} g(-\hat{y}) \right) + \dots$$

It follows from this estimate that the behaviour of the integrand in the integral over the radial variable guarantees the convergence of the initial integral. Thus, the singularity is generated only by the factor $|q|^{-1-2i\eta}$ which is singular in the forward direction.

The representation obtained above correctly expresses the form of the leading singularity of the function f_{11}. However, since the asymptotics (5.4) cease to be correct in the neighbourhood of the forward direction, to determine the exact coefficient for the singularity, it is necessary to use the explicit expression for the Coulomb wave functions. We will not give here the corresponding calculations and show only the final results.

For the functions f_{ij} the following representations hold:

$$f_{11}(\hat{k}', k) = a_{11}|k - k'|^{-1-2i\eta} + b_{11}|k - k'|^{-2i\eta} + \delta f_{11}(\hat{k}', k),$$

$$f_{21}(\hat{k}', k) = a_{21}|k - k'|^{-2i\eta} + \delta f_{21}(\hat{k}', k),$$

$$f_{12}(\hat{k}', k) = a_{12}|k - k'|^{-2i\eta} + \delta f_{12}(\hat{k}', k),$$

where the singular factors are explicitly singled out and the coefficients a_{lj} and δf_{lj} are smooth bounded functions. The factors a_{l1} of the leading singularity of the functions f_{l1} $(l = 1, 2)$ can be represented as

$$a_{l1}(\hat{k}', k) = \frac{(-i)^l}{4\pi}|k|^{2i\eta}e^{-\pi\eta}\Gamma(2 - l + 2i\eta)\tilde{a}_{l1}(\hat{k}', k) \tag{5.29}$$

where the factors $\tilde{a}_{l1}$ can be expressed by means of the multipole moments μ_l of the potential with the help of the following integral representations:

$$\tilde{a}_{l1}(\hat{k}', k) \int d\hat{x}\mu_l(\hat{x})(1 - (\hat{p}, \hat{x})^2)^{i\eta}((\hat{q}, \hat{x}) + i0)^{l-2-2i\eta}. \tag{5.30}$$

Here q is the transfer momentum vector, $q = k - k'$ and p is the sum of the initial and the final momenta, $p = k + k'$. Further, the coefficient b_{11} at the lowest singularity of b_{11} is given by

$$b_{11}(\hat{k}', k) = \frac{i}{4\pi}(2|k|)^{2i\eta}\exp\{2i\arg\Gamma(i\eta))\times$$
$$\times(\mu_1(\hat{p}) + \mu_1(-\hat{p})).$$

Finally, the function a_{12} takes the form

$$a_{12}(\hat{k}', k) =$$
$$i\frac{1}{8\pi|k|}\int_0^{2\pi} d\varphi a_{11}(\hat{k}', k + k'')a_{11}^{(0)}(\widehat{k + k''}k)(\sin^2\varphi)^{-i\eta}\times$$
$$\times(B(\frac{1}{2}, i\eta) + 2\cos\varphi F(1 - i\eta, \frac{1}{2}, \frac{3}{2}, \cos^2\varphi)), \tag{5.31}$$

where $B(x, y)$ is the beta-function and $F(a, b, c, t)$ a hypergeometric function. The integral in (5.31) is taken over the circle $|k"| = |k - k'|$ which lies in the plane perpendicular to k, $\cos\varphi = (\hat{k}'', \hat{q})$. The function $a_{11}^{(0)}$ is given by (5.30) with η equal zero.

Let us stress, that the order of the leading dipole singularity is smaller than the exponent of the Coulomb singularity (5.5) by one. The orders of the singularities corresponding to the multipole components also decrease by one, while the order of decreasing of the potential increases.

Thus, we see that the scattering amplitude f is represented by

$$f(\hat{k}', k) = f_{sing}(\hat{k}', k) + \tilde{f}(\hat{k}', k),$$

where $\tilde{f}$ is a smooth bounded function and $f_{sing}(\hat{k}', k)$ is the singular part. For the latter, we have the representation

$$f_{sing}(\hat{k}', k) = f_c(\hat{k}', k) + \sum_{l=1}^{2} \frac{b_l(\hat{k}', k)}{|k' - k|^{2-l+2i\eta}} \quad k' = |k|\hat{k}', \tag{5.32}$$

where the coefficients b_l are given in terms of the functions a_{lj}, b_{1l}:

$$b_1 = a_{11}, \quad b_2 = a_{12} + a_{21} + b_{11}.$$

In some cases the coefficients b_l can be computed explicitly. For example, for the case of a spherically symmetric potential $\frac{\mu}{|x|^2}$, i.e., $\mu_1(\hat{x}) = \text{const}$, the coefficient b_1 for the leading dipole singularity has the form

$$b_1 = -\frac{\pi\mu}{2}(2|k|)^{2i\eta} \exp\left\{2i\arg\Gamma\left(\frac{1}{2} + i\eta\right)\right\}.$$

For the potential of two fixed Coulomb centres, $\mu_1(\hat{x}) = d\cos\theta$, $\cos\theta = (\hat{x}, \hat{R})$, this coefficient is given by

$$b_1(\hat{k}', k) = id(2|k|)^{2i\eta} \exp\{2i\arg\Gamma\,(1 + i\eta)\}(\hat{x}, \hat{R})$$

$$q = k - k'.$$

The function b_2 for these potentials can be calculated explicitly, as well. However, the corresponding formulae are too cumbersome to be given here.

It must be noted that the representation (5.32) ceases to be valid if the potential $v(x)$ is equal to the sum of multipole parts and does not contain the Coulomb term. Proceeding along the line of the scheme presented above, it can be shown the scattering amplitude in this case has a singularity of the form

$$f_{sing}(\hat{k}', k) = a_{11}^{(0)}(\hat{k}', k)|k - k'|^{-1} + (a_{12}^{(0)}(\hat{k}', k)+$$
$$+a_{21}^{(0)}(\hat{k}', k)\log|k - k'| \tag{5.33}$$

where $a_{jl}^{(0)}$ are smooth bounded functions. The function $a_{11}^{(0)}$ is given by (5.29) and (5.30) with $\eta = 0$. The coefficient $a_{21}^{(0)}$ equals

$$a_{21}^{(0)} = -\frac{1}{4\pi}\int d\hat{x}\mu_2(\hat{x}).$$

The function $a_{12}^{(0)}$, as in the case of charged particles, is expressed by the convolution of the functions $a_{11}^{(0)}$:

$$a_{12}^{(0)}(\hat{k}', k) = -\frac{i}{4\pi|k|}\int_0^{2\pi} d\varphi a_{11}^{(0)}(\widehat{k+k''}, k)a_{11}^{(0)}(\hat{k}, k+k'')$$

where the parameters φ and k'' are defined in (5.31).

5.1.5 The Coulomb potential in R^{3N-3}

For methodological purposes, in this section, we will study the model scattering problem, defined by a spherically symmetric, slowly decreasing potential $q_0|X|^{-1}$ in a (3N - 3)-dimensional space. We will call it a Coulomb potential in R^{3N-3}.

As in three dimensional space, the Schrödinger equation

$$\left(-\Delta_X + \frac{\theta_0}{|X|} - P^2\right)\Psi_N(X, P) = 0 \tag{5.34}$$

can be solved by the method of separation of variables in parabolic coordinates. The wave functions $\Psi_N(X, P)$ are chosen to be the product

$$\Psi_N(X, P) = \exp\{i(X, P)\}\Phi_N(\xi, P) \tag{5.35}$$

where the function Φ depends only on the parabolic coordinate $\xi = |X| - (X, \hat{P})$:

$$\Phi_N(\xi, P) = e^{-\frac{\eta\pi}{2}}\frac{\Gamma\left(\frac{3N-4}{2} + i\eta\right)}{\Gamma\left(\frac{3N-4}{2}\right)}\Phi\left(-i\eta, \frac{3N-4}{2}, i|P|\xi\right), \tag{5.36}$$

$\eta = \dfrac{q_0}{2|P|}$. The wave functions labelled by $(-)$ are expressed in terms of $\Psi(X, P)$ as

$$\Psi_N^{(-)}(X, P) = \Psi_N^{(+)*}(X, -P), \quad \Psi_N^{(+)} \equiv \Psi_N.$$

If $\xi \to \infty$, then, according to the asymptotic formula (5.3), the function Ψ_N becomes equal to the sum of the distorted plane and spherical waves:

$$\Psi_N(X, P) \sim e^{i(X,P)+iW_N(X,P)}(1 + O((|P|\xi)^{-1}))+$$

$$+f_N(\hat{X}, P)\frac{\exp\{i|X|\,|P| + iW_N^{(0)}(X, P)}{|X|^{\frac{3N-4}{2}}} \tag{5.37}$$

where the Coulomb phases are given by

$$W_N(X,P) = \eta \log(|X|\,|P| - (P,X)),$$

$$W_N^{(0)}(X,P) = -\eta \log 2|P|\,|X|,$$

and the amplitude of the spherical wave is

$$f_N(\hat{X},P) =$$

$$= \frac{\Gamma\left(\frac{3N-4}{2} + i\eta\right)}{\Gamma(-i\eta)} \frac{\exp\{-i\eta \log 2\sin^2\frac{\theta}{2}}{\left(2|P|\sin^2\frac{\theta}{2}\right)^{\frac{3N-4}{2}}} e^{-(N-4)\frac{i\pi}{4}}. \qquad (5.38)$$

At $\theta = 0$, $\cos\theta = (\hat{X},\hat{P})$, this amplitude develops a strong singularity. It is easy to see that this singularity is not integrable on the unit sphere $|P| = 1$.

The Green function $R_N(X,X',z)$ can be defined by means of the spectral representation

$$R_N(X,X',z) = \sum_A \frac{\Psi_A(X)\overline{\psi_A(X')}}{z + \varkappa_A^2} +$$

$$+ \frac{1}{(2\pi)^{3N-3}} \int \frac{\Psi_N^{(\pm)}(X,P)\overline{\Psi_N^{(\pm)}(X',P)}}{P^2 - z}\, dP \qquad (5.39)$$

where the sum over the eigenvalues of the operator (5.34)

$$\varkappa_A^2 = -\frac{q_0^2}{4\left(j + \frac{3N-4}{2}\right)^2}, \qquad j = 0,1,2,\ldots$$

differs from zero only for $q_0 < 0$.

The following equations hold:

$$R_N(X,X',Z) =$$

$$\left(-\frac{1}{2\pi y}\frac{\partial}{\partial y}\right)^{\frac{m-1}{2}} G_1(x,y); \quad m = 3N-3, \quad m = 3,9,\ldots,$$

$$R_N(X,X',Z) =$$

$$\left(-\frac{1}{2\pi y}\frac{\partial}{\partial y}\right)^{\frac{m-2}{2}} G_2(x,y); \quad m = 6,12,\ldots,$$

$$G_1(x,y) = \frac{\Gamma(1+i\eta)}{2i}\sqrt{z}(y^2 - x^2)e^{-iz\sqrt{x}} \times$$

$$\times\Phi(1+i\eta,2,-i\sqrt{z}(x+y))\Psi(1+i\eta,2,-i\sqrt{z}(x-y)),$$

$$G_2(x,y)=$$

$$=i\frac{\exp\left\{-i\pi\left(i\eta+\frac{1}{2}\right)\right\}}{\sin\pi\left(i\eta+\frac{1}{2}\right)}\times$$

$$\times\int_{+\infty,\arg(\zeta\pm1)=0}^{(1+)}d\zeta(\zeta+1)^{-i\eta-1/2}(\zeta-1)^{i\eta-1/2}\times$$

$$\times\frac{e^{i\sqrt{z}x\zeta}}{4\sqrt{\pi}}\frac{I_{-1/2}(-i\sqrt{z}(x^2-y^2)^{1/2}(\zeta^2-1)^{1/2})}{\left(-\frac{1}{2}i\sqrt{z}(x^2-y^2)^{1/2}(\zeta^2-1)^{1/2}\right)^{-1/2}},\tag{5.40}$$

where $x=|X|+|X'|$, $y=|X-X'|$, and I_ν denotes the Bessel function of imaginary argument. Thus, in contrast to the two-body case, the Green function for the Coulomb potential in R^{3N-3} is given by a rather complicated integral representation. But, if N is odd $N=2l+1$, this representation can be simplified and one can express $R_N(X,X',z)$ by means of a hypergeometric function.

Let us note that $R_N(X,X',z)$ is smooth if the coordinates X and X' do not coincide $X\neq X'$; when X = X', it has the same pole-like singularity as the free Green function:

$$R_N(X,X',z)\sim\frac{\left(\frac{3N-5}{2}\right)}{4}\frac{(2\pi)^{\frac{-3N+3}{2}}}{|X-X'|^{3N-5}}.\tag{5.41}$$

Let us describe the asymptotic representation of the function $R_N(X,X',z)$. Let us draw the straight line $s(t)=t\hat{X}$ joining the point X and the origin. The set of the points X'_t belonging to the neighbourhood of this line on the opposite side of the origin and satisfying the inequality

$$|X|\,|X'|+(X,X')\leq(|X|+|X'|)^{1+\nu},\quad \nu>2/3$$

will be called the singular domain $\Omega_s(X,X')$ corresponding to the vector X. (see Fig. 19).

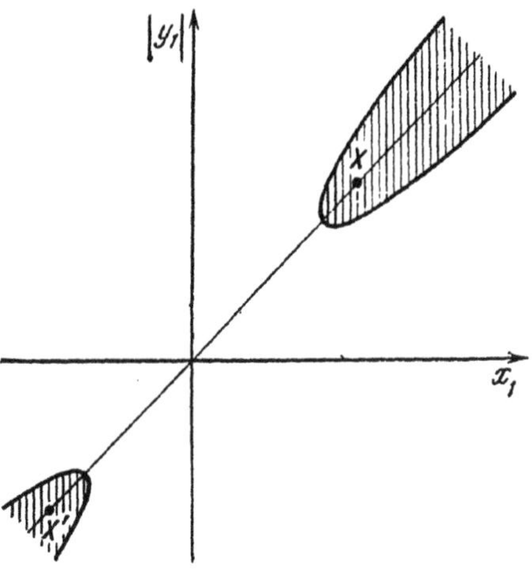

Fig. 19

If the variables X and X' tend to infinity, so that the second argument of the Green function does not belong to the singular domain, then the function $R_N(X, X', z)$ asymptotically takes the eikonal form:

$$R_N(X, X', z) \sim C_0^{(N)} \frac{\exp\{i\sqrt{z}|X - X'|\}}{|X - X'|^{\frac{3N-4}{2}}} \Phi_N(X, X', z) \qquad (5.42)$$

where

$$C_0^{(N)} = \frac{1}{4\pi} \left(\frac{\sqrt{z}}{2\pi}\right)^{\frac{3N-6}{2}} e^{-i\frac{\pi}{4}(3N-6)}$$

and

$$\Phi_N(X, X', z) = \exp\left\{i\frac{q_0}{2\sqrt{z}} \log \frac{|X|\,|X - X'| + (X, X - X')}{|X'|\,|X - X'| + (X', X - X')}\right\}.$$

If we fix one argument, for example, X and the other approaches infinity, the Green function becomes a distorted spherical wave, whose amplitude is given through the wave functions $\Psi_N^{(\pm)}(X, P)$:

$$R_N(X, X', E + i0) \sim$$

$$\sim C_0^{(N)} \Psi_N^{(\pm)}(X, P) \frac{\exp\{\pm i\sqrt{E}|X'| \pm iW_N^{(0)}(X')\}}{|X'|^{\frac{3N-4}{2}}}$$

where $P = \mp\sqrt{E}\hat{X}'$. The asymptotics of the Green function for $|X'| \to \infty$ can be obtained from this formula with the help of the expression

$$R_N(X, X', z) = \overline{R_N(X, X', \bar{z})}.$$

If, finally, the variables X and X' approach infinity while being in a vicinity of the direction $\hat{X}$ and on the same side from the origin, i.e.,

$$(X, X') > 0,$$

$$|X|\,|X'| - (X, X') \leq (|X| + |X'|)^{1+\nu}, \quad \nu > 0, \tag{5.43}$$

then the asymptotics of the Green function R_N for $|X|$, $|X'| \to \infty$ is given by

$$R_0^{(N)}(X, X', z) \exp\left\{i\frac{q_0}{2\sqrt{z}} \log\frac{|X|}{|X'|}\right\} \tag{5.44}$$

where $R_0^{(N)}(X, X', z)$ is the Green function of the operator H_0.

Therefore, we have enumerated the general properties of the Green function $R_N(X, X', z)$. If necessary, more detailed information about behaviour of this function can be obtained from the integral representation (5.40).

5.2 Coordinate Asymptotics of Wave Functions for a System of Three Charged Particles

The main part of this section is devoted to the study of the asymptotics of the wave functions $\Psi_0(X, P)$ corresponding to processes with three free particles in the initial state. The case of functions of the type $\Psi_A(X, p_A)$, $A \neq 0$ describing processes $(2 \to 2)$ and $(2 \to 3)$ is simpler in the technical sense and we will consider it at the end of the chapter. Meanwhile, we disregard the question of justification of the asymptotic formulae and the existence of such functions.

5.2.1 The eikonal approximation

First of all, let us give the heuristic consideration which form the basis of the method used to construct the asymptotic formulae. In the case of neutral particles, we used the method of Fourier transform for the construction of the asymptotics, basing on the study of the properties of kernels of wave operators in momentum representation. This method cannot be used in the case of three charged particles because, as mentioned above, we do not know the properties of the kernels of the resolvent or the T-matrix for positive energies. However, this way of examining the coordinate asymptotics of solutions of differential equations does not exhaust all possibilities. On the contrary, by finding the asymptotics in an independent way, it is also possible to study the properties of the T-matrix in the momentum representation taking the inverse Fourier transform. To this end, let us discuss which methods can be possibly used.

In mathematical physics, there are direct asymptotic methods based on local properties of differential equations: the method of standard equations, the method of semiclassical approximation and its particular case - the method of the eikonal approximation. Now, since the Coulomb potential for $|X| \to \infty$ is becoming arbitrarily small i.e., $V_c(X)E^{-1} \ll 1$, it is natural to use the method of eikonal approximations for construction of the coordinate asymptotics for the wave functions. The reason is that this approximation method applies when the perturbation is small as compared to the energy.

If we make use of this approach we encounter the problem of choosing the trajectories which determine the eikonal approximation. However, these difficulties can be avoided by similarity arguments. Indeed, saw in the example of two charged particles that, preserving the set of asymptotic waves which we had in the case of neutral particles (i.e., plane and spherical waves), the coordinate asymptotics for charged particles in almost all directions can be obtained by simple modification, which reduces in the leading order to the addition of Coulomb phases. These phases depend logarithmically on the distance and can be obtained with the help of the eikonal approximation along the asymptotic trajectories of the particles. We can assume that also for a system of many particles the set of asymptotic waves does not depend whether the particles are charged or not. In the system of three neutral particles, as we have shown in Section 1 of Chapter 4, the coordinate is given by four types of eikonals: plane $(\hat{P}, X)$, spherical $|X|$, single Z_α and double $Z_{\alpha\beta}$. In agreement with our hypothesis, the same set of eikonals should be considered in the case of charged particles. It can also be expected that, as in the case of two particles, the eikonal form of the asymptotics will show irregularities in some singular directions. Even though we are not going discuss this subtle issue now, we will return to it in the next section.

Let us turn to the reformulation of these discussion in technical terms. First of all, we will describe the scheme of construction of the eikonal approximations. As the construction of the latter does not depend on dimensions, we will consider arbitrary number of particles, so that the point X belongs to the space $R^{3(N-1)}$, $N = 3, 4, \ldots$.

Let Z(X) be a solution of the eikonal equation

$$|\nabla Z(X)|^2 = 1.$$

We will look for the solution of Schrödinger equation

$$\left(-\Delta + \sum_{a_{N-1}} n_{a_{N-1}} |x_{a_{N-1}}|^{-1} \right) \Psi(X) = E\Psi(X) \tag{5.45}$$

generated by the eikonal Z and having the form

$$\Psi_z = A_z(X) \exp\{i\sqrt{E} Z + i\tilde{W}_z\} \tag{5.46}$$

where $\exp\{i\tilde{W}_z(X)\}$ is for $\sqrt{E}Z \to \infty$ a slowly oscillating function and $A_z(X)$ is a solution of the continuity equation

$$2(\nabla A_z, \nabla Z) + A_z \Delta Z = 0. \tag{5.47}$$

The function $A_z(X)\exp\{i\sqrt{E}Z\}$ can be shown to be an asymptotic solution of the Schrödinger equation without interaction. The additional phase $\tilde{W}_z$ will be defined so that it satisfies the Schrödinger equation with the Coulomb potential (5.45) for $\sqrt{E}Z \to \infty$.

Substituting the function (5.46) into Schrödinger equation (5.45) we get the following expression for the phase $\tilde{W}_z$:

$$2\sqrt{E}(\nabla \tilde{W}_z, \nabla Z) - \delta W_z = -V_c(X). \tag{5.47'}$$

Here V_c is the sum of the long-range Coulomb potentials:

$$V_c(X) = \sum_{a_{N-1}} n_{a_{N-1}} |x_{a_{N-1}}|^{-1},$$

$$\delta W_Z = i\Delta \tilde{W}_Z - (\nabla \tilde{W}_Z)^2 + 2i\left(\nabla \tilde{W}_Z, \frac{\nabla A_Z}{A_Z}\right) + \frac{\nabla A_Z}{A_Z}.$$

We can look for a formal solution of equation (5.47) in the form of the series in powers of $E^{-k/2}$:

$$\tilde{W}_Z = \sum_{k=0}^{\infty} d_k(X) E^{-k/2-1/2} \tag{5.48}$$

which, in general, has an asymptotic character for $\sqrt{E}Z \to \infty$.

By substituting this expansion into equation (5.47) and comparing the coefficients of equal powers $E^{-k/2-1/2}$, we obtain the following recurrence relations for determination of the functions :

$$2(\nabla d_0, \nabla Z) = -V_0 + A_Z^{-1}\Delta A_Z \tag{5.49}$$

$$2(\nabla d_{k+1}, \nabla Z) = f(d_k). \tag{5.49'}$$

Here

$$f(d_k) = i\Delta d_k - (\nabla d_k)^2 + 2i\left(\nabla d_k, \frac{\nabla A_Z}{A_Z}\right).$$

Let us solve these equations. We denote by $s = s(t)$ the straight line with the directional vector $\hat{K} = \nabla Z$ passing through the considered point X for

$t = z$. Let M be the coordinate of this point on the transversal surface $Z =$ const. The points X_t on the line $s(t)$ are given by $X_t = t\hat{K} + M$, where $M = X - (\hat{K}, X)\hat{K}$. Integrating equation (5.49) along $s(t)$ we represent its solution in the form

$$d_0(X) = C_0(M) + W_Z(X) + \delta f_0(X),$$

where $C_0(M)$ is an arbitrary function of M and

$$W_Z(X) = -\frac{1}{2}\int_{Z_0}^{Z} dt V_c(t\hat{K} + M),$$

$$\delta f_0(X) = \frac{1}{2}\int_{Z_0}^{Z} dt \frac{\nabla A_Z}{A_Z}(t\hat{K} + M).$$

These integrals are taken along a segment of the trajectory $s(t)$ with the endpoint X. In an analogous way we find the coefficients

$$d_{k+1}(X) = \int_{Z_0}^{Z} dt f(d_k(t\hat{K} + M)).$$

Below, by eikonal approximation (or simply eikonal), we mean the approximation of the wave function, corresponding to the trajectory with the direction vector ∇Z; the leading term Ψ_z in the expansions (5.46) and (5.48) is

$$\Psi_Z(X) = f_Z(M)A_Z(X)\exp\{i\sqrt{E}Z + iW_Z(X)\}. \tag{5.50}$$

Let us note that the vector fields $Z(X)$ and $M(X)$ generate an orthogonal curvilinear coordinate system in the 3(N-1)-dimensional space (the variables of the type action - angle); the vector X can be written as the sum

$$X = Z\nabla Z + M,$$

with $(\nabla Z, M) = 0$.

In formula (5.50) the dependence of the eikonal function Ψ_Z on the eikonal variable Z is fixed, but the arbitrary functions of the transversal vector field must be determined from auxiliary conditions. These conditions will be found by using local properties of the Schrödinger equation.

Let us turn to the construction of approximate solutions Ψ_Z of the eikonals of the three-body problem. Below, X will be regarded as a six-dimensional vector, $X \in R^6$.

5.2.2 Plane eikonal Z, $Z = (\hat{P}, X)$

The vector X can be written as the orthogonal sum $X = Z\hat{P}+M$, $(P, M) = 0$, where the vector M corresponds to the points on the hyperplane R^5, defined by the equation $(\hat{P}, X) = 0$. It is easy to see that the wave function in the approximation of the plane eikonal has the form of a distorted plane wave:

$$L_c = \exp\{i(X, P) + iW(X, P)\}. \tag{5.51}$$

The phase $W(X, P)$ is give by the expression:

$$W(X, P) = \sum_\alpha (2|k_\alpha|)^{-1} n_\alpha \log |k_\alpha| \xi^{(\alpha)} \tag{5.52}$$

where $\xi^{(\alpha)} = |x_\alpha| - (x_\alpha, \hat{k}_\alpha)$ are the two-body parabolic variables. Terms of the next orders in the expansion (5.48) are given by the recurrent relations (5.49). One can verify that the k-th coefficient $d_k(X)$, $k \geq 1$ is of order of $O\left((\sum_\alpha |k_\alpha| \xi^{(\alpha)})^{-\frac{k+1}{2}} \right)$ for $|k_\alpha| \xi^{(\alpha)} \to \infty$. The Schrödinger equation for L_c is satisfied up to terms of the order $(\sum_\alpha |k_\alpha| \xi^{(\alpha)})^{-\frac{k+1}{2}}$. It follows that the series (5.48) loses its asymptotic character along the directions where the variables $|k_\alpha| \xi^{(\alpha)}$ are bounded.

Let us note that the function $C(M)L_c$ can also be used as a plane eikonal approximation with the factor $C(M)$ being an arbitrary function of the variable M.

5.2.3 The spherical eikonal $|X|$

As in the two-body case, this eikonal generates the distorted spherical wave:

$$Q_c(X, P) = f(\hat{X})|X|^{-5/2} \exp\{i\sqrt{E}|X| + iW_c(X, P)\} \tag{5.53}$$

where the Coulomb phase has the form of the sum

$$W_0(X, P) = -\frac{|X|}{2\sqrt{E}} \sum_\alpha \frac{n_\alpha}{|x_\alpha|} \log 2\sqrt{E}|X|.$$

The transversal surface $Z(X) = \text{const}$ in this case is a sphere.

5.2.4 The single eikonal Z_α

Let us recall that the eikonal Z_α, describing the single collision of the particle from the pair α, is equal to the sum of the spherical two-body plane eikonals $Z_\alpha = E^{-1/2}(|k_\alpha|\,|x_\alpha| + (p_\alpha, y_\alpha))$. The vector X can be expressed as the orthogonal sum

$$X = Z_\alpha \hat{K}_\alpha + M_\alpha,$$

where the vector $\hat{K}_\alpha$ determining direction of the asymptotic motion is given by

$$\hat{K}_\alpha = E^{-1/2}\{|k_\alpha|\hat{x}_\alpha, p_\alpha\} \tag{5.54}$$

Here the coordinate system associated with pair α is used.

As the orthogonal coordinates on the surface Z_α = const, one can take the spherical coordinates of the vector x_α, the Cartesian coordinates u_α of the projection of vector y_α on the plane (p_α, y_α) = const and the variable

$$\omega_\alpha = E^{-1/2}\{|k_\alpha|(\hat{p}_\alpha, y_\alpha) - |x_\alpha|\,|y_\alpha|\}.$$

Moreover, to the variable ω_α corresponds the direction vector e_{ω_α} given, in the coordinates described above, by

$$e_{\omega_\alpha} = E^{-1/2}\{|k_\alpha|\hat{p}_\alpha, -|p_\alpha|\hat{x}_\alpha\}.$$

The vector field M_α is a function of the coordinates $\hat{x}_\alpha$, u_α, and ω_α satisfying the relation

$$|M_\alpha| = \left(u_\alpha^2 + \omega_\alpha^2\right)^{1/2}.$$

Bellow, we will regard the variables $|x_\alpha|$ and y_α as a point $\tilde{M}_\alpha$ in the four-dimensional space $R^4 = R^1 \oplus R^3$, $\tilde{M}_\alpha = \{|x_\alpha|, y_\alpha\}$. Note that $|\tilde{M}_\alpha| = |X|$. Finally, let us note that it can be shown that the continuity equation is satisfied by the function $|x_\alpha|^{-1}$.

Let us calculate the eikonal phase W_Z. Consider the integral along the straight line $s(t) = t\hat{K}_\alpha + M_\alpha$:

$$W_\alpha^{(\beta)} = -\frac{1}{2\sqrt{E}} \int_{s_\alpha} \frac{n_\beta dt}{|x_\beta(t, M_\alpha)|} \tag{5.55}$$

To start with, let us take $\alpha = \beta$. Because of (5.54), the equation $x_\alpha(t, M_\alpha) = t|k_\alpha|E^{-1/2}$ holds and the integral coincides with the two-body phase

$$W_\alpha^{(\alpha)} = -\frac{n_\alpha}{2|k_\alpha|} \log 2|k_\alpha|\,|x_\alpha|.$$

Next, let $\alpha \neq \beta$. Using equation (3.22), we express $x_\beta(t)$ through the coordinates $x_\alpha(t)$ and $y_\alpha(t)$; taking (5.54) into account, we have

$$x_\beta(t, M_\alpha) = tk_\beta^{(\alpha)} E^{-1/2} + x_\beta^{(\alpha)} \tag{5.56}$$

where

$$k_\beta^{(\alpha)} = c_{\alpha\beta}|k_\alpha|\hat{x}_\alpha + s_{\alpha\beta}p_\alpha, \quad x_\beta^{(\alpha)} = x_\beta + Z_\alpha k_\beta^{(\alpha)} E^{-1/2},$$

and the vector $x_\beta^{(\alpha)}$ is a function of the vector field M_α only, $x_\beta^{(\alpha)} = x_\beta^{(\alpha)}(M_\alpha)$. Substituting (5.56) into the integrand of (5.55) and integrating we obtain

$$W_\alpha^{(\beta)} = -\frac{n_\alpha}{2|k_\beta^{(\alpha)}|} \log\left(2|k_\beta^{(\alpha)}|\,|x_\beta| + (k_\beta^{(\alpha)}, x_\beta)\right). \tag{5.57}$$

By another choice of the of the initial integration point, for the integral (5.55) we can obtain the representation $-W_\alpha^{(\beta)}(x_\beta, -k_\beta^{(\alpha)})$ with the function $W_\alpha^{(\beta)}$ given by (5.57). Moreover, the difference between the functions $W_\alpha^{(\beta)}(x_\beta, k_\beta^{(\alpha)})$ and $-W_\alpha^{(\beta)}(x_\beta, -k_\beta^{(\alpha)})$ depends only on M_α:

$$W_\alpha^{(\beta)}(x_\beta, k_\beta^{(\alpha)}) + W_\alpha^{(\beta)}(x_\beta, k_\beta^{(\alpha)}) = -\frac{n_\beta}{|k_\beta^{(\alpha)}|} \log|k_\beta^{(\alpha)}|\,|x_\beta^{(\alpha)}|.$$

Finally, let us note that the momenta $k_\beta^{(\alpha)}$ can be written as gradients of the corresponding coordinates:

$$k_\beta^{(\alpha)} = \sqrt{E}\nabla_{x_\beta} Z_\alpha.$$

Therefore, to the single eikonal Z_α corresponds the approximate solution of the Schrödinger equation

$$\Psi_\alpha(X, P) = C_\alpha(M_\alpha)\frac{\exp\{i\sqrt{E}Z_\alpha + iW_\alpha\}}{|x_\alpha|}, \tag{5.58}$$

with the Coulomb phase W_α equal to the sum of the integrals $W_\alpha^{(\beta)}$, $W_\alpha = \sum_\beta W_\alpha^{(\beta)}$. The unknown function $C_\alpha(M_\alpha)$ of the transversal vector field will be determined bellow.

5.2.5 The double eikonal $Z_{\alpha\beta}$

The eikonal $Z_{\alpha\beta}$, corresponding to the processes of successive two-body collisions ($\beta \to \alpha$), is given in terms of the kinematic variables by equation (4.46). Correspondingly, the vector X can be expressed by the sum $X = Z_{\alpha\beta}\hat{K}_{\alpha\beta} + M_{\alpha\beta}$, where, in the coordinate system associated with the pair α, the directional vector $\hat{K}_{\alpha\beta}$ is equal $\hat{K}_{\alpha\beta}\{|k_{\alpha\beta}|\hat{x}_\alpha, p_\alpha)$.

It can be shown that the angles $\theta_{\alpha\beta}$, $\varphi_{\alpha\beta}$ given by equation (4.48), the unit vector $\hat{x}_\alpha$, and the variable $\omega_{\alpha\beta}$, $\omega_{\alpha\beta} = (X^2 - Z^2_{\alpha\beta})^{1/2}$ form an orthogonal curvilinear coordinate system on the surface $Z_{\alpha\beta} = \text{const}$. By straightforward calculations of the Lame coefficients, we can check that the Laplace operator in these new variables acts as

$$\Delta f =$$

$$= \frac{A_{\alpha\beta}}{\sqrt{\sin\theta_{\alpha\beta}}|x_\alpha|\,|y_\alpha|}\left(\frac{\partial^2}{\partial Z_{\alpha\beta}} + \frac{\partial^2}{\partial M_{\alpha\beta}}\right)\frac{\sqrt{\sin\theta_{\alpha\beta}}|x_\alpha|\,|y_\alpha|}{A_{\alpha\beta}} f +$$

$$+ O(|x_\alpha|^{-1}\,|y_\alpha|^{-1}|X|), \tag{5.59}$$

where the function $A_{\alpha\beta}$ is given by equation (4.52). It follows from this that, in particular, the function $|x_\alpha|^{-1}\,|y_\alpha|^{-1}A_{\alpha\beta}$ satisfies the continuity equation (5.47) up to the terms of order $O(|x_\alpha|^{-1}\,|y_\alpha|^{-1}|X|)$.

Let us calculate the integral

$$W^{(\gamma)}_{\alpha\beta} = -\frac{1}{2\sqrt{E}}\int_{s_{\alpha\beta}} \frac{n_\gamma\, dt}{|x_\gamma(t, M_{\alpha\beta})|}$$

along the straight line $s_{\alpha\beta}(t) = t\hat{K}_{\alpha\beta} + M_{\alpha\beta}$. For $\gamma = \alpha$, as in the case of the single eikonal, we easily get

$$W^{(\alpha\beta)}_{\alpha} = -\frac{n_\alpha}{2|k_{\alpha\beta}|}\log 2|k_{\alpha\beta}|\,|x_\alpha|. \tag{5.60}$$

If $\gamma \neq \alpha$, we have the following equations

$$x_\gamma(t, M_{\alpha\beta}) = t k_{\gamma,\alpha\beta}E^{-1/2} + x_{\gamma,\alpha\beta} \tag{5.61}$$

where $k_{\gamma,\alpha\beta} = \nabla_{x_\gamma} Z_{\alpha\beta}$ or, explicitly,

$$k_{\gamma,\alpha\beta} = c_{\gamma\alpha}|k_{\alpha\beta}|\hat{x}_\alpha + s_{\gamma\alpha}p_{\alpha\beta}, \quad x_{\gamma,\alpha\beta} = x_\gamma + Z_{\alpha\beta}k_{\gamma,\alpha\beta}E^{-1/2}.$$

By definition, we take $k_{\alpha,\alpha\beta} = k_{\alpha\beta}$. Substituting (5.61) into the integrand of $W^{(\gamma)}_{\alpha\beta}$, we obtain that

$$W^{(\gamma)}_{\alpha\beta} = -\frac{n_\gamma}{2|k_{\gamma,\alpha\beta}|}\log\left(2|k_{\gamma,\alpha\beta}|\,|x_\gamma| + (k_{\gamma,\alpha\beta}, x_\gamma)\right) \tag{5.62}$$

or, up to a term

$$-\frac{n_\gamma}{2|k_{\gamma,\alpha\beta}|}\log 2|k_{\gamma,\alpha\beta}|\,|x_{\gamma,\alpha\beta}|$$

depending only on the transversal variables $M_{\alpha\beta}$, so that we can take the function $-W^{(\gamma)}_{\alpha\beta}(x_\gamma - k_{\gamma,\alpha\beta})$ as the eikonal phase.

Thus, the following asymptotic solution of the Schrödinger equation corresponds to the double eikonal $Z_{\alpha\beta}$:

$$\Psi_{\alpha\beta} = \frac{C_{\alpha\beta}(M_{\alpha\beta})}{|x_\alpha|\,|y_\alpha|} A_{\alpha\beta}(X)\exp\{i\sqrt{E}Z_{\alpha\beta} + iW_{\alpha\beta}\}. \tag{5.63}$$

Here, the Coulomb phase $W_{\alpha\beta}$ is equal to the sum of the integrals (5.60) and (5.62):

$$W = \sum_\gamma W^{(\gamma)}_{\alpha\beta}.$$

Therefore, we constructed the eikonal asymptotic approximations corresponding to the wave functions labelled by (+). To obtain the formulae for the wave functions of the type $\Psi^{(-)}$, one should use the relation $\Psi^{(-)}(X,P) \sim \Psi^{(+)*}(X,-P)$, where as the functions $\Psi^{(+)}$ one takes the eikonal approximations (5.51) - (5.63). Let us now proceed to the study of the asymptotics of Ψ_0.

5.2.6 Asymptotics of the function Ψ_0

We will construct the asymptotic solution of the Schrödinger equation normalized with respect to the distorted plane wave with unit amplitude (5.51). According to the hypothesis presented above, we will look for the asymptotics of this wave function as the sum of the eikonal approximations corresponding to the given eikonals of the three-particle problem:

$$\Psi_0(X,P) \sim L_c + \sum_\alpha \Psi_\alpha + \sum_{\alpha\neq\beta}\Psi_{\alpha\beta} + \tilde{\Phi}_0. \tag{5.64}$$

However, the amplitudes $C_\alpha(M_\alpha)$, $C_{\alpha\beta}(M_{\alpha\beta})$ for the approximation of the single and double eikonals are not known. To find them, we will use the method of sewing the asymptotics.

We will say that the smooth function F satisfies the sewing condition with the eikonal approximation Ψ_Z in the neighbourhood of the direction Ω_0, if it obeys the Schrödinger equation $(H-E)F = 0$ up to terms rapidly decreasing with $|X| \to \infty$ and turns into the corresponding eikonal approximation when X recedes from the direction Ω_0. We will demand the correction term to decrease faster than the product $Z^{-1}F$ of the function by the inverse of the eikonal.

In the scattering theory, the simplest version of the sewing method for solutions of ordinary differential equations is widely used. In this case, in the point, where the sewing is performed, the solutions containing only a finite number of unknown coefficients are compared. These coefficients are then determined by solving a system of algebraic equations. The sewing procedure is much more complicated in the case of partial differential equations where the arbitrariness in the general solution cannot be reduced to a finite number of unknown coefficients.

In such a case, the sewing method is based on the principle of locality which makes it sufficient to construct the solutions locally. In this case, the locality principle is based on the fact that the property of the solution in a vicinity of the manifold Ω_0, where the sewing is performed depends only on the characteristics of the potential in this neighbourhood and on the form of the solution in Ω_0. This circumstances make it possible to solve the sewing problem for the Schrödinger equation by comparing the asymptotic solutions of differential equations with separable variables. These solutions can be obtained by neglecting perturbations with small magnitude and the eikonal asymptotics. This procedure is called the method of standard equation. In the three-body problem (for charged particles) this procedure takes a special form and the arising questions will be discussed in detail in the following sections. In this section, we will show that the lowest terms of the asymptotic expansion of the function Ψ_0 up to the distorted spherical waves can be uniquely determined from the sewing condition in directions Ω_α, where the particles are close to each other.

The calculations will be given on the example of the system where only two particles are charged. This assumption considerably simplifies the constructions. However, all the constructions below can be easily generalized to the case of three charged particles. The necessary modifications will be specified at the end of this section, where the final formulae for the general case will be given as well.

For definiteness, let us assume that the particle 1 is not charged, i.e., $n_2 = n_3 = 0$ and $n_1 \neq 0$. Moreover, in the course of the derivation of the asymptotic formulae, we will assume that the two-body Hamiltonians h_α $(\alpha = 1,2,3)$ have no bound states and the relative momenta of the incoming particles are not equal to zero $k_\alpha \neq 0$ $(\alpha = 1,2,3)$. From the first assumption it follows, in particular, that $n_1 > 0$. The asymptotic formulae for particles with different sign of charge, will be described at the end of this section.

Far away from the singular directions the asymptotics of the term L_c has the form of a distorted plane wave

$$L_c \sim \exp\{i(X,P) + iW\}, \quad W = \frac{n_1}{2|k_1|} \log |k_1| \xi^{(1)}. \tag{5.65}$$

As mentioned above, this expression does not make sense in the direction of the forward scattering of the pair 1, where the variable $\xi^{(1)}$ is bounded. However, it is easy to see that for $|x_\alpha| \to \infty$, $(\alpha = 1,2,3)$ the variables separate and the Schrödinger equation has the solution $\Psi_c^{(1)}(X,P)$ which reduces to a distorted plane wave for $\xi^{(1)} \to \infty$:

$$\Psi_c^{(1)}(X,P) = \exp\{i(X,P)\}\varphi_c(k_1,\xi^{(1)}). \tag{5.66}$$

Here, the function $\varphi_c(k_1,\xi^{(1)})$ is a solution for the scattering problem of two particles and can be expressed through the confluent hypergeometric function $\Phi(a,b,t)$ by the formulae (5.1), (5.2). Let us stress that the function $\Psi_c^{(1)}$ is the exact solution of the three-body Schrödinger equation with the interaction given by the single Coulomb potential $n_1|x_1|^{-1}$.

Thus, the leading terms Ψ_0 in the singular direction where the variable $\xi^{(1)}$ is bounded, are described by the function $\Psi_c^{(1)}$:

$$\Psi_0(X,P) \sim \Psi_c^{(1)}(X,P).$$

For $\xi^{(1)} \to \infty$, the function $\Psi_c^{(1)}$ has the form of the sum of the distorted plane wave L_c and the term $\varphi_c^{(1)}$ corresponding to the single eikonal:

$$\varphi_c^{(1)}(X,P) = f_c(\hat{x}_1, k_1)\frac{\exp\{i\sqrt{E}Z_1 + iW_1\}}{|X_1|}. \tag{5.67}$$

Here, f_c is the two-body Coulomb scattering amplitude given by equation (5.5).

Let us examine Schrödinger equation in the regions Ω_α where the particles are close to each other.

If the point X belongs to the region Ω_α, then asymptotically all potentials except $v_1(x_1)$ become zero for $|X| \to \infty$ and the Schrödinger equation

$$\left(-\Delta_{x_1} - \Delta_{y_1} + v_1^{(s)}(x_1) + n_1|x_1|^{-1} - E\right)\Psi = 0$$

has the solution

$$\Psi = e^{i(p_1,y_1)}\Psi_1(x_1,k_1)$$

which can be found by the method of separation of variables. By $\psi_1(x_1,k_1)$ we denoted the wave function corresponding to the Hamiltonian $h_1 = h_0 + v_1$. For $\xi^{(1)} \to \infty$, this function has the form of a sum of distorted three-dimensional spherical waves and the corresponding scattering amplitude $f_c(\hat{x}_1, k_1)$ is given by the sum of short-range and Coulomb parts (5.22).

By comparing equations (5.58), (5.21) and (5.67), we conclude that far from the directions where $\xi^{(1)}$ is bounded, the term Ψ_0 corresponding to the eikonal $\xi^{(1)}$ is given by formula (5.58) with $C_1(M_1) = f_c(\hat{x}_1, k_1)$.

Next, we will determinate the amplitude $C_\alpha(M_\alpha)$ for $\alpha \neq 1$. Let the point X belong to the region Ω_α, $\alpha \neq 1$. Using formula (3.22) we express the vector x_1 through x_α and y_α and we expand the Coulomb potential $n_1|x_1|^{-1}$ in a series with respect to the small parameter $|x_\alpha|\,|y_\alpha|^{-1}$. We obtain

$$\frac{n_1}{|x_1|} = \frac{n_{1\alpha}}{|y_\alpha|}\sum_{k=0}^{\infty} P_k(\cos\theta_\alpha)\left(\frac{|x_\alpha|c_{1\alpha}}{|y_\alpha|s_{1\alpha}}\right)^k, \quad \cos\theta_\alpha = (\hat{x}_\alpha, \hat{y}_\alpha) \tag{5.68}$$

where $n_{1\alpha} = n_1|s_{1\alpha}|^{-1}$ and $P_k(\zeta)$ are the Legendre polynomials. In the construction of the leading term of the asymptotics of the wave function, we will take into account only the first term of this expansion. Thus, the constructed solution will satisfy the Schrödinger equation in Ω_α up to terms of the order $|x_\alpha|\,|y_\alpha|^{-1}$.

We will seek the solution of Schrödinger equation in the region Ω_α for $\alpha \neq 1$ which turn into the distorted plane wave L_c for $|x_\alpha| \to \infty$. Substituting the sum $\Psi_c^{(1)} + \Phi_\alpha$ into the Schrödinger equation, for the leading term of Φ_α, we obtain:

$$(-\Delta_{x_\alpha} - \Delta_{y_\alpha} + v_\alpha(x_\alpha) + n_{1\alpha}|y_\alpha|^{-1} - E)\,\Phi_\alpha = -v_\alpha(x_\alpha)\Psi_c^{(1)}. \qquad (5.69)$$

To describe a solution of this equation, let us introduce a number of functions.

Let, as above, $\psi_\alpha(x_\alpha, k_\alpha)$ be the wave functions of the two-body Hamiltonian h_α, $h_\alpha = h_0 + v_\alpha$ and let $f_\alpha(\hat{x}_\alpha, k_\alpha)$ be the scattering amplitude for h_α. Further, let $\psi_\alpha^c(y_\alpha, p_\alpha)$ be the wave function of the model two-body Hamiltonian h_α^c corresponding to the effective Coulomb interaction of the pair with the third particle

$$h_\alpha^c f(y_\alpha) = (-\Delta_{y_\alpha} + n_{1\alpha}|y_\alpha|^{-1})\, f(y_\alpha). \qquad (5.70)$$

These functions can be expressed through the confluent hypergeometric function $\Phi(a, b, t)$ by means of the formulae (5.1), (5.2), where the quantities n_1, k_1, and x_1 must be replaced by $n_{1\alpha}$, p_α, and y_α, respectively. Let us denote by $R_\alpha^c(X, X', z)$ the Green function of the energy operator H_α^c generated by the Schrödinger equation (5.69):

$$H_\alpha^c = h_\alpha + h_\alpha^c. \qquad (5.71)$$

Since in this equation the variables separate, the Green function can be represented as the contour integral

$$R_\alpha^c(z) = (2\pi i)^{-1} \oint r_\alpha(\zeta) r_\alpha^c(z - \zeta)\, d\zeta, \qquad (5.71')$$

where $r_\alpha(\zeta)$ and $r_\alpha^c(\zeta)$ denote the two-body Green functions for the operators h_α and h_α^c, respectively.

Let $\chi_1^F(\xi^{(1)})$ be the characteristic function of the singular direction where the variable $\xi^{(1)}$ is bounded:

$$\chi_1^F(\xi^{(1)}) = \begin{cases} 1, & \xi^{(1)} \leq (1 + |x_1|)^\nu, \quad \nu > 0, \\ 0, & \xi^{(1)} \geq (1 + |x_1|)^{\nu'}, \quad \nu < \nu' < 2/3. \end{cases}$$

Using the relations between the functions $\Phi(a, b, t)$ and $\Psi(a, b, t)$:

$$\Phi(a, b, it) = \frac{\Gamma(b)}{\Gamma(b - a)} e^{ia\pi} \Psi(a, b, it) +$$

$$+\frac{\Gamma(b)}{\Gamma(a)}e^{it}e^{ia\pi}\Psi(b-a,b,-it), \quad t>0,$$

we represent the function $\Psi_c^{(1)}$ as the sum

$$\Psi_c^{(1)}(X,P)=(\varphi_c^{(1)}(x_1,k_1)+\varphi_c^{(2)}(x_2,k_2))e^{i(X,P)}, \tag{5.72}$$

where

$$\varphi_c^{(1)}(x_1,k_1)=\chi_1^F(\xi^{(1)})\Psi(-i\eta_1,1,i|k_1|\xi^{(1)})-$$
$$-\frac{\Gamma(1+i\eta_1)}{\Gamma(-i\eta_1)}e^{i|k_1|\xi^{(1)}}\Psi(1+i\eta_1,1,-i|k_1|\xi^{(1)}),$$
$$\varphi_c^{(2)}(x_1,k_1)=(1-\chi_1^F(\xi^{(1)}))\Psi(-i\eta_1,1,i|k_1|\xi^{(1)}).$$

It is easy to see that for $|x_1\to\infty$, the function $\varphi_c^{(1)}$ is of higher order than $\varphi_c^{(2)}$. Far from the singular direction, where $\xi^{(1)}\to\infty$, we have

$$\varphi_c^{(1)}=O(1), \quad \varphi_c^{(2)}=O((\xi^{(1)})^{-1}).$$

In correspondence with this representation, the right-hand side of equation (5.69) can be written as a sum of two terms. First, we will consider the leading order equation:

$$(H_\alpha^c-E)\Phi^{(1)}(X)=-v_\alpha(x_\alpha)\varphi_c^{(1)}(x_1)e^{i(X,P)}, \tag{5.73}$$

and then in the lowest order

$$(H_\alpha^c-E)\Phi^{(2)}(X)=-v_\alpha(x_\alpha)\varphi_c^{(2)}(x_1)e^{i(X,P)}. \tag{5.74}$$

As a consequence of linearity, the solution (5.69) is the sum

$$\Phi_\alpha(X,P)=\Phi^{(1)}(X)+\Phi^{(2)}(X).$$

The solution of the first equation can be represented as the integral

$$\Phi^{(1)}(X)=$$
$$=-\lim\epsilon\downarrow 0\int R_\alpha^c(X,X',E+i\epsilon)v_\alpha(x_\alpha')\varphi_c^{(1)}(x_1')e^{i(X',P)}dX'. \tag{5.75}$$

Since this equation must be satisfied only asymptotically for $|y_\alpha|\to\infty$, the integration over X' can be restricted to the region where $|y_\alpha-y_\alpha'|\le|y_\alpha|^\nu$, $\nu<1$.

To sew the function $\Phi^{(1)}$ with the eikonal approximation (5.58) and to find the coefficients $C_\alpha(M_\alpha)$, let us investigate the asymptotics of the integral (5.75) for $|x_\alpha| \to \infty$. First of all, let us note that in the region Ω_0 the parabolic coordinate $\xi^{(1)}$ can be represented as the sum

$$\xi^{(1)} = \tilde{\xi}^{(1)}_\alpha + c_{1\alpha}(x_\alpha, \hat{y}_\alpha - \hat{k}_1) + O(|x_\alpha|^2|y_\alpha|^{-1}), \tag{5.76}$$

where the variable $\tilde{\xi}^{(1)}_\alpha = |s_{1\alpha}y_\alpha| - s_{1\alpha}(\hat{k}_1, y_1)$ is of the leading order for all direction of the vector y_α as it follows from the inequality $|(x_\alpha, \hat{y}_\alpha - \hat{k}_1)| \leq c\tilde{\xi}^{(1)}_\alpha |x_\alpha||y_\alpha|^{-1}$. This implies that for $|y_\alpha| \to \infty$ the function $\varphi^{(1)}_c$ up to the terms of order $O(|y_\alpha|^{-\nu}$ depends only on the coordinate $\tilde{\xi}^{(1)}$.

Since the potential $v_\alpha(x_\alpha)$ is a rapidly decreasing function, the leading term of the asymptotics of the integral (5.75) for $|x_\alpha| \to \infty$ is generated by that region of integration where the relation $|x'_\alpha| \leq |x_\alpha|^\nu$ holds. Recalling the asymptotic formula (4.17), satisfied when the above mentioned conditions are satisfied and the integral representation for the scattering amplitude (4.18), we obtain

$$\Phi^{(1)}(X, P) \sim \frac{f_\alpha(\hat{x}_\alpha, k_\alpha)}{|x_\alpha|}\tilde{\Phi}^{(1)}(X, P) \tag{5.77}$$

where the function $\tilde{\Phi}^{(1)}$ is given by the integral

$$\tilde{\Phi}^{(1)}(X, P) = \frac{1}{2\pi i}\int_0^\infty dt\, t e^{it|x_\alpha|} u_1(y_\alpha, p_\alpha, t) \tag{5.78}$$

and

$$u_1(y_\alpha, p_\alpha, t) =$$
$$= \lim_{\epsilon \downarrow 0} \int dy'_\alpha r^c_\alpha(y_\alpha, y'_\alpha, E - t^2 + i\epsilon) v_\alpha(x'_\alpha) e^{i(y'_\alpha, p_\alpha)} \varphi^{(1)}_c(\tilde{\xi}^{(1)}_\alpha, k_1).$$

Next, let us r note that since the variables y_α, y'_α do not lie too far apart, $|y_\alpha - y'_\alpha| = O(|y_\alpha|^\nu)$, the function r^c_α asymptotically coincides with the free Green function. Let us assume that $\tilde{\xi}^{(1)}_\alpha \to \infty$. Then, in the leading order, the function $\varphi^{(1)}_c$ will depend only on y_α and can be taken outside the integral which can then be computed explicitly to give

$$u_1(y_\alpha, p_\alpha, t) \sim \frac{e^{i(y_\alpha, p_\alpha)}}{t^2 + p^2_\alpha - E - i0}\varphi^{(1)}_c(\tilde{\xi}^{(1)}_\alpha, k_1).$$

Let us substitute this expression into equation (5.78). For $|x_\alpha| \to \infty$, the integration path along t can be closed in the upper half-plane up to the terms of order $O(|x_\alpha|^{-1-\nu})$, $\nu > 0$. Using the residue theorem, we obtain the formula

$$\tilde{\Phi}^{(1)}(X,P) \sim e^{i\sqrt{E}Z_\alpha}\varphi_c^{(1)}(\tilde{\xi}_\alpha^{(1)}, k_1). \tag{5.79}$$

We note further that in the region Ω_α the variable $\tilde{\xi}^{(1)}$ up to terms of the order $O(|x_\alpha|^2|y_\alpha|^{-1})$ can be expressed through the vector field M_α as

$$\tilde{\xi}_\alpha^{(1)} = \xi_1^{(\alpha)} + O(|x_\alpha|^2|y_\alpha|^{-1})$$

where

$$\xi_1^{(\alpha)} = |u_1^\alpha| - (u_1^\alpha, k_1)\,\mathrm{sign}\, s_{1\alpha}.$$

Here by u_1^α we denote the vector field defined by

$$u_1^\alpha = |k_\alpha|^{-1}(|k_\alpha|y_\alpha - p_\alpha|x_\alpha|).$$

Since $\xi_1^{(\alpha)} \to \infty$, the function $\varphi_c^{(1)}$ can be replaced by its asymptotics. In the leading order we obtain

$$\Phi^{(1)}(X,P) \sim$$

$$\sim \frac{f_\alpha(\hat{x}_\alpha, k_\alpha)}{|x_\alpha|}\exp\{i\sqrt{E}Z_\alpha + i\delta\tilde{W}_1^\alpha + i\log\eta_1\log|M_\alpha|\} \tag{5.80}$$

where the term $\delta\tilde{W}_1^\alpha$ depends only on the angular variables $\hat{M}_\alpha$:

$$\delta\tilde{W}_1^\alpha = \eta_1\log|M_\alpha|^{-1}|s_{1\alpha}k_1|\xi_1^\alpha, \quad \eta_1 = \frac{n_1}{2|k_1|}.$$

Now, let us consider the eikonal approximation Ψ_α on the boundary of the domain Ω_0. We note that the leading order term of the eikonal Z_α in Ω_α is given by $Z_{\alpha\beta} = |p_\alpha|\,|k_\alpha|^{-1}\omega_\alpha$. For the Coulomb phase $W_\alpha^{(1)}$, in the leading order we obtain the representation

$$W_\alpha^{(1)} = -\frac{n_1}{2|k_1^{(\alpha)}|}\log|M_\alpha| + \tilde{W}_\alpha^{(1)}. \tag{5.81}$$

Here

$$\tilde{W}_\alpha^{(1)} = -\frac{n_1}{2|k_1^{(\alpha)}|}\log|M_\alpha|^{-1}k_1^{(\alpha)}\zeta_1^{(\alpha)}$$

and by $\zeta_1^{(\alpha)}$ we denoted the parabolic coordinate

$$\zeta_1^{(\alpha)} = |x_1^{(\alpha 0)}| + (x_1^{(\alpha 0)}, \hat{k}_1^{(\alpha)}), \quad x_1^{(\alpha 0)} = x_1 - |x_\alpha|\,|k_\alpha|^{-1}|k_1^{(\alpha)})|.$$

Let us stress that the phase $W_\alpha^{(1)}$ depends only on the coordinate M_α on the surface $Z_\alpha =$ const and does not change during the motions of the point X along trajectories with the direction vector ∇Z_α. All the dependence on the magnitude of the vector M_α is concentrated in the first term in equation (5.81), and the phase $\tilde{W}_\alpha^{(1)}$ is a function of the angular variables $\hat{M}_\alpha$ only.

Comparing the asymptotic representations (5.58) and (5.80) and bearing in mind equation (5.81) we can see that for $\xi_1^{(\alpha)} \to \infty$, the amplitude $C_\alpha(M_\alpha)$ is given by

$$C_\alpha(M_\alpha) = f_\alpha(\hat{x}_\alpha, k_\alpha)\exp\{ia_\alpha^{(1)}\log|m_\alpha| + i\delta W_\alpha^{(1)}\}, \tag{5.82}$$

where

$$a_\alpha^{(1)} = \frac{n_1}{2|k_1|} + \frac{n_1}{2|k_1^{(\alpha)}|}, \qquad \delta W_\alpha^{(1)} = \delta\tilde{W}_1^{(\alpha)} - \tilde{W}_\alpha^{(1)}.$$

Let us note that the formula (5.82) loses its sense in regions where the phase becomes infinite.

Consider, further, equation (5.74). For $|y_\alpha| \to \infty$, the function $e^{i(P,X)}\varphi_c^{(2)}$ has the following asymptotic form

$$e^{i(P,X)}\varphi_c^{(2)}(X,P) \sim \frac{\tilde{f}_c(\widehat{s_{1\alpha}y_\alpha}, k_1)}{|s_{1\alpha}y_\alpha|}\times$$

$$\times\exp\{i(p_{\alpha 0}^{(1)}, y_\alpha) + i(k_{\alpha 0}^{(1)}, x_\alpha) - i\eta_1\log 2|k_1|\,|s_{1\alpha}y_\alpha|\}, \tag{5.83}$$

where $\tilde{f}_c$ denotes the forward Coulomb scattering amplitude:

$$\tilde{f}_c(\widehat{s_{1\alpha}y_\alpha}, k_1) = (1 - \chi_1(\xi_{1\alpha}^{(1)})f_c(\widehat{s_{1\alpha}y_\alpha}, k_1),$$

$$\xi_{1\alpha}^{(1)} = |s_{1\alpha}y_\alpha| - (\hat{k}_1, s_{1\alpha}y_\alpha),$$

$$p_{\alpha 0}^{(1)} = -s_{\alpha 1}|k_1|\hat{y}_\alpha + c_{\alpha 1}p_1,$$

$$k_{\alpha 0}^{(1)} = -s_{\alpha 1}p_1 + c_{\alpha 1}|k_1|\widehat{s_{1\alpha}y_\alpha}$$

Let us denote by $\varphi_{ac}^{(2)}$ the function which one obtains by multiplying the right-hand side of equation (5.83) by $|s_{1\alpha}y_\alpha|$. Then we can write equation (5.74) as

$$\Phi^{(2)}(X,P) = |s_{1\alpha}y_\alpha|^{-1}(\tilde{\Phi}^{(2)}(X,P) + O(|x_\alpha|\,|y_\alpha|^{-2})$$

where the function $\tilde{\Phi}^{(2)}$ is given by formula (5.75) with the replacement $\varphi_c^{(1)} \to \varphi_{ac}^{(2)}$. For $|x_\alpha| \to \infty$, the asymptotics of the resulting integral can be investigated in the same way as in the calculation of the analogous integral (5.75). The asymptotic formula holds

$$\Phi^{(2)}(X, P) \sim \frac{f_\alpha(\hat{x}_\alpha, k_{\alpha 0}^{(1)}) \tilde{f}_c(\hat{q}_{1\alpha}, k_1)}{|s_{1\alpha} y_\alpha| \, |x_\alpha|} \exp\{i\sqrt{E} Z_{\alpha 1} + i \log \eta_1 \log |M_{\alpha 1}| + i\tilde{W}_1^{\alpha 1}\} \tag{5.84}$$

with the phase $\tilde{W}_1^{\alpha 1}$ given by

$$\tilde{W}_1^{\alpha 1} = -\eta_1 \log 2|k_1 s_{1\alpha}| \sin \Xi_{\alpha 1}, \quad \cos \Xi_{\alpha 1} = \frac{(p_{\alpha 1}, \hat{q}_{1\alpha})}{\sqrt{E}}.$$

Let us now study the eikonal approximation $\Psi_{\alpha 1}$ in the region Ω_α. First of all, let us note that the momenta $k_{\alpha 0}^{(1)}$ and $p_{\alpha 0}^{(1)}$ coincide up to terms of order $|x_\alpha| \, |y_\alpha|^{-2}$ with the momenta k_{a1} and p_{a1} determining the double eikonal Z_{a1}

$$k_{\alpha 0}^{(1)} = k_{a1} + O(|x_\alpha| \, |y_\alpha|^{-2}), \quad p_{\alpha 0}^{(1)} = p_{a1} + O(|x_\alpha| \, |y_\alpha|^{-2})$$

which, in turn, in the leading order reads

$$Z_{a1}^{(0)} = |M_{a1}| \cot \Xi_{a1}.$$

Thus, the logarithmic phase $W_{a1}^{(1)}$ can be represented in the form

$$W_{a1}^{(1)} = \frac{n_1}{2|k_{1,\alpha 1}|} \log |M_{\alpha 1}| + \tilde{W}_{a1}^{(1)} + O(|x_\alpha| \, |y_\alpha|^{-2}) \tag{5.85}$$

where

$$\tilde{W}_{\alpha 1}^{(1)} = -\frac{n_1}{2|k_{1,\alpha 1}|} \log \left(\frac{|k_{1,\alpha 1}|}{|M_\alpha|} (|x_{10}^{(\alpha,1)}| + (\hat{k}_{1,\alpha 1}, x_{10}^{(\alpha,1)})) \right)$$

and the coordinate $x_{10}^{(\alpha,1)}$, which is equal to the value of the vector $x1, \alpha 1$ in the region Ω_α, is expressed in terms of the vector field M_α as

$$x_{10}^{(\alpha,1)} = x_1 - \frac{|x_\alpha|}{|k_{\alpha 1}|} k_{1,\alpha 1}.$$

Let us compare the asymptotic formulae (5.63) and (5.84) bearing in mind relation (5.85). With the help of the locality principle, we find that for $|x_\alpha| \to \infty$, the function $\varphi^{(2)}$ turns into the double eikonal approximation $\Psi_{\alpha 1}$ given by equation (5.63) with the amplitude $C_{\alpha 1}$ equal to

$$C_{\alpha 1} = f_\alpha(\hat{x}_\alpha, k_\alpha) \tilde{f}_c(\hat{q}_{1\alpha}, k_1) \exp\{i a_{\alpha 1}^{(1)} \log |M_{\alpha 1}| + i\delta W_{\alpha 1}^{(1)}\} \tag{5.86}$$

where

$$a_{\alpha 1}^{(1)} = -\frac{n_1}{2|k_1|} + \frac{n_1}{2|k_{\alpha 1}|}, \quad \delta W_{\alpha 1}^{(1)} = \delta \tilde{W}_1^{(\alpha 1)} - \tilde{W}_{\alpha 1}^{(1)}.$$

Note, that in the directions of the configuration space where $(\hat{q}_{1\alpha}, \hat{k}_1) \neq 1$, the amplitude $\tilde{f}_c$ is equal to the two-body Coulomb amplitude (5.5). In the direction where $(\hat{q}_{1\alpha}, \hat{k}_1) = 1$, the amplitude $\tilde{f}_c$ is zero, and this direction coincides with the singular direction where the variable $\xi_1^{(\alpha)}$ is zero.

However, the constructed function $\Phi^{(2)}$ does not fully exhaust the contribution corresponding to the eikonal $Z_{\alpha 1}$. Considering the Schrödinger equation in Ω_α up to terms of order $O(|y_\alpha|^{-1})$, we must take into account the contribution from the asymptotics $\psi_1(x_1, k_1)$ corresponding to the short-range part of the scattering amplitude f_{cs} in (5.22). Demanding that the Schrödinger equation is satisfied up to terms of the order $O(|x_\alpha||y_\alpha|^{-1})$, we conclude that in Ω_α the following equation must hold

$$(H_\alpha^c - E)\Phi_s^{(2)}(X) = -v_\alpha(x_\alpha)\varphi_s^{(2)}(X),$$

where

$$\varphi_s^{(2)}(X) = f_{cs}(\hat{x}_1, k_1)\frac{\exp\{i\sqrt{E}Z_1 + iW_1\}}{|x_1|}.$$

The solution of this equation and its asymptotics can be obtained in the same way as in the case of equations (5.73) and (5.74). By sewing with the approximation of the double eikonal, we find that the function $\Phi_s^{(2)}$ turns into the eikonal approximation $\Psi_{\alpha 1}$, with the amplitude $C_{\alpha 1}$ given by (5.86), where the Coulomb amplitude f_c must be replaced by the short-range part f_{cs}.

Thus, we have shown that the contribution to the asymptotics of the wave function corresponding to the double eikonal $Z_{\alpha 1}$ is equal to the sum of the functions $\Phi^{(2)}$ and $\Phi^{(2)}$:

$$\Psi_{\alpha 1}(X, P) \sim \Phi^{(2)}(X, P) + \Phi_s^{(2)}(X, P).$$

The constructed asymptotic representations lose sense in the directions where the phase becomes infinite. This happens whenever one of the momenta k_1, $k_{\alpha 1}$, or $k_{1,\alpha 1}$ vanishes, or when the coordinate M_α turns into zero. We must note that the whole dependence on the magnitude of the vector M_α

is concentrated in the term $ia^{(1)}_{\alpha 1}\log|M_{\alpha 1}|$. We will denote the set of points where $|M_{\alpha 1}| = 0$ by $\Omega^{(0)}_{\alpha 1}$.

By means of analogous calculations, all the other terms of $\Psi_{\alpha\beta}$ corresponding to the eikonals $Z_{\alpha\beta}$, $b \neq 1$ can be constructed. It can be shown that far from singular directions these terms are given by equation (5.63) with the amplitudes $C_{\alpha\beta}$ equal to

$$C_{\alpha\beta} = f_\alpha(\hat{x}_\alpha, k_{\alpha\beta}) + f_\beta(\hat{q}_{\beta\alpha}, k_\beta)\exp\{ia^{(1)}_{\alpha\beta}\log|M_{\alpha\beta}| + i\delta W^{(1)}_{\alpha\beta}\}. \qquad (5.86')$$

Here for $\alpha = 1$ the variable $a^{(1)}_{\alpha\beta}$ and the phase $\delta W^{(1)}_{\alpha\beta}$ are given by the equations

$$a^{(1)}_{\alpha\beta} = \frac{n_1}{2|k_1|} - \frac{n_1}{2|k_{1\beta}|},$$

$$\delta W^{(1)}_{\alpha\beta} = \frac{n_1}{2|k_1|}\log\frac{|s_{1\beta}k_1|}{|k_\beta|\sin\Xi_{1\beta}}(|k_{1\beta}| + (k_{1\beta}, \hat{k}_1)) - $$

$$- - \frac{n_1}{2|k_{1\beta}|}\log\frac{2k^2_{1\beta}|s_{1\beta}|}{|k_\beta|\sin\Xi_{1\beta}}, \qquad (5.87)$$

and for $\alpha \neq 1$ by

$$a^{(1)}_{\alpha\beta} = \frac{n_1}{2|k_1|} - \frac{n_1}{2|k_{1,\alpha\beta}|}, \qquad (5.88)$$

$$\delta W^{(1)}_{\alpha\beta} = \frac{n_1}{2|k_1|}\log\frac{|s_{1\beta}k_1|}{|k_\beta|\sin\Xi_{\alpha\beta}}(|k_{\alpha\beta}| + (k_{\alpha\beta}, \hat{k}_1)\,\mathrm{sign}\,(s_{\alpha\beta}s_{1\beta})) +$$

$$+ \frac{n_1}{2|k_1^{(\alpha\beta)}|}\log\frac{\hat{k}_1^{(\alpha\beta)}, x_1^{(\alpha\beta)}) + |x_1^{(\alpha\beta)}|}{|s_{1\alpha}|((\hat{k}_{1,\alpha\beta}, \hat{q}_{\alpha\beta}) + 1)} +$$

$$+ \frac{n_1}{2|k_{1,\alpha\beta}|}\log\frac{|k_{1,\alpha\beta}|\,|x^{(0)}_{1,\alpha\beta}| + (k_{1,\alpha\beta}, x^{(0)}_{1,\alpha\beta})}{|M_{\alpha\beta}|}. \qquad (5.89)$$

Here the following new symbols have been used:

$$\cos\Xi_{\alpha\beta} = E^{-1/2}(p_{\alpha\beta}, \hat{q}_{\alpha\beta}),$$

$$k_1^{(\alpha\beta)} = c_{1\beta}|k_\beta|\hat{q}_{\beta\alpha} + s_{1\beta}p_\beta \qquad (5.90)$$

$$x_1^{(\alpha\beta)} = |k_\beta|^{-1}(s_{1\alpha}q_{\beta\alpha} - |s_{\beta\alpha}|\,|k_\beta|^{-1}k_1^{(\alpha\beta)}),$$

$$x^{(0)}_{1,\alpha\beta} = x_1 - |x_\alpha|\,|k_{\alpha\beta}|^{-1}k_{1,\alpha\beta}.$$

As above, we have explicitly separated the term $ia^{(1)}_{\alpha 1}\log|M_{\alpha 1}|$ where all dependence on the magnitude of the vector $M_{\alpha\beta}$ is concentrated. By $\Omega^{(0)}_{\alpha\beta}$ we will denote the set of points where $M_{\alpha\beta} \neq 0$.

Let us note that, similarly to the case of neutral particles, the terms $\Psi_{\alpha\beta}$ change form and turn into distorted spherical waves whenever the point X crosses the singular regions $\Omega_{\alpha\beta}^{(0)}$. The transient phenomena will be considered in the next section.

When one takes charges equal to zero, all Coulomb phases turn into zero, too. Formulae (5.58), (5.63), (5.82), and (5.86) for the eikonal approximations Ψ_α and $\Psi_{\alpha\beta}$ turn into (4.49) and (4.58), the corresponding representations for neutral particles. Here, the recurrent relations (5.49) and (5.49') can be used for the calculations of higher-order corrections, too.

Let us stress that, unlike in the case of neutral particles, the amplitudes C_α and $C_{\alpha\beta}$ depend on the variables $|M_\alpha| = \sqrt{X^2 - Z_\alpha^2}$, $|M_{\alpha\beta}| = \sqrt{X^2 - Z_{\alpha\beta}^2}$. This property of the eikonal approximations is stipulated by the fact that long-range interactions cause the appearance of the Coulomb phase shifts not only after the collision, but also during the process preceding the formation of the asymptotic behaviour. In other words, the phases in the formulae (5.82) and (5.83) correspond to the state of the system at the moment of collision, and the phases W_α and $W_{\alpha\beta}$ of equations (5.58), (5.63) describe the asymptotic motion in the direction ∇Z_α and $\nabla Z_{\alpha\beta}$ after the collisions.

Now, let us consider the general case when all three particles are charged and their charges can have different signs. For $|x_\alpha| \to \infty$ ($\alpha = 1,2,3$) far from the singular directions the coordinate asymptotics of the terms Ψ_α and $\Psi_{\alpha\beta}$ still have the form given above. The amplitudes C_α and $C_{\alpha\beta}$ are given by equations (5.82) and (5.86'), where instead of $a_\alpha^{(1)}$, $a_{\alpha\beta}^{(1)}$, $\delta W_\alpha^{(1)}$, and $W_{\alpha\beta}^{(1)}$ one should substitute the variables a_α, $a_{\alpha\beta}$, δW_α and $\delta W_{\alpha\beta}$ where the Coulomb interactions of all particles are taken into account additively:

$$a_\alpha = \sum_{\gamma\neq\alpha} a_\alpha^{(\gamma)}, \quad \delta W_\alpha = \sum_{\gamma\neq\alpha} \delta W_\alpha^{(\gamma)},$$

$$a_{\alpha\beta} = \sum_{\gamma} a_{\alpha\beta}^{(\gamma)}, \quad \delta W_{\alpha\beta} = \sum_{\gamma} \delta W_{\alpha\beta}^{(\gamma)}. \tag{5.91}$$

The functions $a_\alpha^{(\gamma)}$, $a_{\alpha\beta}^{(\gamma)}$, $\delta W_\alpha^{(\gamma)}$, and $W_{\alpha\beta}^{(\gamma)}$ with different combinations of the indices α, β, γ are defined by equations (5.82), (5.86), (5.86'), and (5.87) – (5.90).

These asymptotic formulae can be obtained by the same way as in the case of two charged particles. What we must only do is to take in terms of the zero-approximation, instead of the function $\Psi_c^{(1)}$ given by equation (5.66), a new function whose asymptotics for $\xi^{(\alpha)} \to \infty$ $(\alpha = 1,2,3)$ is described by equations (5.51), (5.52). In this case instead of a simple representation given by means of the confluent hypergeometric function, we get cumbersome constructions providing the formulae for the new function $\Psi_{a s}$ in the singular directions. The reason why these complicated functions appear is the impossibility of separation of variables in the Schrödinger equation in the case of three charged particles, even asymptotically. As a consequence, for sewing, one uses solutions of various standard equations, depending on the position of the point in the configuration space. If only two particles of the pair α are charged, the variables in the Schrödinger equation (without nuclear potentials) separate in the basis $\{x_\alpha, y_\alpha\}$, as reflected in the representation (5.66). The above mentioned function $\Psi_{a s}$ will be described in the next section.

Thus, we have considered the case where the point X is located in the region Ω_0, where all particles are strongly separated from each another. We have shown that the function (5.64) is an asymptotic solution of the Schrödinger equation, and we have described the eikonal approximations Ψ_α and $\Psi_{\alpha\beta}$. We have seen that the term Ψ_α, corresponding to the process of single scattering, arises as a result of solving the Schrödinger equation (5.69) in the neighbourhood of the direction $x_\alpha = 0$ with a distorted spherical wave as the source. Similarly, the source giving rise to the approximations of the double eikonal in the region Ω_β, $\beta \neq \alpha$, is the function Ψ_α appearing at the rescattering of a plane wave in the region Ω_α. It is clear that the given method of construction of the asymptotics by considering the processes of successive rescatterings in the region Ω_α can be used also in the case of neutral particles.

Now, let us consider the case when the particles are not strongly separated and the two-body subsystems have bound states. In this case, in the regions Ω_α, where the particles are pairwise near to each another, we add to the terms described above the distorted cluster-like spherical waves

$$\sum_j F_{A0}(y_\alpha, P)\psi_A(x_\alpha)\frac{\exp\{i|y_\alpha|\sqrt{P^2+\varkappa_A^2}+iW_A\}}{|y_\alpha|}$$

$$A = (\alpha, j) \tag{5.92}$$

which are associated with the particle capture processes $(3 \to 2)$. The phases

W_A coincide with the two-body phases corresponding to the motion of particles in the effective Coulomb field produced by the bound pair:

$$W_A = -\frac{n_{\alpha\alpha}}{2|p_A|}\log 2|p_A|\,|y_A|, \quad |p_A| = \sqrt{P^2 + \varkappa_A^2}.$$

The parameter $n_{\alpha\alpha}$ characterizing the magnitude of the effective Coulomb interaction up to a multiplicative factor is equal to the sum of charges of the pair α multiplied by the charge of the third particle:

$$n_{\alpha\alpha} = \gamma\sqrt{2\mu_{l,\alpha}}e_l(e_i + e_j), \quad \alpha = (i,j), \quad ,l \neq i \neq j,$$

or, in terms of the coefficients $s_{\beta\alpha}$ and n_β:

$$n_{\alpha\alpha} = \sum_{\beta\neq\alpha} n_\beta |s_{\beta\alpha}|^{-1}.$$

The terms (5.92) decrease rapidly for $|x_\alpha| \to \infty$ and do not contribute to the asymptotics $\Psi_0(X,P)$ in the region Ω_0. To see the reason of their appearance, let us look at equation (5.75) which defines the solution in Ω_α. If the operator h_α has bound states, the Green function (5.71) contains the corresponding terms (4.23). Therefore, along with the functions already described, new terms appear, gathered in the representation (5.92).

Thus, we have described the asymptotics of the wave functions by means of the eikonal approximations. We have also established that the latter lose sense in some directions of the configuration space, called singular directions. These directions can be divided into two groups. We will call principal those singular directions where the eikonals Z, Z_α, and $Z_{\alpha\beta}$ coincide with the spherical eikonal $|X|$. We will show that the amplitude of distorted spherical wave has non-integrable singularities over unit sphere in the principal singular directions.

All the other singular directions and the corresponding singularities will be called the secondary. Analysing the formulae obtained above, we come to the conclusion that secondary singularities appear for the following reasons. First, it happens when the relative momenta of the particles are small after collisions or, as in the case of the eikonals Z_α and $Z_{\alpha\beta}$, at the moment of collision. For example, this is the case in equation (5.86) when $k_{1,\alpha 1} = 0$ or $k_{\alpha 1} = 0$. Second, the eikonal form of the asymptotics loses sense in the cases, when the parabolic coordinates $\xi_1^{(\alpha)}$ are relatively small for one or two pairs of particles after or at the moment of collision.

Finally, let us note that the singular directions correspond to those regions of the configuration space where the semiclassical asymptotics are no longer valid. Indeed, the semiclassical wave functions have the form

$$\Psi_s(X) = A_s(X)e^{iS(X)}$$

where $S(X)$ is the action, given by the Hamilton-Jacobi equation $(\nabla S)^2 = E - V(X)$ and by $A_s(X)$ we denoted the corresponding solutions of the continuity equation $2(\nabla A_s, \nabla S) + A\Delta S = 0$. The semiclassical approximations turn into the eikonal approximation if the ratio of the potential to the kinetic energy tends to zero: $VE^{-1/2} \to 0$. Then the eikonal expression in exponential term $\exp\{\sqrt{E}Z + iW_Z\}$ coincides with the leading term in the expansion of the action in the small parameter $VE^{-1/2}$. Since far from the singular directions, the inequality $|\nabla W_Z| \ll 1$ holds, the variable ∇S is non-zero, because $|\nabla Z|^2 = 1$. In the singular directions the magnitude of ∇W_Z is increasing, therefore the sum $\nabla(Z + W_Z) \sim \nabla S$ can be equal zero. This means that there are turning points in the neighbourhood of whose the semiclassical approximation is inapplicable.

5.2.7 The wave functions $\Psi_A(X, p_A)$

Let us finally describe the coordinate asymptotics of the wave functions $\Psi_A(X, p_A)$ corresponding to the scattering of a bound pair on the third particle.

According to the method employed above, we will seek the wave functions $\Psi_A(X, p_A)$ in the regions Ω_β $(\beta = 1, 2, 3)$ where the particles are not strongly separated as a sum of incoming and scattered waves:

$$\Psi_A(X, p_A) = \psi_A(x_A)\psi_c^{(\alpha)}(y_\alpha, p_\alpha) + \Phi_{\beta A}(X, p_A). \tag{5.93}$$

The function $\psi_c^{(\alpha)}$ in this case describes the asymptotic motion of a bound pair α in the Coulomb field created by the third particle and can be expressed in terms of the confluent hypergeometric function by equations (5.1), (5.2), where new Coulomb parameters should be substituted

$$n_\alpha \to n_{\alpha\alpha}, \quad x_\alpha \to y_\alpha, \quad k_\alpha \to p_\alpha. \tag{5.94}$$

Solving equation (5.69) in the region Ω_β, for the scattered waves $\Phi_{\beta\alpha}$ we obtain the integral representation

$$\Phi_{\beta\alpha}(X, p_A) =$$

$$= \lim_{\epsilon \downarrow 0} \int R^c_\beta(X, X', E_A + i\epsilon)\psi_A(x'_A)\psi^{(\alpha)}_c(y'_\alpha, p_\alpha)\, dX'. \tag{5.95}$$

Here R^c_β denotes the Green function of the model operator H^c_β, given by the equations (5.69) - (5.71) where the new parameters of the Coulomb interaction $n_1|s_{1\beta}| \to n_{\beta\beta}$ should be substituted. The next problem consist of studying the asymptotic behaviour of the integral on the right-hand side of (5.95). This can be done in the same manner as in the case of the integral (5.75). Technically, this problem is even simpler because of the rapid decrease of the function $\psi_A(x'_A)$ for $|X'_A| \to \infty$. This leads to the result that in this case there are no slowly decreasing terms describing rescattering in the regions Ω_β with $\beta \neq \alpha$. As in the case of a system of neutral particles, all slowly decreasing terms reduce to the sum of distorted spherical and cluster waves. We will not present here detailed computation to prove this statement. All necessary technical devices had already been given in the more complicated case of the function Ψ_0. Only the final results will be stated.

First of all, let us consider Ψ_A in the region Ω_α, where the particles of the pair α are sufficiently close to each other, i.e., in the region where elastic scattering takes place. The coordinate asymptotics $\Psi_A(X, p_A)$ contains only terms corresponding to the processes of elastic scattering and to internal rearrangement of the particles of the pair α. If the relative coordinate y_A of the third particle is not parallel to the momentum p_α, the asymptotics Ψ_A is given by

$$\Psi_A(X, p_A) \sim L^c_A(X, p_A) + \sum_j \Phi_{A'A}(X, P_A), \quad A' = (\alpha, j), \tag{5.96}$$

where the first term is a distorted plane cluster wave:

$$L^c_A(X, p_A) =$$

$$\psi_A(x_A)\exp\{i(p_\alpha, y_\alpha) + iw_\alpha^{(0)}(y_\alpha, p_\alpha)\}(1 + O((\xi^\alpha)^{-1})).$$

The summation runs over all internal states of the pair α, and each term equals asymptotically to the distorted spherical cluster wave:

$$\Phi_{A'A}(X, P_A) \sim$$

$$\sim \psi_{A'}(x_A)\frac{\exp\{i|p_{A'}|\,|y_\alpha| + iw_{A'}(y_\alpha, p_\alpha)\}}{|y_\alpha|}F_{A'A}(\hat{y}_\alpha, p_A). \tag{5.97}$$

The Coulomb phases distorting the plane and spherical waves are given by

$$w_\alpha^{(0)} = \frac{n_{\alpha\alpha}}{2|p_\alpha|}\log|p_\alpha|\tilde{\xi}^{(\alpha)},$$

$$w_A(y_\alpha, p_\alpha) = \frac{n_{\alpha\alpha}}{2|p'_A|}\log|p_{A'}|\,|y_\alpha|, \tag{5.98}$$

$$\tilde{\xi}^{(\alpha)} = |y_\alpha| - |y_\alpha, \hat{p}_\alpha|, \quad |p_{A'}^2| = E_A(p_A) + \varkappa_{A'}^2.$$

The amplitudes of the processes of the internal rearrangement $F_{A'A}$ for $A' \neq A$ are smooth functions of the angular variables $\hat{y}_\alpha$. The amplitudes of the elastic scattering turn to infinity in the direction of forward scattering. The following representation holds

$$F_{AA}(\hat{y}_\alpha, p_A) = f_{cA}(\hat{y}_\alpha, p_A) + \tilde{f}_{AA}(\hat{y}_\alpha, p_A) \tag{5.99}$$

where f_{cA} is given by equation (5.5) with the Coulomb parameters changed according to (5.94). The term $\tilde{f}_{AA}$, however, can also have singularities in the forward direction, even though these singularities are weaker than (5.5). The reason is that the effective interaction between the particle and the bound pair is given by the potential (5.68) averaged over the internal state α:

$$v_A^{(c)}(y_\alpha) = \sum_{\beta\neq\alpha}\int dx_\alpha|\psi_A(x_\alpha)|^2|c_{\beta\alpha}x_\alpha + s_{\beta\alpha}y_\alpha|^{-1}. \tag{5.100}$$

For $|y_\alpha| \to \infty$, this potential equals to the sum of the Coulomb and multipole parts:

$$v_A^{(c)}(y_\alpha) = \frac{n_{\alpha\alpha}}{|y_\alpha|} + \frac{1}{|y_\alpha|}\sum_{k=1}^{\infty}\frac{\mu_k(\hat{y}_\alpha)}{|y_\alpha|}, \tag{5.101}$$

where

$$\mu_k(\hat{y}_\alpha) = \sum_{\beta\neq\alpha}\frac{|c_{\alpha\beta}|^k}{|s_{\alpha\beta}|^{k+1}}\int|\psi_A(x_\alpha|^2 P_k(\cos\theta_\alpha)|x_\alpha|^k\,dx_\alpha,$$

$$\cos\theta_\alpha = (\hat{x}_\alpha, \hat{y}_\alpha).$$

Let us denote by $h_A^{(c)}$ the energy operator generated by this potential:

$$h_A^{(c)} f(y) = (-\Delta_y + v_A^{(c)}(y)) f(y)$$

and by $\psi_A^{(c)}(y,p)$ the corresponding wave functions. As it is clear from what was said above, the asymptotics Ψ_A in the direction of forward scattering is in fact described by $\psi_A^{(c)}$. Thus, in equation ' (5.93), the function $\psi_c^{(\alpha)}$ must be replaced by $\psi_A^{(c)}$ and then the term Φ_{AA} will be asymptotically equal to the distorted cluster spherical waves with amplitudes bounded in all directions. But, as we have seen in Section 1, the multipole terms (5.101) create singularities in the scattering amplitude described by equation (5.32). Hence, also the term $\tilde{f}_{AA}$ in (5.99) will have the same singularities in the forward direction.

However, we must note that, provided the bound states $\psi_A(x_\alpha)$ are spherically symmetric, the multipole moments μ_k are equal to zero so that the singular part of the amplitude $F_{A'A}$ coincides with the proper Coulomb part f_{cA}.

Let us consider further the function $\Psi_A(X, p_A)$ in regions Ω_β, $\beta \neq \alpha$, i.e., in the regions corresponding to the rearrangement of the pairs. The incoming wave $\psi_A \psi_c^{(\alpha)}$ is rapidly decreasing in these regions, so that the leading terms of the asymptotics reduce to the distorted cluster spherical waves:

$$\Psi_A(X, p_A) = \sum_j \Phi_{BA}(X, p_A), \quad B = (\beta, j), \qquad ,(5.102)$$

where

$$\Phi_{BA} \sim \psi_B(x_\beta \frac{\exp\{i|p_B|\,|y_\beta| + iW_B(y_\beta, p_B)\}}{|y_\beta|} F_{BA}(\hat{y}_\beta, p_A).$$

The Coulomb phases are determined by equation (5.98) with $p_B^2 = E_A(p_A) + \varkappa_B^2$.. The amplitudes F_{BA} are smooth bounded functions in all scattering directions $\hat{y}_\beta$.

Finally, let us describe the asymptotic behaviour of the wave function $\Psi_A(X, p_A)$ in the region Ω_0, where all particles are far from one another. Suppose that for all $\beta = 1, 2, 3$ the conditions $|x_\beta||X|^{-1} \geq \delta > 0$ hold. In this case, $\Psi_A(X, p_A)$ has the form of a distorted six-dimensional spherical wave with a bounded amplitude:

$$\Psi_A(X, p_A) \sim F_{0A}(\hat{X}, p_A) \frac{\exp\{i\sqrt{E_A}|X| + iW_0(X, E_A)\}}{|X|^{5/2}}. \tag{5.103}$$

For small values of $|x_\beta|\,|X|^{-1}$, i.e., by approaching the regions Ω_β, this amplitude has a rather complicated behaviour. To verify this, let us consider, first of all, the Green function (5.71) which, according to equation (5.95) gives a solution of the Schrödinger equation in Ω_β. In analogy to equation (4.24), the asymptotics of this function can be expressed by means of the asymptotics (5.14) of the functions r_β and r_β^c using the saddle-point method. Here the critical point is determined by the equation

$$\frac{\partial}{\partial\zeta}(|x_\beta - x'_\beta|\sqrt{\zeta} + |y_\beta - y'_\beta|\sqrt{z-\zeta}) = 0$$

and is given by

$$\zeta_0 = \frac{|x_\beta - x'_\beta|^2}{|X - X'|^2} z.$$

Consequently, the exponent becomes equal to $i\sqrt{z}|X - X'|$. If $|X| \to \infty$ and $|x_\beta|\,|X|^{-1} \geq \delta > 0$, the asymptotics of the Green function takes the form of a distorted spherical wave in R^6. The solution (5.95) will have the same form, and can be sewed with (5.103). If x_β becomes too small, so that $\sqrt{\zeta_0} = O(|x_\beta|^{-3/4})$, the Green function $r_\beta(x_\beta, x'_\beta, z)$ in (5.71) has an anomalous asymptotic behaviour of the kind (5.8) or (5.9) and the arguments presented above are no longer valid. In this case, we must modify the asymptotic representation (5.103) in accordance with (5.8) and (5.9). On the basis of this considerations, one can show that if the particles of the pair β have charges of different sign $n_\beta < 0$, then the amplitude F_{0A} becomes infinite for $\theta_\beta = 0$, $\sin\theta_\beta = |x_\beta|\,|X|^{-1}$:

$$F_{0A}(\hat{X}, p_A) \sim (\sin\theta_B)^{-1/2} \exp\{i\eta_B(\log|\eta_B| - 1)\}\tilde{f},$$

$$\eta_B = \frac{n_\beta}{2\sqrt{E_B}\sin\theta_\beta}, \tag{5.103'}$$

and in the case of Coulomb repulsion it decreases as some power

$$F_{0B} \sim (\sin\theta_B)^N, \quad N \gg 1.$$

With this we finish the description of the asymptotics of the function $\Psi_A(X, p_A)$. We see that it is constructed almost as simply as the corresponding function for three neutral particles. It is only necessary to replace the plane and spherical waves by their distorted analogies. Of course, specific difficulties arise in a system of charged particles when we try to describe the scattering in the forward direction. However, all singularities have a two-body nature and correspond to the effective interaction of a pair with free particle.

5.3 The Asymptotics of Ψ_0 in Forward Direction

In this and the next sections we proceed with our investigation of the wave function $\Psi_0(X, P)$. We have seen above that the eikonal formulae describe the asymptotics of $\Psi_0(X, P)$ incorrectly in those directions of the configuration space where the phases turn into infinity. For the construction of the asymptotics of $\Psi_0(X, P)$ in such singular regions we sew the solutions of partial differential equations by means of the method of parabolic equations. According to this method it is necessary to choose characteristic parameters, called local parabolic coordinates in the neighbourhood of singular directions, drop the leading order terms and solve the remaining equation exactly. This solution, satisfying the sewing conditions, is determined uniquely by comparing its asymptotics far from the singular directions with the corresponding eikonal approximation.

5.3.1 Distorted plane waves

We will perform the investigation of the asymptotics of $\Psi_0(X, P)$ in two stages. First, we will study the long-range effects of the Coulomb potentials, and then the processes of rescattering caused by the interaction of the particles at short distances. Let us note in this context that, as it was shown in the last section, the rescattering of the particles takes place at the expense of those parts of the potential which are concentrated in the regions Ω_α $(\alpha = 1, 2, 3)$ where the particles are strongly interacting. On the other hand, the particles are moving almost freely in the region Ω_0 and their paths are only slightly disturbed by a weak Coulomb field. Therefore, it is appropriate to divide the two-body potentials into two parts separating the small long-range Coulomb background:

$$v_\alpha(x_\alpha) = \hat{v}_\alpha(X) + v_\alpha^{(a)}(X)$$

such that the term $v_\alpha^{(a)}$ corresponding to the background is different from zero only in the region Ω_0, where $|x_\alpha| > a(1 + |y_\alpha|)^\nu$, and equals the pure

Coulomb potential there:

$$v_\alpha^{(a)}(X) = (1 - \hat{\chi}_\alpha(X))\frac{n_\alpha}{|x_\alpha|},$$

$$\hat{v}_\alpha(X) = v_\alpha^{(s)}(X) + \hat{\chi}_\alpha(X)\frac{n_\alpha}{|x_\alpha|}. \tag{5.104}$$

By $\hat{\chi}_\alpha(X)$ we denote the characteristic function of the region Ω_α, which is equal to zero beyond boundaries of some neighbourhood of this region:

$$\hat{\chi}_\alpha(X) = \begin{cases} 1, & |x| < a(1+|y_\alpha|)^\nu, \quad \nu' > \nu \\ 0, & |x| > a'(1+|y_\alpha|)^{\nu'}, \quad a' > a \end{cases} \tag{5.104'}$$

In terms of such a function we can take arbitrary function which need not necessarily to be equal to zero beyond the boundaries of Ω_α; it must just tend towards zero sufficiently fast for $|x_\alpha|\,|y_\alpha|^{-\nu} \to \infty$, e.g., as $\exp\{-a|x_\alpha|\,|y_\alpha|^{-\nu}\}$. We will call $\hat{v}_\alpha$ the short-range part of the potential $v_\alpha(x_\alpha)$ and $v_\alpha^{(a)}$, the long-range part. By H_a we will denote the energy operator, generated by the long-range parts of the potentials:

$$H_a = H_0 + V^{(a)}, \quad V^{(a)} = \sum_\alpha V_\alpha^{(a)}.$$

This operator can be interpreted as a Hamiltonian of the system of three particles moving on the background of smooth Coulomb perturbations. Let us stress that the magnitude of the perturbation $v^{(a)}$ can be made arbitrarily small by the appropriate choice of the screening constant a.

We will proceed according to the following plan: To start with, we will construct a smooth bounded function $\Psi_{as}(X,P)$ defined in the whole domain of definition of the variables, coinciding with the distorted plane waves far from the singular directions, and satisfying the Schrödinger equation for the operator H_a up to terms rapidly decreasing for $|X| \to \infty$. As a result, we will obtain an approximate wave function for H_a which describes correctly the effect of the long-range Coulomb potentials. This function can be chosen as the zeroth approximation for the wave function $\Psi_0(X,P)$ mentioned in the last section. Carrying on with the calculation given there, we can get the eikonal representations of the terms Ψ_α and $\Psi_{\alpha\beta}$, corresponding to the rescattering of the particles. The next task will be to investigate the asymptotics of these terms in the singular regions and this question will be investigated in Section 4.

The approximate wave function $\Psi_{as}(X, P)$ depends on the coordinates X and momenta P of the particles. We will regard this function as a function of one variable $\{X, P\}$ taking values on the domain $R_{cm} = R^6 \oplus R^6$. We divide this domain into parts in which the function $\Psi_{as}(X, P)$ is given by different analytic expressions. The problem is to extend the definition of this function to those regions where eikonal formulae cannot be used. The latter lose sense not only for bounded values of the parabolic coordinates $\xi^{(\alpha)}$, i.e., in the directions of forward scattering, but also at the region of relatively small momenta satisfying the conditions (5.6). For these reasons, we must impose restrictions not only on the coordinate but also on the momentum variables.

Bearing this in mind, we introduce the following characteristic regions. The region of finite relative momenta $\mathcal{D}_\alpha^{(+)}$ is the set of points in R_{cm} satisfying the inequalities

$$|k_\alpha| > (a_0 + \xi^{(\alpha)})^{-1/2+\nu}, \quad \nu > 0. \tag{5.105}$$

The set of points satisfying the opposite inequality will be called the region of zero momenta of the pair α and denote by $\mathcal{D}_\alpha^{(-)}$. The intersection of the regions $\mathcal{D}_\alpha^{(+)}$, where all pairs have non-zero momenta, will be denoted by $\mathcal{D}^{(+)}$, $\mathcal{D}^{(+)} = \cup_\alpha \mathcal{D}_\alpha^{(+)}$.

The region of forward scattering of the pair α is the set of points $\mathcal{D}_F^{(\alpha)}$ where the inequalities

$$\xi^{(\alpha)} < |x_\alpha|^\nu, \quad 0 < \nu < 1/2. \tag{5.105'}$$

are satisfied.

By $\mathcal{D}_F$ we will denote the three-body forward scattering region where the variables $\xi^{(\alpha)}$ are restricted by these inequalities for all α. By the region of forward scattering we will also mean the set of points in the configuration space Ω_F which satisfy the above inequalities for fixed momentum variables.

The set of points from $\mathcal{D}^{(+)}$ satisfying the inequalities opposite to (5.105) and (5.105') will be called the eikonal region and denoted by $\mathcal{D}_e$. Hence, in the eikonal region, both the initial momenta k_α and the parabolic coordinates $\xi^{(\alpha)}$ must be relatively large.

Let us now turn to description of the function $\Psi_{as}(X,P)$. We assume, first, that the point $\{X,P\}$ is located in the eikonal region $\mathcal{D}_e$. In this case, we define $\Psi_{as}(X,P)$ by means of the asymptotic expansion (5.46), (5.48) for $Z=(X,\hat{Z})$, where only a finite number N, $N\gg 1$ of terms must be taken:

$$\Psi_{as}(X,P)=\Psi_e^{(N)}(X,P). \tag{5.106}$$

We write the function $\Psi_e^{(N)}$ as the product of a plane wave and a phase factor:

$$\Psi_e^{(N)}(X,P)=\exp\{i(X,P)\}\tilde{\Phi}_e^{(N)}(X,P),$$

given by

$$\tilde{\Phi}_e^{(N)}(X,P)=\exp\{iW_{as}(X,P)+i\delta W_N(X,P)\}.$$

If the particles are strongly separated $X\in\Omega_0$, the Coulomb phase $W_{as}(X,P)$ is equal to the sum (5.52). If the point of observation is located in Ω_α, $X\in\Omega_\alpha$, the corresponding term in (5.52) should be replaced by

$$\tilde{W}^{(\alpha)}=\frac{1}{2\sqrt{E}}\int_{Z_0}^{Z}v_\alpha^{(\alpha)}(t\hat{P}+M)dt-W^{(\alpha)}(X_0,P),$$

$$W^{(\alpha)}=\frac{n_\alpha}{2|k_\alpha|}\log|k_\alpha|\xi^{(\alpha)}, \tag{5.106'}$$

where $X_0=\{Z_0,M\}\equiv\{x_\alpha^{(0)},y_\alpha^{(0)}\}$ is an arbitrary point in Ω_0 and $(x_\alpha^{(0)},k_\alpha)<0$. Then the phase $\tilde{W}^{(\alpha)}$ coincides with $W^{(\alpha)}$ if X lies in $\Omega_{0\alpha}=R^6\setminus\Omega_\alpha$. The correction term δW_N is given through the eikonal phase W_{as} by the formulae (5.49), (5.49'). Let us note that the phase $\tilde{W}^{(\alpha)}$ is chosen such that the function $\Psi_e^{(N)}$ remains continuous when the point X passes from the proper Coulomb region Ω_0 to the region Ω_α where the Coulomb potential is screened by the function $\hat{\chi}_\alpha(X)$.

Let us denote by $A_{as}(X,P)$ the result of action of the operator H_0-P^2 to the function $\Psi_{as}(X,P)$:

$$(H_{as}-P^2)\Psi_{as}(X,P)=A_{as}(X,P). \tag{5.107}$$

Note that the order of the function $A_{as}(X,P)$ for $|X|\to\infty$ characterizes the degree to which the approximation of the function $\Psi_{as}(X,P)$ to the correct solution of Schrödinger equation is accurate.

Equations (5.48), (5.106) imply that in the eikonal region $\mathcal{D}_e$ the function $A_{as}(X,P)$ is smooth, rapidly decreasing, and together with its derivatives is subjected to the estimate

$$|A_{as}(X,P)| \leq C\left(a\sum_{\alpha}|k_\alpha|\xi^{(\alpha)}\right)^{-N}, \quad N \gg 1. \tag{5.108}$$

Therefore, the eikonal formulae describe the behaviour of the wave function in the region Ω_0 with high accuracy.

If the point $\{X,P\}$ leaves the region $\mathcal{D}_e$, the eikonal formula (5.106) loses its sense. In this case, we describe $\Psi_{as}(X,P)$ by means of special functions.

Let $S^{(c)}_{cm}$ be a surface in R_{cm}, consisting of the points X and P located on the unit spheres $|X| = 1$, $|P| = 1$. Let us denote by $\mathcal{D}_s$ the set of point on $S^{(c)}_{cm}$ determining the directions of the coordinate and momentum variables at the points where the eikonal approximation ceases to be correct. It is clear that in the limit $|X| \to \infty$ this set has measure zero. In this sense one can say that the eikonal approximation works almost in all directions of the configuration space.

5.3.2 The forward scattering

Assume the spatial variable X lies in the region Ω_0. Let us construct $\Psi_{as}(X,P)$ in the forward scattering region $\mathcal{D}_F$. Because of the fact that in the region Ω_0 the interaction potentials are purely Coulomb, we define, in analogy with (5.66), the zeroth approximation for $\Psi_{as}(X,P)$ in a factorized form:

$$\Psi^{(0)}_F = \exp\{i(X,P)\}\prod_{\alpha}\varphi^{(c)}_{\alpha}(\xi^{(\alpha)},k_\alpha), \tag{5.109}$$

where the two-body functions $\varphi^{(c)}_{\alpha}$ are expressed through the confluent hypergeometric function by equations (5.1), (5.2). Here, the function $\Psi^{(0)}_F$ smoothly turns into the eikonal approximations (5.51) determining for $\xi^{(\alpha)} \to \infty$ the leading terms of the asymptotics of $\Psi_0(X,P)$.

However, this function does not obey the sewing conditions as it does not describes with sufficient accuracy the behaviour of the solution of Schrödinger equation in the neighbourhood of the direction of forward scattering. This is reflected in the slow decrease of the function A_F defined by

$$A_F(X,P) = (H_a - P^2)\Psi_F^{(0)}(X,P).$$

Indeed, differentiating the product (5.104) and bearing in mind that the function $\exp\{i(k_\alpha, x_\alpha)\}\varphi_\alpha^{(c)}(\xi^{(\alpha)}, k_\alpha)$ satisfies the two-body Schrödinger equation with the potentials $n_\alpha|x_\alpha|^{-1}$ we find

$$A_F(X,P) = 2\exp\{i(P,X)\}(1 + P_{231} + P_{321})c_{21}\times$$

$$\times\varphi_3^{(c)}(\xi^{(3)}, k_3)\left(\nabla_{x_1}\varphi_1^{(c)}(\xi^{(1)}, k_1), \nabla_{x_2}\varphi_2^{(c)}(\xi^{(2)}, k_2)\right), \quad (5.110)$$

where the first derivatives of the functions $\varphi^{(c)}$ are present. The symbols P_{231} and P_{321} denote the operators of cyclic permutation of the particle labels. Let us note at this point that if only one pair of particles is charged, e.g., (12), the function A_F becomes zero, since then $\varphi_1^{(c)} = \varphi_2^{(c)} = 1$ and the asymptotics Ψ_0, in agreement with equation (5.66), describes the function $\Psi_F^{(0)}$ correctly.

Since the function $\varphi^{(c)}$ depends on x_α through the parabolic coordinate $\xi^{(\alpha)}$, the scalar product of the gradients can be rewritten as

$$\left(\nabla_{x_1}\varphi_1^{(c)}, \nabla_{x_2}\varphi_2^{(c)}\right) = \varphi_1^{(c)\prime}\varphi_2^{(c)\prime}\left(\nabla_{x_1}\xi^{(1)}, \nabla_{x_2}\xi^{(2)}\right), \quad (5.111)$$

where the derivative of the function $\varphi_\alpha^{(c)}(\xi^{(\alpha)}, k_\alpha)$ with respect to the first argument is denoted by $\varphi_\alpha^{(c)\prime}$. Calculating the derivatives of the parabolic coordinates we get the expression

$$\left(\nabla_{x_1}\xi^{(1)}, \nabla_{x_2}\xi^{(2)}\right) = \frac{\xi^{(1)}(\xi^{(2)}\cos\theta_{21} - \sqrt{\xi^{(2)}\zeta^{(2)}}(\hat{u}_2, k_1)}{|x_1|\,|x_2|} -$$

$$-\frac{\sqrt{\xi^{(1)}\zeta^{(1)}}}{|x_1|\,|x_2|}(\hat{u}_1, e_1^{(1)})\sin\theta_{21} + \frac{\sqrt{\xi^{(1)}\zeta^{(1)}\xi^{(2)}\zeta^{(2)}}}{|x_1|\,|x_2|}(\hat{u}_1, \hat{u}_2), \quad (5.112)$$

where u_α denotes the projection of the vector x_α on the plane perpendicular to the vector k_α and the variables $\zeta^{(\alpha)}$, $\cos\theta_{21}$ and the vector $e_1^{(1)}$ are given by

$$e_1^{(1)} = \frac{\hat{k}_2}{\sin\theta_{21}} - \hat{k}_1\cot\theta_{21},$$

$$\cos\theta_{21} = (\hat{k}_2, \hat{k}_1),$$

$$\zeta^{(\alpha)} = |x_\alpha| + (x_\alpha, \hat{k}_\alpha), \quad \alpha = 1,2,3.$$

Recalling the properties of the function $\varphi^{(c)}$ from Section 1, we get the inequality

$$|A_F(X,P)| \leq \sum_{\alpha\neq\beta} c_\alpha(k_\alpha)c_\beta(k_\beta)|u_\alpha|\,|u_\beta|(1+a+|x_\alpha|)^{-1}\times$$

$$\times(1+|x_\beta|)^{-1}(1+|k_\alpha|\xi^{(\alpha)})^{-1}(1+|k_\beta|\xi^{(\beta)})^{-1}, \tag{5.113}$$

showing that for bounded $\xi^{(\alpha)}$ ($\alpha = 1,2,3$), the function A_F decreases not faster than the Coulomb potential and hence cannot be neglected as a higher-order correction. If $\xi^{(\alpha)} \to \infty$ ($\alpha = 1,2,3$), this function is of the order $O(|X|^{-1-\nu})$, $\nu > 0$, and can be regarded as a small perturbation.

Thus, we must "repair" $\Psi_F^{(0)}$ in the direction of forward scattering to obtain a function satisfying the Schrödinger equation at least up to terms of order $O(|X|^{-1-\nu})$, $\nu > 0$. To this end, let us consider the Schrödinger equation in the neighbourhood of the direction of forward scattering. Let us introduce parabolic coordinates ξ and ζ:

$$\xi = \frac{|X|-Z}{2}, \quad \zeta = \frac{|X|+Z}{2},$$

where by Z we denote the component of the vector X in direction of the vector P. By M we denote the projection of the vector X onto the plane R^5 perpendicular to P, so that $X^2 = Z^2 + M^2$, $M^2 = \zeta\xi$. Expanding the long-range potential $v^{(a)}(X)$ in the neighbourhood of the forward scattering direction in powers of the small parameter $\xi\zeta^{-1}$, we obtain the representation

$$v^{(a)}(X) = V_P(X) + O(\xi\zeta^{-2}),$$

where $V_P(X)$ is the spherically symmetric Coulomb potential in R^6: $V_P(X) = q_0(P)/|X|$. Here the effective charge $q_0(P)$ varies depending on the initial direction of the flux of particles according to the equation

$$q_0(P) = |P|\sum_\alpha \frac{n_\alpha}{|k_\alpha|}.$$

By H_P we denote the model Hamiltonian generated by this potential, $H_P = H_0 + V_P$.

Let us define the function $\Psi_{as}(X,P)$ in the neighbourhood of the forward scattering direction as the sum

$$\Psi_{as}(X,P) = \Psi_F^{(0)}(X,P) + \Phi_F(X,P). \tag{5.114}$$

We will seek the function Φ_F as to satisfy the Schrödinger equation up to terms of the order $|X|^{-1-n}$, $\nu > 0$. Substituting this sum into the Schrödinger equation and discarding terms of the order $\xi\zeta^{-2}$, for Φ_F we obtain the inhomogeneous Schrödinger equation with a spherically symmetric potential $V_P(X)$:

$$(-\Delta_X + q_0(P)|X|^{-1} - P^2)\,\Phi_F(X,P) = -A_F(X,P). \tag{5.115}$$

Here the function $A_F(X,P)$ is given by equation (5.107). The solution of this equation, to be sewed with the eikonal approximation (5.51), can be represented in the form

$$\Phi_F(X,P)\lim_{\epsilon\downarrow 0}\int R_P(X,X',P^2+i\epsilon)\times$$
$$\times\chi_F(X')A_F(X',P)\,dX' \tag{5.116}$$

where $R_P(X,X',z)$ is the Green function of the model Hamiltonian H_P, and χ_F is the characteristic function of the singular direction; χ_F is equal one in Ω_F and smoothly goes to zero when X leaves the region of the forward scattering.

By construction, the sum $\Psi_F = \Psi_F^{(s)} + \Phi_F$ satisfies the Schrödinger equation in the neighbourhood of the direction of forward scattering up to terms of order $\xi\zeta^{-2}$. Therefore, in order to check that the sewing conditions are satisfied, it is sufficient to show that this function takes the eikonal form when $\xi \to \infty$.

5.3.3 The asymptotics of the function Φ_F

Let us study the asymptotic behaviour of the function Φ_F for $\xi \to \infty$. First of all, we express the integrand $A_F(X,P)$ in equation (5.116) in parabolic coordinates. To this end, we use equations (5.110-113) and consider the equality $\xi^{(\alpha)} = \xi|P|\,|k_\alpha|^{-1}\cos^2\theta_\alpha + O(\xi\zeta^{-2})$ where by $\cos\theta_\alpha$ we denote the ratio $\cos\theta_\alpha = |u_\alpha||M|^{-1}$. As the result, we obtain

$$A_F(X,P) = \frac{\xi}{\xi+\zeta}e^{i(X,P)}V_F(\xi,\hat{M}), \tag{5.117}$$

where the function $V_F(\xi, \hat{M})$ is given by

$$V_F(\xi, \hat{M}) = (1 + P_{321} + P_{123})c_{21} \cos\theta_1 \cos\theta_2(\hat{u}_1, \hat{u}_1)\times$$

$$\times\varphi_1^{(c)\prime}(\tilde{\xi}^{(1)}, k_1)\varphi_2^{(c)\prime}(\tilde{\xi}^{(2)}, k_2)\varphi_3^{(c)}(\tilde{\xi}^{(3)}, k_3)(1 + O(\xi\zeta^{-2});$$

$$\tilde{\xi}^{(\alpha)} = \xi\frac{|P|}{|k_\alpha|}\cos^2\theta_\alpha.$$

Note that both the integration variable X' and the first argument of the Green function $R_P(X, X', E + i0)$ are located in the neighbourhood of the direction of forward scattering Ω_F. Therefore, when $|X| \to \infty$, the kernel $R_P(X, X', E + i0)$ can be replaced by the asymptotic representation (5.42). By integrating by parts with respect to the variable $q = |X - X'|$, we can prove that the integral over the region located in the neighbourhood of X for $q \leq |X|^\nu$, $\nu < 1/2$ is a smooth function of X. Formula (5.117) implies that this function can be expressed as a product of a plane wave and a factor tending to zero as $O(|X|^{-\nu'})$, $\nu' > 0$ and, therefore, is a higher order correction to the eikonal approximation $L_c(X, P)$.

Let us consider the remaining integral over the exterior of the sphere of radius $|X|^\nu$ centered at X. Note that the main contribution to this integral comes from the points where the relation $|Z - Z'| \geq |M - M'|$ holds, with Z and Z' being the projections of the points X and X' on the axis P and M, M' are the projections of the points X and X' on the plane perpendicular to the vector P. Using this relation, let us expand the difference $|X - X'|$ into a series in the small parameter $|Z - Z'|^{-1}|M - M'|$:

$$|X - X'| = |Z - Z'| + \frac{1}{2}\frac{|M - M'|^2}{|Z - Z'|} + O\left(\frac{|M - M'|^2}{|Z - Z'|^3}\right)$$

and then express the leading terms in the parabolic coordinates. We can easily see that the leading terms of A_F for $|X| \to \infty$ are generated by those points Z', for which the inequality $Z' \leq Z - Z^\nu$, $\nu < 1$, holds. Changing the variables $q = \sqrt{\frac{\zeta'}{|X|}}\hat{M}'$, $q \in R^5$, we come to the asymptotic equality

$$\Phi_F(X, P) \sim e^{i(X,P)}\tilde{\Phi}(\xi, M) \tag{5.118}$$

where the function $\tilde{\Phi}$ is given by the integral

$$\tilde{\Phi} = e^{i|P|\xi}\lim_{\epsilon\downarrow 0}\int_0^\infty d\xi'\xi'^{5/2}\int_{|q|\leq 1}\exp\left\{-i|P|\sqrt{\xi\xi'}(q, \hat{M})-\right.$$

$$-i\frac{q_0(P)}{2|P|}\log\frac{|q|}{2}+i\frac{|P|q^2\xi'}{2}-\epsilon\xi'\Big\}\,V_F(\xi',\hat{M}')\epsilon(q). \tag{5.119}$$

Here $\epsilon(q)$ denotes the cutoff function equal to one in the interior of unit sphere $|q| \leq 1$ and equal to zero in some neighbourhood of its boundary.

For small ξ the function $\Phi_F(\xi, M)$ does not go to zero for $|X| \to \infty$. Therefore, we can say that this function "repairs" the zeroth approximation $\Psi_F^{(0)}$ in the neighbourhood of the direction of forward scattering. To find the asymptotics of Φ_F for $x \to \infty$, let us make the change of the variable $\sqrt{\xi}q = q'$ in the integral (5.119). After simple calculations we get the asymptotic representation

$$\Phi_F(X,P) \sim F_{as}(\theta,\hat{M})|X|^{-5/2}\exp\{i|X|\,|P|+iW_0(X,P)\}. \tag{5.120}$$

Therefore, when receding from the singular direction, Φ_F takes the form of a spherical wave. The amplitude of this wave has a strong singularity in the direction of forward scattering:

$$F_{as}(\theta,\hat{M}) = \frac{f_0(\theta,M)}{(1-\cos\theta)^{5/2+iq_0/2|P|}}, \qquad \cos\theta = (\hat{X},\hat{P}). \tag{5.121}$$

The coefficient f_0 is a smooth function of the angular variables $\hat{M}$ and in the leading order does not depend on the angle between X and P:

$$f_0(\theta,M) \sim \lim_{\epsilon_1\downarrow 0,\epsilon_2\downarrow 0}\int_0^\infty d\xi'\xi'^{5/2}\int dq' e^{-\epsilon_1\xi'-\epsilon_2|q'|}\times$$

$$\times V_F(\xi',\hat{M})\exp\left\{-i|P|\sqrt{\xi'}(q',M)-i\frac{q_0(P)}{2|P|}\log|q'|\right\}. \tag{5.122}$$

At this point we finish the construction of the function $\Psi_{as}(X,P)$ in the direction of forward scattering. We established that this function satisfies the Schrödinger equation up to terms of order $|X|^{-1-\nu}$, $\nu > 0$ and in the direction of forward scattering in the leading order has the form of the product

$$\Psi_{as}(X,P) = e^{i(X,P)}\tilde{\Phi}_{as}(X,P), \tag{5.123}$$

where the function $\tilde{\Phi}_{as}(X,P)$ is equal to the sum

$$\tilde{\Phi}_{as}(X,P) = \prod_\alpha \varphi_\alpha^{(c)}(\xi^{(\alpha)},k_\alpha) + \tilde{\Phi}_F(X,P).$$

As the point X departs from this direction, Ψ_{as} turns into the sum of the eikonal approximations, corresponding to the plane and spherical eikonals. Here the function A_{as} given by equation (5.107), obeys the estimate

$$A_{as}(X,P) \leq \left(1 + \sum_{\alpha} |k_\alpha|\,|x_\alpha|\right)^{-1-\nu}, \quad 0 < \nu \leq 1. \tag{5.123'}$$

Let us further describe the function Ψ_{as} in the remaining singular regions. If the point X leaves the forward scattering region, but has not yet reached the eikonal domains, we let Ψ_{as} to be equal to the product of a plane wave and the two-body functions $\varphi_\alpha^{(c)}$, i.e.,

$$\Psi_{as}(X,P) = \Psi_F^{(0)}(X,P). \tag{5.124}$$

If for some pairs α, the momenta or coordinates fall in the eikonal region, the corresponding function $\varphi_\alpha^{(c)}$ will be replaced by the leading asymptotic term:

$$\varphi_\alpha^{(c)} \sim (\xi^{(\alpha)})^{i\eta_\alpha}, \quad \eta_\alpha = \frac{n_\alpha}{2|k_\alpha|}.$$

The function $A_{as}(X,P)$ is smooth and obeys the estimate (5.113) if the momenta of the pairs are sufficiently large, $\{X,P\} \in \mathcal{D}^{(+)}$. For small relative momenta, when $\{X,P\} \in \mathcal{D}^{(-)}$, the estimating functions c_α and c_β depend on the momentum variables $|k_\alpha|$ and $|k_\beta|$. A detailed description of A_{as} this case can be obtained by means of the asymptotic representation of $\Phi(a,b,x)$ for large values of the parameter a. We will not give here the detailed formulae for $|k_\alpha| \to 0$. We only note that according to equations (5.8) and (5.9), the functions c_α decrease if the particles of the pair α have the same sign of charge $n_\alpha > 0$, and they increase as $|k_\alpha|^{-1/2}$ in the case of attractive Coulomb potentials $n_\alpha < 0$.

Thus, we fully discussed the case, when the spacial variable is located in the region Ω_0, corresponding to the strongly separated particles.

5.3.4 Plane waves in Ω_α

If the point X moves into the domain Ω_α where the particles of the pair α are close together, we must use for the sewing of the asymptotic solutions of the new model Hamiltonians describing the behaviour of the solution in the limit of small distances. As we have seen in the last section, the parameter with respect to which one should expand the potential $v^{(a)}(X)$ is in this case the ratio $|x_\alpha|\,|y_\alpha|^{-2}$. The asymptotic equation

$$v^{(a)}(X) = v_\alpha^{(a)}(x_\alpha, y_\alpha) + \tilde{v}_\alpha^{(c)}(y_\alpha) + O(|x_\alpha|\,|y_\alpha|^{-2})$$

holds in Ω_α up to terms of the order $|x_\alpha|\,|y_\alpha|^{-2}$ where the potential $v_\alpha^{(a)}$, describing the interaction in the pair α, depends on the variable y_α as on a parameter, and the term $\tilde{v}_\alpha^{(c)}(y_\alpha)$ corresponds to the effective interaction of the pair with the third particle:

$$\tilde{v}_\alpha^{(c)}(y_\alpha) = \sum_{\beta\neq\alpha} v_\beta^{(a)}(x_\beta, y_\beta)|_{x_\alpha=0}.$$

For sufficiently large $|y_\alpha|$, the potential $\tilde{v}_\alpha^{(c)}$ coincides with the proper Coulomb potential $n_{\alpha\alpha}/|y_\alpha|$ from equation (5.101).

Using this remark, let us define the model three-body Hamiltonian, corresponding to the leading terms of the asymptotics of the potential $\tilde{v}_\alpha^{(c)}$ in the region Ω_α:

$$H_{as}^{(0)} = H_0 + V_{as}^{(\alpha)}, \quad V_{as}^{(\alpha)} = V_\alpha^{(a)} + \tilde{V}_\alpha^{(c)}. \tag{5.125}$$

To describe the structure of the Green function of this operator, let us consider the two-body Hamiltonians $h_\alpha^{(a)}$ and $h_\alpha^{(c)}$, generated by the potentials $v_\alpha^{(a)}$ and $\tilde{v}_\alpha^{(c)}$, respectively.

The operator $h_\alpha^{(a)}$ is given by

$$(h_\alpha^{(a)} f)(x_\alpha) = \left(-D_{x_\alpha} + v_\alpha^{(a)}(x_\alpha, p_\alpha)\right) f(x_\alpha).$$

We will regard the variable y_α as a parameter determining the cutoff radius $a(1+|y_\alpha|)^\nu$ of the Coulomb potential for $|x_\alpha| > a(1+|y_\alpha|)^\nu$. The wave functions of this operator can be studied with the help of the modified equation of the perturbation theory (5.20). Let us note that the magnitude of the potential $n_\alpha|x_\alpha|^{-1}$ outside the ball $|x_\alpha| \le a(1+|y_\alpha|)^\nu$, where the operator $h_\alpha^{(a)}$ is equal to the free Hamiltonian h_0, tends to zero for $|y_\alpha| \to \infty$. We make the scale transformation $x_\alpha \to |y_\alpha|^\nu a x_\alpha$. As a result, the cutoff radius becomes finite and the scattered energy $k_\alpha'^2$ and the effective charge n_α' attain new values:

$$k_\alpha'^2 = a^2|y_\alpha|^{2\nu}k_\alpha^2, \quad n_\alpha' = a|y_\alpha|^\nu n_\alpha, \tag{5.126}$$

going to infinity when $|y_\alpha| \to \infty$. Equations (5.20) – (5.24) imply that the wave functions of this Hamiltonian $\psi_\alpha^{(a)}(X, k_\alpha)$ can be written in form of the sum

$$\psi_\alpha^{(a)}(X, k_\alpha) = e^{i(k_\alpha, x_\alpha)}(\varphi_c(x'_\alpha, k'_\alpha) + \tilde{\varphi}_\alpha^{(a)}(X', k'_\alpha)),$$

$$X' = \{x'_\alpha, y_\alpha\},$$

where the first term coincides with the Coulomb wave function, the parameters of which are given by equations (5.126), and the second corresponds to the short-range potential. The latter equals to the potential $\frac{n_\alpha}{|x_\alpha|}\chi_\alpha(x_\alpha, y_\alpha)$ from equation (5.104) taken with opposite sign, i.e., it represents a compensation potential cutting off the Coulomb potential at short distances, $0 \le |x_\alpha| \le a(1 + |y_\alpha|)^\nu$. According to the equations (5.20) – (5.23) the asymptotics of the function $\psi_\alpha^{(a)}$ for $|k_\alpha|\,|y_\alpha|^\nu \xi^{(\alpha)} \to \infty$ has the form of the sum of distorted plane and spherical waves; the amplitude of the latter has a small value $O(|y_\alpha|^{-\nu})$ outside the forward scattering direction, $f_\alpha^{(a)}(\hat{x}_\alpha, y_\alpha, k_\alpha) = O(|y_\alpha|^{-\nu})$,

$$\psi_\alpha^{(a)} \sim$$

$$\sim \exp\{i(k_\alpha, x_\alpha) + iW_\alpha\} + f_\alpha^{(a)}(\hat{x}_\alpha)\frac{\exp\{i|k_\alpha|\,|x_\alpha| + iw_\alpha^{(0)}\}}{|x_\alpha|}. \tag{5.127}$$

If the relative momentum k_α decreases so that $|k_\alpha|\,|y_\alpha|^\nu \to 0$, the Coulomb parameters turn into the semiclassical range. The asymptotics of $\psi_\alpha^{(a)}$ will then be given by the exact solution $\psi_\alpha(x_\alpha, k_\alpha)$ which has in this case a rather complicated form. We will not describe here the full asymptotic behaviour, we just refer to equation (5.7).

If the particles of the pair α have charges of opposite signs, the operator $h_\alpha^{(a)}$ has discrete spectrum. But when the parameter $a(1 + |y_\alpha|)^\nu$ increases, the lower bond of the eigenvalues $-\varkappa_i^{(a)}$ tends to zero, so that starting with some sufficiently large $|y_\alpha|$, no point, even arbitrarily close to the origin, can belong to the discrete spectrum of the operator $h_\alpha^{(a)}$.

The operator $\tilde{h}_\alpha^{(a)}$, given by the equation

$$(\tilde{h}_\alpha^{(a)} f)(y_\alpha) = \left(-\Delta_{y_\alpha} + \tilde{v}_\alpha^{(a)}(y_\alpha)\right) f(y_\alpha)$$

corresponds, as it was mentioned above, to the effective interaction of the pair α with the third particle. Depending on the total charge of the pair α, the Coulomb part of the potential $\tilde{v}_\alpha^{(a)}$ determines attraction or repulsion. If the charges compensate each another, this potential is short ranging. The construction of wave functions for such operators was described in Section 1.

Let us represent the model operator $H_{as}^{(a)}$ in the form of the sum $H_{as}^{(\alpha)} = h_\alpha^{(a)} + \tilde{h}_\alpha^{(c)}$,

$$(H_{as}^{(\alpha)} f)(x_\alpha, y_\alpha) = \left(-\Delta_{x_\alpha} + v_\alpha^{(a)}(x_\alpha, y_\alpha^{(0)})\right) f(x_\alpha, y_\alpha) +$$

$$+ \left(-\Delta_{y_\alpha} + \tilde{v}_\alpha^{(c)}(y_\alpha)\right) f(x_\alpha, y_\alpha),$$

and we will regard the variable $y_\alpha^{(0)}$, present in the definition of the operator $h_\alpha^{(a)}$, fixed. We will assume y_α belongs to a cylindrical layer of the width $(|y_\alpha^{(0)}| + a)^\nu$ containing the point $y_\alpha^{(0)}$:

$$|y_\alpha - y_\alpha^{(0)}| \leq |a + |y_\alpha^{(0)}||^\nu.$$

When sewing the solutions in terms of this operator, we must take into account the fact that the asymptotics of the function $\psi_\alpha^{(a)}(x_\alpha, y_\alpha^{(0)}, k)$ for different $y_\alpha^{(0)}$ satisfying the inequality above, coincide, up to distorted spherical waves with infinitely small amplitudes of the order $|y_\alpha|^{-\nu}$. Let us note that the resolvent of the operator $H_{as}^{(\alpha)}$, $R_{as}^{(\alpha)} = (H_{as}^{(\alpha)} - z)^{-1}$ can be expressed explicitly through the resolvents of the two-particle Hamiltonians $h_\alpha^{(a)}$ and $\tilde{h}_\alpha^{(c)}$ by means of relation (5.71').

Knowing the properties of the Green function of the model Hamiltonian $H_{as}^{(\alpha)}$, we can construct the function Ψ_{as} in Ω_α operating in the same manner as in the case of large distances between the particles, studied above. The necessary changes reduce to the replacement of the purely Coulomb functions $\varphi_\alpha^{(c)}(\xi^{(\alpha),k_\alpha}$ by the functions $\varphi_\alpha^{(a)}(X, k_\alpha) = e^{-i(x_\alpha, k_\alpha)}\psi_\alpha^{(c)}(X, k_\alpha)$ and the model operator H_P by the operators $H_{as}^{(\alpha)}$.

It can be shown that the function A_{as}, given by equation (5.107), for $X \in \Omega_\alpha$ obeys the estimates (5.108) and is of order $O(|X|^{-1-n})$ depending on the values of the momentum variables. The computations necessary for the proof of this statement are analogous to those used in Section 1 for the construction of the asymptotics of the terms $\Phi^{(1)}(X)$. Let us note that, in contrast to the terms $\Phi^{(1)}$, the asymptotics of the function $\Psi_{as}(X, P)$ does not contain terms corresponding to the eikonal approximations Ψ_α and $\Psi_{\alpha\beta}$. Indeed, as we have seen in Section 2, such terms were obtained as the result of rescattering of plane or spherical waves in the regions Ω_α and Ω_β, where the two-particle potentials were substantially different from zero. In the case of the operator $H_{as}^{(\alpha)}$, the interaction $V_{as}^{(\alpha)}$ vanishes in this regions. Therefore, there are no secondary rescattered waves, as well. We will not go into further details of these considerations. The necessary computational technique was described above. Let us also note that since for $|y_\alpha| \to \infty$ the operators $h_\alpha^{(a)}$ have no non-zero eigenvalues, the coordinate asymptotics of the function Ψ_{as} does not contain the cluster waves (5.92) corresponding to the scattering of the particle on a bound pair.

Thus, we have described the function $\Psi_{as}(X, P)$ in the whole domain of definition R_{cm}. The transition from one analytic formula to another is realized with the help of the characteristic functions of the regions introduced above. It is important that the various constructions of $\Psi_{as}(X, P)$ the leading terms coincide in the adjacent regions.

5.4 Asymptotics of the Function Ψ_0 in Singular Directions $\Omega_\alpha^{(0)}$ and $\Omega_{\alpha\beta}^{(0)}$

In this section we will investigate the coordinate asymptotics of the wave functions $\Psi_0(X,P)$ in the neighbourhood of the singular directions $\Omega_\alpha^{(0)}$ and $\Omega_{\alpha\beta}^{(0)}$.

5.4.1 The direction $\Omega_\alpha^{(0)}$

Let us consider the eikonal approximation of Ψ_α given by equations (5.58), (5.82) and (5.91). For time being, we will assume that the vectors k_α and p_α are not parallel, $(\hat{k}_\alpha, \hat{p}_\alpha) \neq \pm 1$. In this case, the main singular direction does not intersect the secondary one.

Let q_α be the three-dimensional vector field, defined by

$$q_\alpha = \omega_\alpha p_\alpha + m_\alpha.$$

By $\tilde{\Omega}_\alpha^{(0)}$ we denote the neighbourhood of the direction $\Omega_\alpha^{(0)}$ given by the inequality

$$|X| - Z_\alpha \leq |X|^\nu, \quad \nu < \frac{1}{2},$$

and by $\partial\tilde{\Omega}_\alpha^{(0)}$ the boundary of $\tilde{\Omega}_\alpha^{(0)}$. Expand the function Ψ_α on the boundary in a series of spherical harmonics $Y_l^m(\hat{q}_\alpha)$:

$$\Psi_\alpha(X,P) = \sum_{l,m} \Psi_{\alpha,l}^m(|q_\alpha|, Z_\alpha) Y_l^m(\hat{q}_\alpha) \frac{\tilde{f}_c^{(\alpha)}(\hat{x}_\alpha, k_\alpha)}{|x_\alpha|}. \tag{5.128}$$

Here $\tilde{f}_c^{(\alpha)}$ denotes the scattering amplitude for the pair α, the Coulomb part of which is multiplied by cutoff function $\chi_\alpha(\xi^{(\alpha)})$ which is equal to zero in a neighbourhood of the direction of forward scattering:

$$\tilde{f}_c^{(\alpha)}(\hat{x}_\alpha, k_\alpha) = f_c^{(\alpha)}(\hat{x}_\alpha, k_\alpha)\chi_\alpha^F(\xi^{(\alpha)}) + f_{cs}^{(\alpha)}(\hat{x}_\alpha, k_\alpha). \tag{5.128'}$$

Recall that the function $\chi_\alpha^F(\xi^{(\alpha)})$ was used in equation (5.72) above. The function are given by $\Psi_{\alpha,l}^m$ integrals

$$\Psi_{\alpha,l}^m(|q_\alpha|, Z_\alpha) = \int d\hat{q}_\alpha Y_l^m(\hat{q}_\alpha)\Psi_\alpha(X,P). \tag{5.129}$$

Substituting the asymptotic representation (5.58) into equation (5.129), we conclude that the leading term of the asymptotics of $\Psi^m_{\alpha,l}$ for $Z_\alpha \to \infty$ has the form

$$\Psi^m_{\alpha,l} \sim B^m_{\alpha,l} \exp\{i\sqrt{E}Z_\alpha + iW_\alpha + ia_\alpha \log|q_\alpha|\}, \qquad (5.130)$$

where $B^m_{\alpha,l}$ is a constant dependent only on the numbers l and m:

$$B^m_{\alpha,l} = \int d\hat{q}_\alpha Y^m_l(\hat{q}_\alpha) \exp\{\delta W_\alpha\}.$$

Let us define the continuation of the function Ψ_α into the singular region $\tilde{\Omega}^{(0)}_\alpha$ in the neighbourhood of parabolic coordinates with the help of the expansion (5.128) where instead of the functions $\Psi^m_{\alpha,l}$ we substitute the expression $\tilde{\Psi}^m_{\alpha,l} = f^m_{\alpha,l} \exp\{i\sqrt{E}Z_\alpha + iW_\alpha\}$ with the unknown functions $f^m_{\alpha,l}$. To find these functions, we will use the method of the parabolic equation.

In the neighbourhood of $\Omega^{(0)}_\alpha$ let us introduce parabolic coordinates corresponding to the eikonals Z_α and $|X|$: $\xi_\alpha = |X| - Z_\alpha$, $\zeta_\alpha = |X| + Z_\alpha$, so that $|q_\alpha| = \sqrt{\xi_\alpha\zeta_\alpha}$. These coordinates reflect correctly a scale of changes of variables in the singular region. Let us substitute the function Ψ_α in the form (5.129) into the Schrödinger equation (5.45) and use the representation of Laplace operator in parabolic coordinates. Disregarding terms of order $\xi_\alpha\zeta_\alpha^{-2}$, we get the following parabolic equation for $f^m_{\alpha,l}$

$$\xi_\alpha^{-1/2}\frac{\partial}{\partial\xi_\alpha}\xi_\alpha^{3/2}\frac{\partial}{\partial\xi_\alpha}f + 4i\sqrt{E}\left(\zeta_\alpha\frac{\partial}{\partial\zeta_\alpha} - \xi_\alpha\frac{\partial}{\partial\xi_\alpha}\right)f + \\ + \frac{l(l+1)}{4\xi_\alpha}f = 0. \qquad (5.131)$$

The solution of this equations can be obtained by separation of variables

$$f^m_{\alpha,l} = C^m_{\alpha,l}\xi_\alpha^{l/2}\zeta_\alpha^{i\mu/\sqrt{E}}\Phi\left(\frac{l}{2} - i\frac{\mu}{\sqrt{E}}, l + \frac{3}{2}, i\sqrt{E}\xi_\alpha\right), \qquad (5.132)$$

the coefficient $C^m_{\alpha,l}$ and the constant μ can be determined from the sewing condition of $f^m_{\alpha,l}$ with the eikonal asymptotics (5.130). Remembering the asymptotic formula for the confluent hypergeometric function (5.3), we get

$$C^m_{\alpha,l} = \frac{\Gamma\left(\frac{l}{2} + \frac{3}{2} + i\frac{a_\alpha}{2}\right)}{\Gamma\left(l + \frac{3}{2}\right)}(i\sqrt{E})^{\frac{l}{2} - i\frac{a_\alpha}{2}} B^m_{\alpha,l}, \quad \mu = \frac{a_\alpha\sqrt{E}}{2}. \qquad (5.133)$$

Let us further note that the asymptotics of the solution $f^m_{\alpha,l}$ contains also a term corresponding to the spherical eikonal.

On the basis of (5.3), we find that the functions $f^m_{\alpha,l}$ which describe the asymptotics of Ψ_α in $\tilde{\Omega}^{(0)}_\alpha$, generate a distorted spherical wave (5.53) with the singular amplitude $F^{(s)}_\alpha$ given by the expression

$$F^{(s)}_\alpha(X,P) = \tilde{f}^m_{\alpha,l}(\hat{x}_\alpha, k_\alpha)\frac{A^{(s)}_\alpha(\hat{y}_\alpha, |x_\alpha|\,|X|^{-1}}{Q_\alpha|^{3+ia_\alpha}}. \tag{5.134}$$

Here the factor $A^{(s)}_\alpha$ not depending on $\hat{x}_\alpha$ is expressed as follows

$$A^{(s)}_\alpha = \frac{|P|}{|k_\alpha|}(i|P|)^{-\frac{a_\alpha}{2}} \exp\left\{-i\sum \frac{n_\alpha |X|}{|P|\,|x_\alpha|} \log \frac{|x_\alpha|}{|X|}\right\} \times$$

$$\times \sum_{l,m} B^m_{\alpha,l} \frac{\Gamma\left(\frac{l}{2}+\frac{3}{2}+i\frac{a_\alpha}{2}\right)}{\Gamma\left(l+\frac{3}{2}\right)} Y^m_l(\hat{q}_\alpha) \tag{5.135}$$

and by Q_α we denote the reduced difference between the spherical and singular eikonals, $Q_\alpha = (1-|X|^{-1}Z_\alpha)^{1/2}$.

Thus, we have shown that the asymptotics of the term Ψ_α in the singular directions $\Omega^{(0)}_\alpha$ is described by means of the function $\tilde{\Psi}_\alpha$, given by equations (5.128) – (5.133). Let us study properties odf this function in more details.

Under the assumption about the directions of the vector k_α and p_α made above, the function Ψ_α depends smoothly on the angles $\hat{q}_\alpha$, so that the coefficients $B^m_{\alpha,l}$ decrease as an arbitrary power l^{-N}, $N \gg 1$. Therefore, the series (5.129) and (5.135) converge rapidly and represent smooth bounded functions, too. If $(\hat{n}_\alpha, \hat{p}_\alpha) = \pm 1$, then for $X \in \tilde{\Omega}^{(0)}_\alpha$ the parabolic coordinates $\xi^{(\alpha)}_\beta$, $\alpha \neq \beta$ describing the state of the particles of the pair β at the moment of collision of the pair α are small. The asymptotics of the function Ψ_α is then a rapidly changing function of the angles $\hat{q}_\alpha$ and its derivatives with respect to $\hat{q}_\alpha$ grow as $|q_\alpha|$. Formulae (5.128) – (5.135) can be obtained also in this case, however now it is difficult to investigate the convergence of the series, since the derivatives of the function Ψ_α with respect to $\hat{q}_\alpha$ are unbounded for $|q_\alpha| \to \infty$. In this situation, it is necessary to use integral representations of the form (5.116). They will be given in the next section.

On the unit sphere $S^{(5)}$ let us introduce coordinates associated with the variables k_α and p_α. The surface element is then given by

$$d\hat{X} = |X|^{-4}\sqrt{X^2 - y^2_\alpha}\, dx_\alpha \wedge dy_\alpha. \tag{5.136}$$

Next, we check that the difference between the eikonals $|X|$ and Z_α which determines singularities of the amplitude of the distorted and the spherical waves $F^{(s)}_{\alpha}$, is proportional to the difference between the initial and final momenta

$$Q_{\alpha} = |p_{\alpha} - p'_{\alpha}|(\tilde{Q}_{\alpha} + O(|p_{\alpha} - p'_{\alpha}|)),$$

where

$$p'_{\alpha} = E^{-1/2}|X|^{-1}y_{\alpha}, \quad \tilde{Q}_{\alpha} = \left(1 + \frac{p_{\alpha}}{E}(\hat{y}_{\alpha}, \hat{p}_{\alpha})\right).$$

Equation (5.134) implies then that the singularity of $F^{(s)}_{\alpha}$ has the form of the polar singularity

$$|p_{\alpha} - p'_{\alpha}|^{-3-ia_{\alpha}}.$$

By virtue of equation (5.136) such a singularity is unintegrable over the unit sphere $S^{(5)}$. Therefore, when integrating over $S^{(5)}$, the amplitude $F^{(s)}_{\alpha}$ cannot be regarded as an ordinary function and we will associate with it a distribution. The definition of this distribution will be given in the next chapter.

5.4.2 The direction $\Omega^{(0)}_{\alpha\beta}$

Now we will investigate the asymptotics of the terms in $\Psi_{\alpha\beta}$ in the singular directions $\Omega^{(0)}_{\alpha\beta}$. Let us denote by $\Omega^{(+)}_{\alpha\beta}$ ($\Omega^{(-)}_{\alpha\beta}$) the regions in configuration space where the variable $\omega_{\alpha\beta} = E^{-1/2}(|x_{\alpha}|\,|p_{\alpha\beta}| - |k_{\alpha\beta}|(\hat{p}_{\alpha\beta}, y_{\alpha})$ is positive, $\omega_{\alpha\beta} > 0$ (respectively negative $\omega_{\alpha\beta} < 0$). It is not difficult to see that in the singular direction $\Omega^{(0)}_{\alpha\beta}$ this variable is equal to zero. It can be shown that the introduced regions coincide with the decomposition of configuration space made in Chapter 4 with sign of the function $k^2_{\beta} - k^2_{\beta\alpha}$ taken into account.

It can be shown by means of the modified compact integral equations which will be presented in Section 5 that, as in the case of neutral particles, the term $\Psi_{\alpha\beta}$ changes the form from the wave corresponding to the double eikonal, into a distorted spherical wave when the point X moves from the region $\Omega^{(-)}_{\alpha\beta}$ into the region $\Omega^{(-)}_{\alpha\beta}$. We will bear this circumstances in mind when constructing the asymptotics of $\Psi_{\alpha\beta}$.

Let us denote by $\tilde{\Omega}^{(0)}_{\alpha\beta}$ the neighbourhood of the direction $\Omega^{(0)}_{\alpha\beta}$ where the inequality $|M_{\alpha\beta}| < Z^{\nu}_{\alpha\beta}$, $0 < \nu < 1/2$ is satisfied. To describe the transient behaviour of $\Psi_{\alpha\beta}$, let us introduce parabolic coordinates by the formulae

$\xi_{\alpha\beta} = |X| - Z_{\alpha\beta}$, $\zeta_{\alpha\beta} = |X| + Z_{\alpha\beta}$ and let us consider the function $\tilde{\Psi}_{\alpha\beta}$ given by

$$\tilde{\Psi}_{\alpha\beta}(X,P) = f_{\alpha\beta}(\xi_{\alpha\beta},\zeta_{\alpha\beta}) \frac{f_c^{(\alpha)}(\hat{x}_\alpha, k_\alpha) f_c^{(\beta)}(\hat{q}_{\beta\alpha}, k_\beta)}{|x_\alpha|\,|y_\alpha|} \times$$
$$\times A_{\alpha\beta}(\hat{X},P)\exp\{\sqrt{E}Z_{\alpha\beta} + iW_{\alpha\beta}\}, \tag{5.137}$$

where the function $f_{\alpha\beta}$ will be found with the help of the sewing condition for sewing $\tilde{\Psi}_{\alpha\beta}$ with the eikonal approximation $\Psi_{\alpha\beta}$ constructed above.

Let us substitute the expression (5.137) into the Schrödinger equation (5.45). Using Laplace operator in parabolic coordinates (5.59) and neglecting terms of order $\xi_{\alpha\beta}\zeta_{\alpha\beta}^{-2}$, we get for $f_{\alpha\beta}$ the following equation with separating variables:

$$\xi^{1/2}\frac{\partial}{\partial\xi}\xi^{1/2}\frac{\partial}{\partial\xi}f + i\sqrt{E}\left(\zeta\frac{\partial}{\partial\zeta} - \xi\frac{\partial}{\partial\xi}\right)f = 0,$$
$$\xi \equiv \xi_{\alpha\beta}, \quad \zeta \equiv \zeta_{\alpha\beta}. \tag{5.138}$$

After the separation, the equation for the variable ξ reduces to the confluent hypergeometric equation. Its solution describing the variation of the asymptotic form of the term $\tilde{\Psi}_{\alpha\beta}$ can be obtained if we choose the function $\Psi(a,b,t)$ as a solution of the confluent hypergeometric equation. Then in the region of shadow $\Omega_{\alpha\beta}^{(+)}$ the function $f_{\alpha\beta}$ should be defined as

$$f_{\alpha\beta}(\xi_{\alpha\beta},\zeta_{\alpha\beta}) =$$
$$= c_{\alpha\beta}\zeta_{\alpha\beta}^{i\mu}e^{i\sqrt{E}\xi_{\alpha\beta}}\Psi\left(\frac{1}{2}+i\mu, \frac{1}{2}, -i\sqrt{E}\xi_{\alpha\beta}\right), \tag{5.139}$$

where the constant μ and the amplitude $C_{\alpha\beta}$ are found by sewing the eikonal approximation of $\Psi_{\alpha\beta}$ for $x \to \infty$ in the region $\Omega_{\alpha\beta}^{(+)}$, which will be carried out below.

Describing the transition through the singular direction $\Omega_{\alpha\beta}^{(0)}$, we will assume that the variable $x_{\alpha\beta}$ vary in the complex plane with a cut along the positive semi-axis Π_0. We will suppose that the upper edge of the cut corresponds to the region $\Omega_{\alpha\beta}^{(+)}$, so that the argument of $\xi_{\alpha\beta}$ in equation (5.139) is fixed by the condition $\arg\xi_{\alpha\beta} = 0$ for $\xi_{\alpha\beta} > 0$. The passage on the lower edge of the cut, where $\arg\xi_{\alpha\beta} = 2\pi$ for $\xi_{\alpha\beta} > 0$ is in correspondence with the motion of the point X into the region $\Omega_{\alpha\beta}^{(+)}$. The function $f_{\alpha\beta}$ becomes then equal to

$$f_{\alpha\beta}(\xi_{\alpha\beta},\zeta_{\alpha\beta}) = C_{\alpha\beta}\zeta_{\alpha\beta}^{i\mu}e^{i\sqrt{E}\xi_{\alpha\beta}}\times$$

$$\times \left(\frac{2\sqrt{\pi}}{\Gamma(1+i\mu)} e^{-i\sqrt{E}\xi_{\alpha\beta}} \Phi \left(-i\mu, \frac{1}{2}, i\sqrt{E}\xi_{\alpha\beta} \right) - \right.$$

$$\left. - \Psi \left(\frac{1}{2} + i\mu, \frac{1}{2}, -i\sqrt{E}\xi_{\alpha\beta} \right) \right). \tag{5.140}$$

Taking the asymptotics of the functions $\Phi(a, b, t)$ and $\Psi(a, b, t)$ for $t \to \infty$ into account, we find from the sewing conditions

$$\mu = \frac{a_{\alpha\beta}}{2},$$

$$C_{\alpha\beta} =$$

$$\frac{1}{2\pi} E^{i\frac{a_{\alpha\beta}}{4}} \Gamma \left((1 + i\frac{a_{\alpha\beta}}{2} \right) \Gamma \left(\frac{1}{2} + i\frac{a_{\alpha\beta}}{2} \right) \exp \left\{ i\delta W_{\alpha\beta} - \pi \frac{a_{\alpha\beta}}{4} \right\}. \tag{5.141}$$

Let us give some corollaries of the formulae (5.139) – (5.141). In the region $\Omega^{(-)}_{\alpha\beta}$ the asymptotics of $\tilde{\Psi}_{\alpha\beta}$ can be expressed as the sum

$$\tilde{\Psi}_{\alpha\beta}(X, P) = \Psi_{\alpha\beta}(X, P) + \Phi^{(-)}_{\alpha\beta}(X, P),$$

where the eikonal approximation $\Psi_{\alpha\beta}$ has been described in Section 1 and $\Phi^{(-)}_{\alpha\beta}$ is the distorted spherical wave (5.53) with the singular amplitude

$$\Phi^{(-)}_{\alpha\beta}(\hat{X}, P) =$$

$$= C_0(E) \frac{\tilde{f}^{(\alpha)}_c(\hat{x}_\alpha, k_\alpha(p'_\alpha, p_\beta) \tilde{f}^{(\beta)}_c(\hat{k}_\beta(p'_\alpha, p_\beta, k_\beta)}{|s_{\alpha\beta}|^2 (k^2_\beta(p'_\alpha, p_\beta) - k^2_\beta - i0)^{1+ia_{\alpha\beta}}} A^{(s)}_{\alpha\beta} \tag{5.142}$$

Here the following notation is used:

$$A^{(s)}_{\alpha\beta} = \left(\frac{A_{\alpha\beta}|s_{\alpha\beta}|E^{1/4}}{2\sqrt{2}|k'_\alpha|\,|p'_\alpha|} \right)^{ia_{\alpha\beta}} \Gamma \left(1 + i\frac{a_{\alpha\beta}}{2} \right) \Gamma \left(\frac{1}{2} + i\frac{a_{\alpha\beta}}{2} \right) \times$$

$$\times \exp \left\{ i\delta W_{\alpha\beta} + \frac{a_{\alpha\beta}\pi}{2} \right\} \exp \left\{ -i \sum_\alpha \frac{n_\alpha}{\sqrt{E}} \frac{|X|}{|x_\alpha|} \log \frac{|x_\alpha|}{|X|} \right\}.$$

Moreover, as above, we used the notation $p'_\alpha = \sqrt{E}|X|^{-1}$ and took into account the equation

$$1 - \frac{Z_{\alpha\beta}}{|X|} \simeq -(k^2_\beta(p'_\alpha, p_\beta) - k^2_\beta) \frac{A_{\alpha\beta}|s_{\alpha\beta}|}{2\sqrt{2}|k'_\alpha|\,|p'_\alpha|},$$

$$|k'_\alpha| = \sqrt{E - {p'_\alpha}^2}.$$

In the region $\Omega_{\alpha\beta}^{(+)}$ the term in $\tilde{\Psi}_{\alpha\beta}$ corresponding to the double eikonal $Z_{\alpha\beta}$ is equal to zero and the asymptotics of $\tilde{\Psi}_{\alpha\beta}$ has the form of the distorted spherical wave:

$$\tilde{\Psi}_{\alpha\beta}(X,P) \sim F_{\alpha\beta}^{(+)}(\hat{X},P)|X|^{s/2}\times$$

$$\times \exp\{i|X|\,|P| - iW_0(X,P)\}.$$

Here the amplitude $F_{\alpha\beta}^{(+)}$ is not equal in magnitude to the amplitude $F_{\alpha\beta}^{(-)}$ as it was in the case of neutral particles. These amplitudes, calculated at points symmetric with respect to the boundary $\Omega_{\alpha\beta}^{(0)}$ are connected by the relation

$$F_{\alpha\beta}^{(+)} = \exp\{i\pi(1+ia_{\alpha\beta})\}F_{\alpha\beta}^{(-)}. \tag{5.143}$$

Let us note that the singular denominator in (5.142) can be understood as the distribution$(t - i0)^{-1-ia}$. Then one can explain the appearance of the additional factor $\exp\{i\pi + ia_{\alpha\beta})\}$ as the result of the analytical continuation of this function from the right half-plane, where $\arg t = 0$ for $t > 0$, into the region of negative t, $\arg t = \pi$. One moves around the point $t = 0$ in the positive direction. Remember that in a system of neutral particles the singularity of the amplitude of the spherical wave, corresponding to double collisions, has the character of a pole $(t - i0)^{-1}$, so that the transition into the left half-plane corresponds to the change of the sign of the amplitude.

5.4.3 The parabolic equation in forward scattering region

Thus we have constructed the asymptotics of the function $\Psi_0(X,P)$ in the neighbourhood of the singular directions $\Omega_{\alpha}^{(0)}$ and $\Omega_{\alpha\beta}^{(0)}$ by solving the model equations with separating variables. Here, contrary the case of forward scattering considered in the previous section, we will find the answer for Ψ_α not in the form of an integral representation, but as a series in spherical functions. Let us find analogous representations also for the neighbourhood of the direction of forward scattering.

Let us substitute Ψ_{as} in the form of the product

$$\Psi_{as}(X,P) = e^{i(X,P)}\Phi_{as}(X,P)$$

into the Schrödinger equation written in parabolic coordinates ξ and ζ. For Φ_{as} we obtain the equation

$$-\frac{4}{\xi+\zeta}\xi^{1/2}\frac{\partial}{\partial\xi}\xi^{1/2}\frac{\partial}{\partial\xi}\Phi-\frac{8}{\xi+\zeta}\left(\frac{\partial}{\partial\xi}-\frac{\partial}{\partial\zeta}\right)\Phi-$$

$$-\frac{4i\sqrt{E}}{\xi+\zeta}\left(\zeta\frac{\partial}{\partial\zeta}-\xi\frac{\partial}{\partial\xi}\right)\Phi+\frac{q_0(p)}{\xi+\zeta}\Phi-\frac{1}{\xi\zeta}\Delta_{\hat{X}}\Phi=0 \tag{5.144}$$

satisfied up to terms of order $\xi\zeta^{-2}$. Here by $\Delta_{\hat{X}}$ we denote the Laplace – Beltrami operator on the unit sphere $S^{(4)}$ in the space R^5. Let $Y_J(\hat{M})$ be the spherical functions on $S^{(4)}$, defined as the normalized eigenfunctions of the operator $\Delta_{\hat{X}}$:

$$-\Delta_{\hat{X}}Y_J(\hat{M})=L_JY_J(\hat{M}).$$

Here, J is a multi-index, $J=\{j,m_1,m_2,m_4\}$, labelling these functions and L_J are the corresponding eigenvalues

$$L_J=j(j+2),\quad j=1,2,\ldots$$

We will seek the functions Φ_{as}, as above, in the form of a series in spherical functions:

$$\Phi_{as}(X,P)=\sum_J f_j(\xi,\zeta)Y_J(\hat{M}), \tag{5.145'}$$

where the functions f_J satisfy the equation (5.144) with the operator $\Delta_{\hat{X}}$ replaced by the factor $L_J=j(j+2)$. To find the desired solution of this equation, we give its asymptotics for $\xi\to\infty$ which is determined by the sewing condition with the eikonal approximation L_c.

To this end, let us expand the function $\Phi_{as}(X,P)$ on the coordinate surface $\xi=|X|^\nu$ into a series (5.145'), so that the coefficients f_J are given by the integrals

$$f_J(\xi,\zeta)=\int dMY_J^*(\hat{M})\prod_\alpha\varphi_c^{(\alpha)}(x_\alpha,k_\alpha) \tag{5.145}$$

and let us investigate the asymptotics of f_J for $\xi\to\infty$. In the leading order we get the expressions

$$f_J(\xi,\zeta)\sim\exp\left\{i\frac{q_0(P)}{2|P|}\log\xi\right\}B_J \tag{5.146}$$

which do not depend on ξ. Moreover, if the eigenvalues are not too large,

$$L_J=O(|X|^\nu),\quad \nu<1/2, \tag{5.146'}$$

the coefficient B_J can be calculated from equation (5.145) with the function $\prod_\alpha \varphi_c^{(\alpha)}$ replaced by its asymptotics

$$\exp\left\{i\sum_\alpha \log\frac{|P|}{|k_\alpha|}\cos^2\theta_\alpha\right\}, \quad \cos\theta_\alpha = \frac{|u_\alpha|}{|M|}.$$

Since the asymptotic boundary conditions (5.146) are invariant with respect to translations in ζ, we must take the solutions of equation (5.144) which do not depend on ζ in the leading order. Up to terms of order $\xi\zeta^{-2}$ such solutions are

$$f_J(\xi) = C_J\xi^{\mu_J/2}\Phi\left(\left(\frac{\mu_J}{2}+i\frac{q_0(P)}{2|P|}, \mu_J+\frac{5}{2}, i\sqrt{E}\xi\right\}, \tag{5.147}$$

where the constants μ_J are given by

$$\mu_J = \sqrt{\frac{9}{4}+l_J^2}-\frac{3}{2}.$$

Taking into account the asymptotic form of the function $\Phi(a,b,t)$ for $t\to\infty$, from the sewing conditions we find

$$C_{J-} = B_J\frac{\Gamma\left(\frac{\mu_J}{2}+\frac{5}{2}+i\frac{q_0(P)}{2|P|}\right)}{\Gamma\left(\frac{5}{2}+\mu_J\right)}(i\sqrt{E})^{\mu_J-i\frac{q_0(P)}{2|P|}}.$$

Finally, let us note that the function $\Psi_{\alpha\beta}$ contains asymptotically also a distorted spherical wave with the amplitude which is singular for $(\hat{X},\hat{P})=1$; this amplitude is given as a series in hyperbolic functions

$$F_{as} = \frac{1}{(1-\cos\theta)^{\frac{5}{2}+i\frac{q_0(P)}{2|P|}}}f_0$$

$$f_0 = \sum_J B_J\frac{\Gamma\left(\frac{\mu_J}{2}+\frac{5}{2}+i\frac{q_0(P)}{2|P|}\right)}{\Gamma\left(\frac{\mu_J}{2}-i\frac{q_0(P)}{2|P|}\right)}\times$$

$$\times(i\sqrt{E})^{\mu_J-i\frac{q_0(P)}{2|P|}-\frac{5}{2}}Y_J(\hat{M}). \tag{5.147'}$$

$$\cos\theta = (\hat{X},\hat{P}).$$

This expression represents an expansion of the singular amplitude (5.121) with respect to the functions $Y_J(M)$, obtained above as solution of the inhomogeneous equation (5.115) in terms of the Green function of the operator H_P.

In derivation of formulae (5.147) and (5.147'), we assumed that the eigenvalues L_J satisfy inequalities (5.146'). This means that in expansions (5.145) and (5.147) we must use effectively only those terms which obey the stated conditions. However, it can be shown that the series (5.145) and (5.147) converge slowly, so that it becomes necessary to consider also terms which do not satisfy the conditions. Therefore, it is difficult to use these series to study local properties of the functions Φ_{as} and f_c. For this purposes, the integral representations (5.116) and (5.119) are more suitable. The expansions (5.145) and (5.147) are helpful, first of all, when working with integrals of the kind

$$\int \Psi_0(X,P) f(P)\, d\hat{P}$$

where f are smooth functions. In this case, the series obtained after averaging converges rapidly, so that the conditions (5.146') can be regarded satisfied. We will consider such integrals in the next chapter when computing the matrix elements of the Coulomb scattering matrix.

5.4.4 The eigenfunctions

In conclusion to this section, we will describe asymptotics of the eigenfunctions belonging to the discrete spectrum. To construct these asymptotics, we use the method of the eikonal expansions.

In the case of the discrete spectrum it is convenient to represent the asymptotic solution of the Schrödinger equation which corresponds to the eikonal Z in the form

$$\Psi_Z(X) A_Z f_Z \tilde{W}_Z \exp\{-\sqrt{E} Z\}.$$

Here, we use notations A_Z, f_Z from Section 1, and $\tilde{W}_Z$ is given by

$$\tilde{W}_Z = \exp\left\{(2\sqrt{E})^{-1} \int_{s(t)} V_c(t\nabla Z + M)\, dt\right\}.$$

In analogy with the case of a system of neutral particles, we will consider two types of eikonals: the spherical eikonal $|X|$ and the cluster eikonals

$$Z^{(\alpha)} = E^{-1/2}(E_\alpha^{1/2}|x_\alpha| + (E - E_\alpha)^{1/2}|y_\alpha|),$$

$$E_\alpha = \max_j \varkappa_A^2, \quad A = \{\alpha, j\}.$$

The eikonal approximation, corresponding to the spherical eikonal, has the form

$$\Psi_0(X, P) = f_0(\hat{X})(2\sqrt{E}|X|)^{\rho_0}\frac{\exp\{-\sqrt{E}|X|\}}{|X|^{5/2}}, \tag{5.148}$$

where the exponent ρ_0 is equal to the sum of two-particle exponents

$$\rho_0 = -\sum_\alpha \frac{n_\alpha}{2\sqrt{E}}\frac{|X|}{|x_\alpha|}.$$

The following approximations correspond to the cluster eikonal:

$$\Psi^{(\alpha)}(X, E) = F^{(2)}(\hat{x}_\alpha, \hat{y}_\alpha, \omega_\alpha)\tilde{W}^{(\alpha)}\frac{\exp\{-\sqrt{E}Z^{(\alpha)}\}}{|x_\alpha|\,|y_\alpha|}. \tag{5.149}$$

The function $\tilde{W}^{(\alpha)}$ above is given by the formula

$$\tilde{W}^{(\alpha)} = (2\sqrt{E}|x_\alpha|)^{-\frac{n_\alpha}{2\sqrt{E_\alpha}}}\prod_{\beta\neq\alpha}|k_{\beta\alpha}\zeta_{\beta\alpha}|^{-\frac{n_\beta}{2|k_{\beta\alpha}|}}$$

where

$$k_{\beta\alpha} = c_{\beta\alpha}\sqrt{E}\hat{x}_\alpha - s_{\beta\alpha}\sqrt{E - E_\alpha}\hat{y}_\alpha,$$

$$\zeta_{\beta\alpha} = |x_\beta| + (\hat{k}_{\beta\alpha}, x_\beta).$$

Here, the functions W_0 and $W^{(\alpha)}$ determine increasing or decreasing factors, depending on the relation between the charges.

Note that in agreement with general formulae (5.50), the angular coordinates $\hat{x}_\alpha$, $\hat{y}_\alpha$ and the variable $\omega_\alpha = E^{-1/2}(\sqrt{E - E_\alpha}|x_\alpha| + \sqrt{E_\alpha}|y_\alpha|)$ define an orthogonal coordinate system on the transversal surface $Z^{(\alpha)} = \text{const}$.

Let $\Omega_\alpha^{(\pm)}$ be the region where $\operatorname{sign}\omega_\alpha = \pm 1$, and let $\Omega^{(0)}$ be the transient region where the inequality $|\omega_\alpha| \leq (1 + |X|)^\nu$, $0 < \nu < 1/2$, is satisfied. We have encountered such a decomposition of the configuration space already in the case of neutral particles. Remember that the region Ω_α, where the particles of the pair α are separated weakly, is contained in the region $\Omega_\alpha^{(-)}$.

The eigenfunction $\Psi(X)$, corresponding to the bound state energy $-E$, can be expressed as the sum

$$\Psi(X) = \sum_{\alpha} \Phi_{\alpha}(X) \tag{5.150}$$

where the terms Φ_α have the form of the eikonal approximations. Let us describe these terms in more details.

In the region Ω_α, the function $\Phi_\alpha(X)$ is equal to the product of the eigenfunction of the ground state of the operator h_α and the cluster spherical wave (5.97), where E is equal to the eigenvalue, i.e.,

$$\Phi_\alpha(X) \sim g_\alpha(\hat{y}_\alpha)\psi_\alpha(x_\alpha)|y_\alpha|^{-\eta_\alpha} \frac{\exp\{-\sqrt{E-E_\alpha}|y_\alpha|\}}{|y_\alpha|}. \tag{5.151}$$

Here $g_\alpha(\hat{y}_\alpha)$ is a smooth function of the angular variables $\hat{y}_\alpha$ and by η_α we denoted the Coulomb parameter

$$\eta_\alpha = \frac{n_{\alpha\alpha}}{2\sqrt{E-E_\alpha}}.$$

When $|x_\alpha| \to \infty$, the two-body eigenfunction takes the following asymptotic form:

$$\psi_\alpha(x_\alpha) \sim f_\alpha(\hat{x}_\alpha)(2\sqrt{E_\alpha}|x_\alpha|)^{-\frac{n_\alpha}{2\sqrt{E_\alpha}}} \frac{\exp\{-\sqrt{E_\alpha}|x_\alpha|\}}{|x_\alpha|}.$$

The function Φ_α turns into the eikonal approximation (5.149), where the amplitude $F^{(\alpha)}$ is given by

$$F^{(\alpha)} = \left(-\frac{\sqrt{E}\omega_\alpha}{\sqrt{E_\alpha}}\right)^{\tilde{\eta}_{\alpha\alpha}} \left(\sqrt{\frac{E_\alpha}{E}}\right)^{\eta_\alpha} f_\alpha(\hat{x}_\alpha)g_\alpha(\hat{y}_\alpha)\times$$

$$\times \prod_{\beta\neq\alpha} \left(|s_{\beta\alpha}k_{\beta\alpha}|(1-(\hat{k}_{\beta\alpha}, y_\alpha)\,\mathrm{sign}\, s_{\beta\alpha})\right)^{\frac{n_\beta}{2|k_{\beta\alpha}|}}.$$

Here we introduced the notation

$$\tilde{\eta}_{\alpha\alpha} = \frac{n_{\alpha\alpha}}{2\sqrt{E-E_\alpha}} + \sum_{\beta\neq\alpha} \frac{n_\beta}{2|k_{\beta\alpha}|}$$

This amplitude, as in the case of the wave function $\Psi_0(X,P)$, is determined from the sewing conditions of the eikonal approximation with the function (5.151).

Similarly to the system of neutral particles, the function Φ_α has in the region $\Omega_\alpha^{(+)}$ the form of the eikonal approximation (5.148), corresponding to the spherical eikonal. Therefore, the next problem is to sew together all the indicated asymptotic regimes and to describe the asymptotics of Φ_α in the transient region $\Omega_\alpha^{(0)}$. This can be done by the method of standard equation.

Let us introduce the parabolic coordinates $\xi = |X| - Z^{(\alpha)}$, $\zeta = |X| + Z^{(\alpha)}$ into the region $\Omega_\alpha^{(0)}$. On the boundary between the regions $\Omega_\alpha^{(-)}$ and $\Omega_\alpha^{(+)}$, $\xi = 0$ and in the transient region $\xi < |X|^\nu$. We will look for the function F_α in the transient region $\Omega_\alpha^{(0)}$ having the form of the product

$$\Phi_\alpha = F^{(\alpha)}(M_\alpha)W^{(\alpha)}\frac{\exp\{-\sqrt{E}Z^{(\alpha)}\}}{|x_\alpha|\,|y_\alpha|}u_\alpha(\xi,\zeta), \tag{5.152}$$

where the function $u_\alpha(\xi,\zeta)$ depends only on the variables ξ and ζ. Substituting this expression into the Schrödinger equation (5.45) and neglecting small corrections of order $\xi\zeta^{-2}$, we get for the function $u_\alpha(\xi,\zeta)$ the following equation:

$$\left(\xi^{1/2}\frac{\partial}{\partial\xi}\xi^{1/2}\frac{\partial}{\partial\xi} - \sqrt{E}\left(\xi^{1/2}\frac{\partial}{\partial\xi} + \zeta^{1/2}\frac{\partial}{\partial\zeta}\right)\right)u_\alpha(\xi,\zeta) = 0. \tag{5.153}$$

The sewing condition provides the following boundary condition for this equation: the function (5.152) must have the asymptotics (5.148) if $\xi \to \infty$ and X lies in $\Omega_\alpha^{(+)}$; and it becomes the cluster spherical wave (5.149) if $\xi \to \infty$ on the other side of the boundary.

Note further that equation (5.153) has a solution of the form

$$u_\alpha(\xi,\zeta) = C_\alpha\zeta^\mu F_\alpha(-\sqrt{E}\xi)$$

where the function F_α satisfies the confluent hypergeometric equation

$$\xi F_\alpha'' + \left(\frac{1}{2} - \xi\right)F_\alpha' - \mu F_\alpha = 0.$$

The boundary conditions indicated above give the following values of the constants:

$$\mu = \frac{1}{2}\tilde{\eta}_{\alpha\alpha}, \qquad C_\alpha = -\frac{\Gamma(1+\mu)\Gamma\left(\frac{1}{2}+\mu\right)}{2\pi E^{\mu/2}},$$

and determine the solution

$$F_\alpha(-\sqrt{E}\xi) = e^{-\sqrt{E}\xi}\Psi\left(\frac{1}{2}+\mu, \frac{1}{2}, \sqrt{E}\xi\right).$$

When $x \to \infty$, we can replace the function $\Psi(a, b, t)$ by its asymptotics, $\Psi(a, b, t) \sim t^{-a}$. Then the solution (5.152) turns into the approximation of the spherical eikonal (5.148).

To obtain the function (5.152) in the region $\Omega_\alpha^{(+)}$, consider, as above, the coordinate ξ and regard it as the variable on the complex plain with the cut Π_0. We will assume that for X from $\Omega_\alpha^{(+)}$, the variable ξ belongs to the upper edge of the cut, $\xi = \lambda + i0$; to the region $\Omega_\alpha^{(-)}$ corresponds the lower edge of the cut. Carrying the analytical continuation along the contour around the point $\xi = 0$ in the positive direction, and bearing in mind the formula (5.140), we get the approximation of the cluster eikonal (5.149).

Thus, we have described the asymptotic form of the functions Φ_α. The asymptotics of the eigenfunction $\Psi(X)$, as in the case of neutral systems, is given by the leading terms in the sum (5.150).

5.5 Compact Equations in Configuration Space

In this section, we will describe the integral equations for the resolvent and the wave functions in the configuration space. As it was noted in Chapter 3, the basic difficulty one need to overcome to obtain an equation of the Fredholm type in a system of charged particles, results from the non-trivial action of the Coulomb potential on large distances. In particular, for that reason, the methods based on perturbation theory, cannot be used anymore.

However, let us note that an adequate mathematical techniques going beyond the framework of the perturbation theory and permitting description of the long-range corrections, was developed during the construction of the coordinate asymptotics of the wave function. In fact, in the first two sections, we found the singular part of the operator $(H - z)^{-1}$ corresponding to the long-range parts of the potential, i.e., we solved the problem mentioned in Section 5 of Chapter 4 in the context of the investigation of singularities of the kernels of equations (3.29) for positive energies. Therefore it remains to study the smooth part of this operator by means of the integral equations of the type of perturbation theory. Below, we will make this remark more precise and formulate the corresponding procedure.

5.5.1 Integral equations for components of the resolvent

Let us regard the operator H_a, defined on page 237 as an unperturbed Hamiltonian. We define the modified components of the $\hat{\mathrm{T}}$-matrix $\hat{M}_{\alpha\beta}$ corresponding to the perturbation $\hat{V}_\alpha$, by the relation

$$\hat{M}_{\alpha\beta} = \hat{V}_\alpha \delta_{\alpha\beta} - \hat{V}_\alpha R(z) \hat{V}_\beta.$$

In analogy with equation (3.57), we introduce the following components of the resolvent:

$$\hat{R}_{\alpha\beta}(z) = -R_a(z) \hat{M}_{\alpha\beta} R_a(z);$$

here by $R_a(z)$ we denote the resolvent of the operator H_a, $R_a(z) = (H_a - z)^{-1}$.

It is obvious that the resolvent of the operator H is equal to the sum

$$R(z) = R_a(z) + \sum_{\alpha,\beta} \hat{R}_{\alpha\beta}.$$

The integral equation for the kernels of the operators $M_{\alpha\beta}$ and $R_{\alpha\beta}$ can be obtained in the same way as in the case of neutral particles. We will not repeat here already known reasoning. Let us just present the final result. Denote by $\hat{H}_\alpha$ the operator

$$\hat{H}_\alpha = H_\alpha + \hat{V}_\alpha = H_0 + V_\alpha + \sum_{\beta\neq\alpha} V_\beta^{(0)}$$

and by $\hat{R}_\alpha(Z)$ its resolvent, $\hat{R}_\alpha(Z) = (\hat{H}_\alpha - z)^{-1}$. The following modified equations for the components $\hat{R}_{\alpha\beta}$ holds:

$$\hat{R}_{\alpha\beta}(z) = \delta_{\alpha\beta}(\hat{R}_\alpha(z) - R_a(z)) - \hat{R}_\alpha(z)\hat{V}_\alpha \sum_{\gamma\neq\alpha} \hat{R}_{\gamma\beta}(z). \qquad (5.154)$$

These equations are best suited for the investigation of the Coulomb Green function.

To study the modified integral equations, it is necessary to know the properties of the kernels of the operators R_a and $\hat{R}_\alpha$. However, unlike the analogous operators of the system of three neutral particles, the explicit form of these kernels is not known and they must be investigated independently. We will solve the given problem by constructing the asymptotic solutions of the inhomogeneous Schrödinger equations for Green functions and by subsequent turning to the integral equations of the perturbation theory related to these equations. In other words, we will correctly describe the singular Coulomb part of the Green function already in the zeroth approximation, so that the remaining perturbation will have a short-range character.

5.5.2 The operator $R_a(z)$

Let us start with the construction of the approximate Green function. We will call the function $G_{as}(X, X', z)$ the asymptotic Green function for the operator H_s, if it satisfies the following conditions:

1. The function $G_{as}(X, X', z)$ is smooth and bounded for $X \neq X'$ and for all z from the complex plane with the cut P_0. The estimate

$$|G_{as}(X, X', z)| \leq C(1 + |X - X'|)^{-5/2}$$

holds. If the variables X and X' coincide, G_{as} has a pole-like singularity

$$G_{as}(X, X', z) \sim (4\pi^3)^{-1}|X - X'|^{-4}.$$

2. The kernel $G_{as}(X, X', z)$ is a symmetric function $G_{as}(X, X', z) = \overline{G_{as}(X, X', z)}$ and satisfies the inhomogeneous Schrödinger equation with δ-type singularity up to rapidly decreasing terms for $|X| \to \infty$ ($|X'| \to \infty$). The following equation holds

$$(-\Delta_{X'} + V^{(0)}(X') - z)G_{as}(X, X', z) =$$

$$= \delta(X - X') - G_{as}(X, X', z)V_{as}(X, X', z), \qquad (5.155)$$

where the slowly oscillating function $V_{as}(X, X', z)$ decreases more rapidly than $|X'|^{-3-\epsilon}$, $\epsilon > 0$, uniformly with respect to the variable X.

3. For $|X| \to \infty$ and $|X'| < (1 + |X|)^{\nu}$, $\nu < 1/2$, the asymptotics of $G_{as}(X, X', z)$ is given by the eikonal approximation (5.50), corresponding to the spherical eikonal

$$G_{as}(X, X', z) \sim C_z|X|^{-5/2}\exp\{\sqrt{z}|X| + iW_{as}(X, z)\}f(X', \hat{X}). \qquad (5.155')$$

If the function $G_{as}(X, X', z)$ having the listed properties, is known, we can find the equation of perturbation theory to which Fredholm alternative is applicable. Indeed, equations (5.155) for the function G_{as} can be rewritten in the operator form:

$$G_{as}(z)(Ha - z) = I - G_{as}V_{as},$$

where by G_{as} and $G_{as}V_{as}$ we denote the operators defined by the kernels $G_{as}(X, X', z)$ and $G_{as}(X, X', z)V_{as}(X, X', z)$. Multiplying this expression from the right by the resolvent of the operator H_a, we obtain the desired equation of perturbation theory:

$$R_a(z) = G_{as}(z) - G_{as}(z)V_{as}R_a(z). \qquad (5.156)$$

This equation can be interpreted as a modification of equation (2.9), in whose the kernel $V_{as}(X, X', z)$ plays the role of a short-range potential.

The proof that this equation is of Fredholm type is based on the fact that the kernel $(G_{as}V_{as})(X, X', z)$ decreases sufficiently rapidly in second variable. This implies that the properties of the integral

$$f(X) = \int G_{as}(X, X', z)V_{as}(X, X', z)g(X')\,dX'$$

determining the action of the operator $G_{as}V_{as}$, depend on the properties of the kernel $G_{as}(X, X', z)$ only regarded as a function of X. In particular, if $g(X)$ is continuous and decreasing as $(1+|X|)^{-5/2+\nu}$, $\nu > 0$, then the function $f(X)$ is uniformly bounded and equicontinuous, and it decreases as $|X|^{-5/2}$. Thus, the set of functions $g(X)$ can be chosen so that the operator $G_{as}V_{as}$ improves their properties. Hence, this operator is absolutely continuous.

Let us note that the norm of the operator $G_{as}V_{as}$ can be made arbitrarily small at the by selecting the cutoff parameter a in equation (5.104'). This constant can be chosen so that the corresponding homogeneous equation

$$f = G_{as}V_{as}f$$

has no non-trivial solutions for z lying beyond the boundaries of an arbitrarily small neighbourhood of the zero. Therefore, if we consider a fixed point z, we can assume that the homogeneous equation (5.156) has no non-trivial solutions.

We must note that the requirement that the function $V_{as}(X, X', z)$ in equation (5.155) decreases can be weaken. In particular, one can assume that this function is of order $O(|X|^{-1})$ for the values of the arguments close together $X \sim X'$. This does not change the essence of the matter - the operator $G_{as}V_{as}$ remains completely continuous. Further below we will construct the asymptotic Green function $G_{as}(X, X', z)$ which indeed satisfies this weaken condition.

Thus, our principal problem is to construct the asymptotic Green function $G_{as}(X, X', z)$.

As it was shown in Section 2, the asymptotics of solutions of the Schrödinger equation for the function $\Psi_{as}(X,P)$, corresponding to the operator H_a, is determined by the straight trajectories of free particles, given by the plane eikonal $(X,\hat{P})$. Therefore, as in the case of neutral particles, it is natural to take as a solution of the inhomogeneous Schrödinger equation (5.155), the eikonal approximation corresponding to the eikonal $|X - X'|$ which describes a straight propagation of a spherical wave from the point X' to the point X. As a result, for $|X - X'| \to \infty$ we get the asymptotic solution of this equation in the form of the product of a free Green function and a Coulomb phase factor:

$$G_{as}(X,X',z) = C_z \frac{\exp\{i\sqrt{z}|X - X'|\}}{|X - X'|^{5/2}} \tilde{G}_{as}(X,X',z). \tag{5.157}$$

The latter is given by

$$\tilde{G}_{as}(X,X',z) = \exp\{iW_{as}(X,X',z) + i\delta W_N(X,X',z)\}.$$

where the function W_{as} is determined by an integral along the straight line joining the points X and X':

$$W_{as}(X,X',z) = \int_{(X',\hat{P}_{XX'})}^{(X,\hat{P}_{XX'})} V^{(0)}(t\hat{P}_{XX'} + M_{XX'})\, dt,$$

$$P_{XX'} = X - X', \quad M_{XX'} = X - \hat{P}_{XX'}(X,\hat{P}_{XX'}),$$

and the additional phase δW_N can be found from the recurrent relation (5.49'), where a finite number N of terms must be taken. Let us note that if both the points X and X' belong to the region Ω_0, so that the potential $V^{(0)}(X)$ is equal to the sum of Coulomb potentials, the function W_{as} reduces to the sum of the two-body phases (5.14'), where we must replace the energy factor z by the variables $z^{-1/2}|x - X'|\,|x_\alpha - x'_\alpha|^{-1}$:

$$W_{as}(X,X',z) =$$

$$= \frac{|X - X'|}{2\sqrt{z}} \sum_\alpha \frac{n_\alpha}{|x_\alpha - x'_\alpha|} \log \frac{|x_\alpha - x'_\alpha|\,|x_\alpha| + (x_\alpha - x'_\alpha, x_\alpha)}{|x_\alpha - x'_\alpha|\,|x'_\alpha| + (x_\alpha - x'_\alpha, x'_\alpha)}.$$

As in the case of wave functions, the eikonal formulae lose sense in some singular directions. These directions can be characterized on the basis of the classification of the asymptotic behaviour of the function $\Psi_0(X,P)$ given in Section 3 if we associate the expressions $q_\alpha = \sqrt{z}\frac{|x_\alpha - x'_\alpha|}{|X-X'|}$ with the energy variables k_α and the parabolic variables $\xi^{(\alpha)} = |x_\alpha| - (x_\alpha, \hat{k}_\alpha)$ with the analogous coordinates, constructed with help of the vector $-\tilde{x}'_\alpha$:

$$\xi_{x'_\alpha}(x_\alpha) = |x_\alpha| + (x_\alpha, \tilde{x}'_\alpha).$$

In particular, the validity of the eikonal formulae is limited by the conditions on coordinate variables

$$\xi_{x'_\alpha}(x_\alpha) \geq \left(|x_\alpha| + |x'_\alpha|\right)^\nu, \quad \nu > 0 \quad (\alpha = 1,2,3) \tag{5.158}$$

i.e., the points x_α and x'_α should not lie on the opposite sides of a straight line passing through the origin; further, it is limited by the following condition on the energy variables

$$|q_\alpha| > \left(\xi_{x_\alpha}(x'_\alpha) + \xi_{x'_\alpha}(x_\alpha)\right)^{\mu'}, \quad \mu' > -1/2. \tag{5.159}$$

If these conditions are satisfied, the function $V_{as}(X,X',z)$ decreases as an arbitrary power $|X-X'|^{-N}$ for $|X-X'| \to \infty$. Let us note that we have already encountered such conditions when describing the validity limits for the eikonal formulae (5.14) and (5.15) for Green functions $r_c(x,x',z)$ and $R_N(X,X',z)$.

For construction of the asymptotic resolvent in the singular regions the method of the model equation should be used. However, in order not to repeat already known reasoning, we will use knowledge of the asymptotic wave functions $\Psi_{as}(X,P)$ and so we will give the asymptotic Green function by means of an approximate spectral integral, similar to the representation (5.39):

$$W_{as}(X,X',z) = \frac{1}{(2\pi)^6}\int dP'\chi_c(P',z)\frac{\Psi_{as}(X,P')\Psi^*_{as}(X',P')}{P'^2 - z}. \tag{5.160}$$

Here by $\chi_c(P',z)$ we denoted a smooth cutoff function equal to one in vicinity of the point $P'^2 = \operatorname{Re} z$ and smoothly going to zero at some distance from this point. The asymptotics of this integral can be computed by the scheme used in the proof of the formula (4.119). First of all, one must integrate over the angular variables and make use of the saddle-point method, then calculate the remaining integral over the radial variable with the help of the residue theorem. So, for $z = E + i0$, one can obtain the representation (5.157), where the kernel $\tilde{G}_{as}$ is given by the integral

$$\tilde{G}_{as}(X, X', E + i0) =$$

$$-\frac{1}{2}\left(\frac{2}{\pi}\right)^{5/2} \exp\left\{-i\sqrt{E}\xi_{XX'} - \sum_{\alpha} i\frac{|X - X'|}{|x_\alpha - x'_\alpha|}\sqrt{E}n_\alpha \log\frac{\zeta_{XX'}}{2}\right\} \times$$

$$\times \lim_{\epsilon \downarrow 0} \int dq \tilde{\Phi}_F(\xi, M) \times$$

$$\times \exp\{-\epsilon|q| - 2i\sqrt{E}q^2 - 2i\sqrt[4]{E}\sqrt{\xi_{XX'}}(q, \hat{M}_{XX'})\}, \tag{5.161}$$

where

$$q = \sqrt{\xi}\hat{M}, \quad q \in R^3,$$

$$\hat{M}_{XX'} = \hat{X}(\hat{X}, X - X') + \hat{X}'(\hat{X}', X' - X),$$

$$\xi_{XX'} = |X| + |X'| - |X - X'|, \quad \zeta_{XX'} = |X| + |X'| + |X - X'|.$$

Finally, let us consider the case when the difference $|X - X'|$ remains bounded. We assume that the inequality $|X - X'| \leq (|X| + |X'|)^\nu$, $\nu < 1/6$ is satisfied.

Let the points X and X' belong to the region Ω_0. We expand the interaction potential $\sum_\alpha \frac{n_\alpha}{|x_\alpha|}$ in a series with respect to the small parameter $\frac{X - X'}{|X| + |X'|}$. The leading term of this expansion is the Coulomb potential in R^6, $q_0|X|^{-1}$, where the effective charge is given by

$$q_0 = (|X| + |X'|) \sum_\alpha \frac{n_\alpha}{|x_\alpha| + |x'_\alpha|}.$$

In accordance with this, we take $G_{as}(X, X', z)$ equal to the model Green function (5.40) for this potential:

$$G_{as}(X, X', z) = R_3(X, X', z).$$

If the points X and X' belongs to Ω_α, the leading asymptotic terms of the potential are defined by equation (5.125). Therefore, in this case we choose for the model Green function the kernel $R_{as}^{(\alpha)}(X, X', z)$ of the resolvent of the operator $H_{as}^{(\alpha)}$, given by equation (5.125).

Thus, we described the function G_{as} in the whole domain of definiteness. By construction, this kernel satisfies all conditions listed above and consequently it can be taken as the asymptotic Green function for the operator

$R_a(z)$. Note that this construction implies the following asymptotic formula similar to equation (4.79)

$$G_{as}(X, X', E + i0)\ |X'| \tilde{\to} \infty$$

$$\sim C_E \Psi_{as}(X, P') \frac{\exp\{i\sqrt{E}|X| + iW_{as}(X)\}}{|X|^{5/2}}, \qquad (5.162)$$

where $P' = -\sqrt{E}\hat{X}'$ and $\Psi_{as}(X, P')$ is the approximate wave function described in Section 3.

5.5.3 The Green Function of the Operator $\hat{H}_\alpha$

An analogous method, based on the equations of perturbation theory with an approximate Green function, can be used in the investigation of the resolvent $\hat{R}_\alpha(z)$. Before constructing the approximate Green function, let us consider some functions which will be used in the course of its description. Let $G_\alpha^{(1)}(X, X', z)$ be the function given by the integral

$$G_\alpha^{(1)}(X, X', z) = \int dy''_\alpha R_\alpha(X, X'', z) R_\alpha(X'', X', z), \qquad (5.163)$$

where we put $x''_\alpha = 0$. By definition this function satisfies the Schrödinger equation (5.45) if the variables X and X' lie in the region Ω_0. Let us consider the asymptotics of this function for $|X| \to \infty$ or $|X'| \to \infty$. According to (5.157), the leading term of the asymptotics is generated by the critical point of the exponential curve $\exp\{i\sqrt{z}(|X = X''| + |X'' - X'|)\}$ with the additional condition $x''_\alpha = 0$. In the basis corresponding to the pair α, this point is given by

$$X^{(0)} = \left\{0, \frac{|x'_\alpha| y_\alpha - |x_\alpha| y'_\alpha}{|x_\alpha| + |x'_\alpha|}\right\}.$$

At this point the exponent is equal to the single eikonal $Z_\alpha(X, X')$. Therefore, the function $G_\alpha^{(1)}(X, X', z)$ is closely related to the eikonal approximation corresponding to the eikonal $Z_\alpha(X, X')$. Indeed,

$$G_\alpha^{(1)}(X, X', z) \sim C_z \exp\{i\sqrt{z} Z_\alpha(X, X')\} \tilde{G}_\alpha^{(1)}(X, X', z), \qquad (5.164)$$

where, as above, by C_z we denoted the quantity $C_z = -\frac{1}{2} e^{i\pi/4} (2\pi)^{-5/2} z^{3/4}$. This notation will be also used below.

If the points X, $X^{(0)}$, and X' do not belong to the singular regions, defined by conditions (5.158), (5.159), the function $\tilde{G}_{\alpha}^{(1)}(X,X',z)$ is defined by the eikonal formula

$$\tilde{G}_{\alpha}^{(1)}(X,X',z) = \frac{|x_\alpha| + |x'_\alpha|}{|x_\alpha|\,|x'_\alpha|} \frac{\exp\{iW_{\alpha\alpha}(X,X',z)\}}{Z_\alpha^{5/2}(X,X')}. \quad (5.165)$$

The phase $W_{\alpha\alpha}$ is equal to the sum of the eikonal phases contained in the asymptotics of the function G_{as}:

$$W_{\alpha\alpha}(X,X',z) = W_{as}(X,X^{(0)},z) + W_{as}(X^{(0)},X',z).$$

Let us note that similarly to equation (5.86'), in the last formula we can separate the phase $\tilde{W}_{\alpha\alpha}$ dependent on the eikonal variable Z_α, and the term $\delta W_{\alpha\alpha}$, being a function of the point $M_{\alpha\alpha}$ only on the transversal surface $Z_\alpha = \text{const}$:

$$M_{\alpha\alpha} = X - (X - \hat{K}_\alpha)\hat{K}_\alpha, \quad \hat{K}_\alpha = \nabla_Z Z_\alpha(X,X').$$

If $X \in \Omega_0$, the phase $\tilde{W}_{\alpha\alpha}$ can be cast into the following form:

$$\tilde{W}_{\alpha\alpha} = -\sum_{\alpha}\left(\frac{n_\beta}{2\sqrt{z}|k_{\beta\alpha}|}\log(|k_{\beta\alpha}|\,|x_\beta| - (k_{\beta\alpha},x_\beta)) + \right.$$

$$\left. + \frac{n_\beta}{2\sqrt{z}|k'_{\beta\alpha}|}\log(|k_{\beta\alpha}|\,|x'_\beta| - (k'_{\beta\alpha},x_\beta))\right).$$

Here the momentum variables $k_{\beta\alpha}$ and $k'_{\beta\alpha}$ are given by the equations

$$k_{\beta\alpha} = \nabla_{x_\beta} Z_\alpha(X,X'), \quad k'_{\beta\alpha} = -\nabla_{x'_\beta} Z_\alpha(X,X')$$

and the phase $\delta W_{\alpha\alpha}$, depending on the transversal coordinate $M_{\alpha\alpha}$, is equal to the sum

$$\delta W_{\alpha\alpha}(X,X',z) = \delta W_{\alpha\alpha}^{(1)}(X,X',z) + \delta W_{\alpha\alpha}^{(2)}(X,X',z),$$

where

$$\delta W_{\alpha\alpha}^{(1)}(X,X',z) = \sum_{\beta\neq\alpha}\frac{n_\beta}{2\sqrt{z}|k_{\beta\alpha}|}\log(|k_{\beta\alpha}|\,|x_\beta^{(0)}| - (k_{\beta\alpha},x_\beta^{(0)}))$$

and

$$\delta W_{\alpha\alpha}^{(2)}(X,X',z) = -\sum_{\beta\neq\alpha}\frac{n_\beta}{2\sqrt{z}|k'_{\beta\alpha}|}\log(|k'_{\beta\alpha}|\,|x_\beta^{(0)}| - (k'_{\beta\alpha},x_\beta^{(0)})).$$

By $x_\beta^{(0)}$ we denote the coordinate of the point $X^{(0)}$ in the basis corresponding to the pair β: $X^{(0)} = \{x_\beta^{(0)}, y_\beta^{(0)}\}$.

The eikonal formulae (5.164), (5.165) lose sense in the singular directions indicated above. In these directions the phase $W_{\alpha\alpha}$ turns into infinity. In particular, this happens for $|M_{\alpha\alpha}| = 0$. In this case, the function $G_\alpha^{(1)}$ is given by means of the integral representations (5.163). These can be cast into the form (5.161), if we make the change of the variable $\tilde{y} = Z_\alpha^{-1} y''_\alpha$ in the integral (5.161) and reduce the domain of integration over $\tilde{y}$ to the neighbourhood of the critical point $y^{(0)}$. We will not describe here the cumbersome formulae obtained as the result of this procedure.

Thus, we see that the function $G_\alpha^{(1)}$ satisfies all conditions required for an approximate solution of the Schrödinger equation: it has asymptotically eikonal form in non-singular directions and satisfies the sewing conditions in the singular regions. With help of this function, let us now construct the asymptotic Green function for the operator $\hat{H}_\alpha$.

First, let us assume that the points X and X' lie in the region Ω_0. We will then define the function $G_\alpha(X, X', z)$ by the following equation:

$$G_\alpha(X, X', z) = R_a(X, X', z) + \hat{G}_\alpha(X, X', z).$$

Here R_a is the Green function of the asymptotic Hamiltonian H_a and $\hat{G}_\alpha$ is a smooth function, generated by the perturbation $\hat{V}_\alpha$. Let us define function by the equation

$$\hat{G}_\alpha(X, X', z) = f_\alpha^{(\pm)}(\hat{x}_\alpha, k_\alpha) G_\alpha^{(1)}(X, X', z),$$

$$k_\alpha = -\tilde{x}_\alpha \frac{|x_\alpha| + |x'_\alpha|}{Z_\alpha(X, X')} \mathrm{Re}\sqrt{z}, \tag{5.166}$$

where the function $G_\alpha^{(1)}$ was defined above, and the function $f_\alpha^{(\pm)}$ describes effects of the perturbation $\hat{V}_\alpha(X)$. This function is expressed through the two-body scattering amplitude. Indeed, let us assume that

$$(\hat{x}_\alpha, \hat{x}'_\alpha) \neq -1.$$

Then $f_\alpha^{(\pm)}$ coincides with the scattering amplitude for the energy operator h_a (5.22) with the label (+) corresponding to positive values of $\mathrm{Re}\sqrt{z}$, and (-) to negative ones. If $(\hat{x}_\alpha, \hat{x}'_\alpha) = -1$, the Coulomb part of the scattering amplitude f_c turns into infinity and in the neighbourhood of the directions x_α and x'_α it must be replaced by the short-range part of the scattering amplitude for the model operator $h^{(\alpha)}_{as}$, defined in the last section, i.e. the function $f_\alpha = f_c + f_{cs}$ is replaced by the function $\tilde{f}_{cs} = f^{(a)}_\alpha - f_c$ defined according to (5.127).

Let us note that in almost all directions in the configuration space, the kernel $G^{(1)}_\alpha$ is described by the eikonal formulae (5.164) and (5.165). Therefore, the leading asymptotic terms can be completed by eikonal corrections so that the Schrödinger equation (5.45) for $\hat{G}_\alpha$ will be satisfied up to terms of an arbitrary power Z_α^{-N}, $N \gg 1$. We will assume below that this procedure is completed. Thus, we define the function $\hat{G}_\alpha$ from equation (5.166) in non-singular directions by the sum (5.48) with a finite number of terms N, $N \gg 1$.

By precisely the same relation one can define the function $\hat{G}_\alpha(X, X', z)$ in the case, when the points X and X' are situated in the regions $\Omega_\beta(a, n)$ for $\beta \neq \alpha$. Finally, let these points pass into the region Ω_α, where the short-range perturbation $\hat{V}_\alpha$ is not zero. For those directions of the vectors X and X' not getting into the singular regions, defined by the relations (5.158) and (5.159), we put

$$\hat{G}_\alpha(X, X', z) = R^{(c)}_\alpha(X, X', z), \tag{5.167}$$

where the model Green function $R^{(c)}_\alpha$ is defined above by equation (5.71'). Then the Schrödinger equation will be satisfied up to terms of order $|x_\alpha|\,|y_\alpha|^{-2}$. If the points X and X' get into singular regions, we add to the right-hand side of equation (5.167) terms which, similarly to (5.114), guarantee the sewing of the solution in the singular region with the eikonal formulae. This can be done, e.g., if we give $\hat{G}_\alpha$ by means of the approximate spectral integral (5.160), where we must take as the asymptotic wave functions the solutions $\Psi_{as}(X, P)$ constructed in Section 3.

If, finally, one of the points, say x'_α, falls outside the limits of the region Ω_α, the kernel $\hat{G}_\alpha$ can again be defined by the eikonal formulae (5.164), (5.165), where, however, we must replace the ratio $f_\alpha|x_\alpha|^{-1}$ by the wave function of the operator h_α. Here the argument x_α of the function $G^{(1)}_\alpha$

should to be put equal to zero:

$$\hat{G}_\alpha(X,X',z) = C_z\psi_\alpha^{(\pm)}(x_\alpha,k_\alpha)\left[\frac{\exp\{iW_{\alpha\alpha}(X,X',z)\}}{Z_\alpha^{5/2}(X,X')}\right]_{x_\alpha=0}$$

Let us note that the model Green function $R_\alpha^{(c)}(X,X',z)$ and, consequently, the asymptotic Green function $\hat{G}_\alpha(X,X',z)$ contains a non-trivial part corresponding to the discrete spectrum of the operator h_a. The kernel of the operator $P_A R_\alpha^c$ has the form

$$(P_A R_\alpha^c)(X,X',z) = \psi_A(x_A)\psi_A^*(x_A')r_\alpha^c(y_\alpha,y_\alpha',z-\varkappa_A^2), \tag{5.168}$$

where by r_α^c we denote the Green function of the operator h_α^c. Let us recall that the latter describes the scattering of the pair α on Coulomb centre.

Therefore, we described the asymptotic Green functions in the whole domain of definiteness. By construction, this function along with the kernel $G_{as}(X,X',z)$ has the properties listed above in conditions 1 and 2. Let us check which form equation (5.155) takes. Consider the expression

$$\left((-\Delta + v_\alpha(x_\alpha) + \sum_{\beta\neq\alpha} v_\beta^{(0)}(X) - z\right)\hat{G}_\alpha(X,X',z) =$$

$$= \delta(X-X') + \hat{A}_\alpha(X,X',z), \tag{5.169}$$

which can be regarded as the definition of the kernel $\hat{A}_\alpha(X,X',z)$. As in the case of the kernel $G_{as}(X,X',z)$, the function $\hat{A}_\alpha(X,X',z)$ is rapidly decreasing in the second variable. However, together with terms present in the representation (5.155'), new terms will appear in this case in the region Ω_α, the existence of whose is induced by the cluster Green functions (5.168). therefore, we have the representation

$$\hat{A}_\alpha(X,X',z) =$$

$$G_{as}(X,X',z)\hat{V}_{as}(X,X',z) + \sum_A G_A(X,X',z)\hat{v}_A(y_A,y_A',z), \tag{5.170}$$

where by G_A we denote the kernel (5.168). The functions $\hat{V}_{as}$ and $\hat{v}_A$ are rapidly decreasing in variables X' and y_α' uniformly with respect to X and y_A, respectively.

As in the case of the operator $R_a(z)$, the asymptotic Green function $\hat{G}_\alpha$ can be taken as the zeroth approximation for the kernel of resolvent $\hat{R}_\alpha(X, X', z)$. In such a way we obtain the following equation of perturbation theory:

$$\hat{R}_\alpha(z) = -\hat{A}_\alpha(z)\hat{R}_\alpha(z) + \hat{G}_\alpha(z), \tag{5.171}$$

where by $\hat{A}_\alpha(z)$ we denote the operator defined by the kernel $\hat{A}_\alpha(X, X', z)$. This equation can be investigated along the same lines as the analogous equation (5.165). The only difference is that the asymptotics of the solution, along with terms decreasing like $(1 + |X|)^{-5/2}$, contains also terms with the asymptotics of the cluster type corresponding to the terms (5.168). One can check that equation (5.171) is compact. We will not go into details of the proof of this statement, they have been discussed in length in the case of equations (4.12) and (4.70). We will present here only some implications of equation (5.171).

Let us note that the constant a from equation (5.104') determines the magnitude of the perturbations $\hat{V}_{as}$ and $\hat{v}_A$ which can be made arbitrarily small. Consequently, for arbitrary fixed z not coinciding with the threshold values $z \neq 0 - \varkappa_A^2$, $(A = \{\alpha, i\},\ i = 1, 2, \ldots N_\alpha)$, the constant a can be chosen so that there are no non-trivial solutions of the homogeneous equation (5.171).

Bellow, we will need the asymptotic form of the function $\hat{R}_\alpha(X, X', z)$ for z lying on the cut $[-\varkappa_A^2, \infty)$ and $|X'| \to \infty$. According to equations (5.166), (5.162), and (5.168) the following relation is holds

$$\hat{R}_\alpha(X, X', E \pm i0) \sim$$

$$\sim C_E \hat{F}_{0\alpha}(X, P') \frac{\exp\{\pm i\sqrt{E}|X| + iW_0(X)\}}{|X|^{5/2}} +$$

$$+ \frac{1}{4\pi} \sum_j \psi_A(x'_A) F_A(X, p'_\alpha) \frac{\exp\{\pm i\sqrt{E + \varkappa_A^2}|y_\alpha| + iW_A(y_\alpha)\}}{|y_\alpha|}. \tag{5.172}$$

Here the sum runs over all eigenvalues of the operator h_a and by $F_{0\alpha}$ and $F_{A\alpha}$ we denote the bounded amplitudes of the spherical waves, which are functions of X and $P' = \mp\sqrt{E}\hat{X}'$ or $p_\alpha = \mp\sqrt{E}|X'|^{-1}y'_\alpha$, correspondingly.

We will see bellow that these amplitudes are closely related to the wave functions of the operator $\hat{H}_\alpha$.

Therefore, we have completed construction of the Green functions $R_a(X, X', z)$ and $\hat{R}_\alpha(X, X', z)$, appearing in the equations (5.154), and now we can proceed to investigation of the properties of the modified resolvent $\hat{R}_{\alpha\beta}$ and of the resolvent itself.

5.5.4 The Green function $R(X, X', z)$

We will follow the procedure used in the case of neutral particles. First of all, let us construct the functional space where we could study the integral equation (5.154). To do this, consider the behaviour of the iteration kernels $\hat{R}^{(n)}_{\alpha\beta}$. These kernels are equal to the sum (4.70'), where the operators $\hat{R}^{(n)}_{\alpha\alpha_1\ldots\alpha_n\beta}$ are defined by equations analogous to (4.71')

$$\hat{R}^{(n)}_{\alpha\alpha_1\ldots\alpha_n\beta} = (-1)^{n+1}\hat{R}_\alpha \hat{V}_\alpha \cdots \hat{R}_{\alpha_n}\hat{V}_{\alpha_n}(\hat{R}_\beta - \hat{R}_\alpha).$$

Let us note that the properties of these kernels $\hat{R}^{(n)}_{\alpha\alpha_1\ldots\alpha_n\beta}$ can be studied by means of the same technique as in the case of neutral particles. The most significant from the point of view of applicability of this technique is the fact that the functions $\hat{V}_\alpha(X)$, like the short-range two-particle potentials $v_\alpha(x_\alpha)$, are concentrated in the region Ω_α where the particles of the pair α are weakly separated. Therefore, the asymptotics of the kernels $\hat{R}^{(n)}_{\alpha\alpha_1\ldots\alpha_n\beta}$, given by integrals of the type (4.71), is determined by the same factors as those considered in Section 4 of the previous chapter. The specific features of Coulomb interactions exhibits themselves only in the fact that in the rapidly oscillating exponents additional phases appear logarithmically depending on the coordinates. However, these phases are by no means an obstacle for use of the saddle-point method, because the factors induced by them are slowly oscillating functions in the sense of the definition (4.117).

It follows from representation (5.172) that the kernels $\hat{R}^{(n)}_{\alpha\alpha_1\ldots\alpha_n\beta}$ are of the type $\mathcal{D}_{\alpha\beta}$ (4.72). The components of these kernels have properties similar to those listed in Section 4 of the previous chapter. One can check, for example, that the asymptotics of the component $F(X, X', z)$ has the form of the eikonal approximation corresponding to the eikonals $|X|$, $Z_\alpha(X, X')$, $Z_{\alpha\beta}(X, X')$ etc. Moreover, for the same reason as it was in the case of neutral particles, the kernel $\hat{R}^{(n)}_{\alpha\beta}$ has different asymptotic forms pending on the position of the points X and X' in configuration space. This kernel is equal to the eikonal approximation constructed from the eikonal $Z_{\alpha\beta}(X, X')$, if the points X and X' can be joined by a straight line with two breaking points on the manifolds $x_\alpha = 0$ and $x_\beta = 0$ and reduces to a product of distorted spherical waves in the opposite case.

It can be shown that starting with a sufficiently large number n, the properties of the components of the kernels $\hat{R}^{(n)}_{\alpha\beta}$ are getting stabilized and they fall into the class $\tilde{\mathcal{D}}_{\alpha\beta}$, defined in connection with the investigation of equation (4.70)

By using decomposition of unity (4.23'), the integral equations (5.154) can be reduced to an integral equation of second kind in this class, in analogy with equations (4.77). It is clear that the proof of compactness of these equations can be given in the same way as in the case of the equations (4.77). Let us note that the singular points of equations (5.154) coincide with the points of the discrete spectrum of the operator H, if they are not equal to the threshold values,

$$E_i \neq 0, -\varkappa_A^2, \quad A = \{a, i\}, \quad \alpha = 1, 2, 3, \quad i = 1, 2, \ldots N_\alpha,$$

and differ from the singular points of equations (5.156) and (5.171).

We will not continue with discussion of technical questions anymore, these problems have been considered in detail above. Instead, we will describe some implications of equations (5.154).

The properties of the Green function $R(X, X', z)$ can be described with the help of the proposition formulated at the end of Section 4 of Chapter 4. The only difference is that, instead of (4.78), we have here the representation

$$R(z) = R_a(z) + \sum_\alpha (\hat{R}_\alpha(z) - R_a(z)) +$$

$$+ \sum_{k=1}^{4} \sum_{\alpha,\beta} \hat{R}^{(k)}_{\alpha\beta}(z) + \sum_{\alpha,\beta} \tilde{R}_{\alpha\beta}(z) \tag{5.173}$$

where the kernels $\tilde{R}_{\alpha\beta}(z)$ belong to the class $\mathcal{D}_{\alpha\beta}$ and the properties of the kernels $\hat{R}_\alpha$, R_a, and $\hat{R}^{(k)}_{\alpha\beta}(z)$ $k \leq 4$ have been discussed above.

Let us characterize the asymptotics of the Green function $R(X, X', z)$ for real z in more detail. To this end, we introduce new classes of functions.

Let $Q_B^{(\pm)}(y_B, E)$ be distorted spherical waves in R^3

$$Q_B^{(\pm)}(y_B, E) = |y_B|^{-1} \exp\left\{ \pm i\sqrt{E + \varkappa_B^2}|y_B| \mp \right.$$

$$\left. \mp i \frac{n_{\beta\beta}}{2\sqrt{E + \varkappa_B^2}} \log 2\sqrt{E + \varkappa_B^2}|y_B| \right\}, \tag{5.174}$$

and $Q_0^{(\pm)}(X, E)$ the distorted spherical waves in R^6

$$Q_0^{(\pm)}(X, E) =$$

$$= \frac{1}{|X|^{5/2}} \exp\left\{ \pm i\sqrt{E}|X| \mp i \frac{|X|}{2\sqrt{E}} \sum_\alpha \frac{n_\alpha}{|x_\alpha|} \log 2\sqrt{E}|X| \right\}.$$

By $S_{E,c}^{(\pm)}$, $B_{E,c}^{(\pm)}$, $\Phi_{E,c}^{(\pm)}$ we denote the classes of functions defined by the formulae (4.27) - (4.29) of Section 2, Chapter 4, where instead of the spherical waves in R^3 and R^6,one should use the distorted spherical waves $Q_B^{(\pm)}$ and $Q_0^{(\pm)}$.

The following proposition follows from the results of this section:

Let E does not coincide with the singular points of equation (5.154) and let $|X - X'| \geq \delta > 0$. Then the limits of the kernels $R(X, X', z)$ exist for $\epsilon \downarrow 0$, and they are functions of the class $S_{E,c}^{(\pm)}$ as functions of X'. Therefore, for $|X'| \to \infty$ we have the representation

$$R(X, X', E + i0) \underset{|X'| \to \infty}{\sim} C_B F_0^{(\pm)}(X, P) Q_0^{(\pm)}(X', P) +$$
$$+ \frac{1}{4\pi} \sum_A F_A(X, p_\alpha) \Psi_A(x_A') Q_A^{(\pm)}(y_A', P), \tag{5.175}$$

where $P = \mp\sqrt{E}\hat{X}'$ and $p_\alpha = \mp\sqrt{E}|X'|^{-1} y_\alpha'$. By virtue of the symmetry of the kernel $R(X, X', z)$, an analogous representation holds for $|X| \to \infty$, as well.

If the variables X and X' tend to infinity independently, then the asymptotics of the Green function is more complicated. In this case, we must consider iterations of the equations separately. Recall that the first three terms in equation (5.173) are described by eikonal formulae which we considered above. The last term in equation (5.173), equal to the sum of kernels of the class $\tilde{\mathcal{D}}_{\alpha\beta}$, is a function of the class $S_{E,c}^{(\pm)}$ in both X and X'.

Let us note that the representation (5.176) is closely related to the analogous representation (2.46) for the kernel of the resolvent in momentum space. Namely, after performing the Fourier transform, the singular denominators generate the distorted spherical waves

$$
\begin{aligned}
(p_A^2 - \varkappa_A^2 - E \mp i0)^{-1+i\eta_A} &\rightarrow Q_A^{(\pm)}(y_A, E), \\
(P^2 - E \mp i0)^{-1+i\eta_A} &\rightarrow Q_0^{(\pm)}(X, E).
\end{aligned}
\tag{5.176}
$$

Therefore, we justified the representation (2.46) which was obtained above on the basis of heuristic reasoning. Moreover, we can now study in detail the structure of the functions $\tilde{R}_{AB}$ (2.46) contained in definition of the singular factors as well. It is clear that the slowly decreasing terms of the asymptotics $\hat{R}_{\alpha\beta}^{(n)}$ generate secondary singularities of this functions similar to equation (3.31) as it was in the case of neutral particles. We will not present here a more concrete amplification of this remark for the general case. If necessary, these singularities can be found by studying the Fourier transform which defines the transition to momentum representation. Below, we will describe such singularities in the case when the variables p_A, p'_B, and z lie on the energy shell $E_A(p_A) = E_B(p'_B) = z$.

5.6 Boundary Conditions for Wave Functions

In this section, we give formal definitions of the wave functions $\Psi_0(X, P)$ and $\Psi_A(X, p_A)$, and, on the basis of the Schrödinger equation and the differential equations for the components, we will describe their boundary conditions.

5.6.1 Definition of the wave functions

In Sections 1-4 of this chapter we have been busy investigating the asymptotic forms of the wave functions, regarding them as formal solutions of the Schrödinger equation. After having studied the properties of the kernel of the resolvent, we can give a more precise definition of the wave functions which is convenient to be employed in justification of the scattering problems. We will make use of the prescription formulated at the end of Chapter 4. Namely, let us consider the asymptotics of the Green functions $R(X, X', z)$ for $|X'| \to \infty$ (5.176). Obviously, the amplitudes of the distorted cluster and spherical waves satisfy the Schrödinger equation in X

$$\left(-\Delta_X + \sum_\alpha v_\alpha(x_\alpha) - E\right) F_A(X, p_A) = 0. \tag{5.177}$$

Therefore, as in the case of neutral particles, we can take the representation (5.176) as the definition of wave functions. In agreement with the normalization chosen above, we put

$$\Psi_0^{(\pm)}(X, P) = F_0^{(\pm)}(X, P), \tag{5.178}$$

$$\Psi_A^{(\pm)}(X, p_A) = F_A^{(\pm)}(X, p_A), \quad A \neq 0. \tag{5.179}$$

As expected, the asymptotic formulae constructed in Sections 1 – 4, describe the behaviour of the wave functions at infinity correctly. This follows immediately from the definitions of the formal solutions and their relation to the asymptotic Green functions.

Similarly, one can define the wave functions of the operators H_a and $\hat{R}_\alpha$. According to equation (5.162), the wave functions of the operator H_a are defined as the amplitudes of the distorted spherical waves. We will denote this functions by the same symbol as the asymptotic solution (5.51) $L_c(X, P)$. The asymptotic Green functions for the operator $\hat{H}_\alpha$ (5.72) contains both six-dimensional spherical waves and cluster spherical waves generated by the eigenvalues of the operator h_α. We denote by $L_{\alpha c}(X, P)$ the wave functions defined by equation (5.178) and by $L_{Ac}(X, p_A)$ the wave functions of the type (5.179).

Next, let us determine the components of the wave function by means of the asymptotic representation for the component of the resolvent kernel

$$\hat{R}(X, X', E + i0) \sim C_B F_{\alpha\beta}^{(\pm)}(X, P) Q_0^{(\pm)}(X, E) +$$
$$+ \frac{1}{4\pi} \sum_j F_{\alpha B}^{(\pm)}(X, p_B) \psi_B(x'_B) Q_B^{(\pm)}(y'_B, E). \tag{5.180}$$

In analogy to (5.178) and (5.179), we put

$$\Phi_{\alpha 0}^{(\pm)}(X, P) = \sum_\beta F_{\alpha\beta}^{(\pm)}(X, P),$$

$$\Phi_{\alpha B}^{(\pm)}(X, p_B) = F_{\alpha B}^{(\pm)}(X, p_B).$$

Then the wave functions are given by the sum

$$\Psi_0^{(\pm)}(X, P) = \sum_\alpha \Phi_{\alpha 0}^{(\pm)}(X, P) + L_c(X, P);$$

$$\Psi_B^{(\pm)}(X, p_B) = \sum_\alpha \Phi_{\alpha B}^{(\pm)}(X, p_B). \tag{5.180'}$$

Representation (5.180) is convenient for derivation of compact equations which are satisfy by the components (5.180'). These equations can be obtained from the integral equations (5.154) by the already known procedure. We must take the limit $X' \to \infty$ in equations (5.154) for the kernels $\hat{R}_{\alpha\beta}(X, X', E \pm i0)$, compare the corresponding asymptotic terms on both sides, and then drop the distorted spherical or cluster waves. In this way, we get the following equations:

$$\Phi_{\alpha 0} = L_{\alpha c} - L_c - \hat{R}_\alpha(E + i0)\hat{V}_\alpha \sum_{\gamma \neq \alpha} \Phi_{\gamma 0},$$

(5.181)

$$\Phi_{\alpha B} = \delta_{\alpha\beta} L_{Bc} - \hat{R}_\alpha(E_B + i0)\hat{V}_\alpha \sum_{\gamma\neq\alpha} \Phi_{\gamma B}.$$

If the point E is not singular, these equations determine the components of wave functions uniquely.

The integral equations for the kernels L_c, $L_{\alpha c}$, and L_{Ac} can be obtained in the same way from the integral equations (5.156) and (5.171). They coincide in the form with equations presented above and differ only in free terms. It is clear that as these free terms one should take the asymptotic solutions, constructed in Sections 1 – 3.

Alternatively, the wave functions can be obtained with the help of the differential formalism as solutions of the Schrödinger equation in a class of smooth functions with a given asymptotic form. Let us describe this approach on the example of the function Ψ_A.

Let us denote by $\tilde{S}_A$ the set of functions having the form of the sum

$$f_A(X) = \psi_A(x_\alpha)\psi_A^{(c)}(y_\alpha, p_\alpha) + \tilde{f}_A(X), \tag{5.182}$$

where the function $\psi_A^{(c)}$ corresponds to the potential (5.100), and the functions $\tilde{f}_A$ belong to the class $S_{E,c}$ for $E = p_\alpha^2 - \varkappa_A^2$. The following proposition holds.

The Schrödinger equation (5.177) *can be solved uniquely in the class* $\tilde{S}_A$. *The solution coincides with the wave function* $\Psi_A(X, p_A)$.

We will not give here the justification of the differential formulation, it almost coincides with that presented in Chapter 4.

Let us note that the first term in the sum (5.182), fixed for all functions of the class $\tilde{S}_A$, can be given with various accuracies. For example, it can be replaced by the function $\psi_A(x_A)\psi_c^{(\alpha)}(y_\alpha, p_\alpha)$ given by equation (5.93). Then the singularities, mentioned in Section 1 and corresponding to the multipole moment of the effective potentials $v_A^{(c)}$ which earlier were concentrated in the first term of (5.183), will have the amplitudes of the distorted spherical wave for the second, variable, term.

The wave functions can also be determined with the help of differential equations for the components. To find the latter, let us apply the operator $\hat{H}_\alpha - E$ to the homogeneous equation (5.181).

We get the expression

$$\left(-D_X + v_\alpha(x) + \sum_{\gamma\neq\alpha} v_\gamma^{(0)}(X) - E\right) \Phi_{\alpha B}(X,P) =$$

$$= -\hat{v}_\alpha(X) \sum_{\gamma\neq\alpha} \Phi_{\gamma B}(X). \tag{5.183}$$

By summing them, we see that the sum $\sum_\gamma \Phi_{\gamma B}$ satisfies Schrödinger equation (5.177).

Let us denote by $\tilde{B}_\alpha$ the set of the vector functions $\Phi = \{f1, f2, f3\}$, whose components we write in the form of the sum

$$f_\gamma(X) = \delta_{\alpha\gamma}\psi_A(x_\alpha)\psi_A^{(c)}(y_\alpha, p_\alpha) + \tilde{f}_\gamma(X), \tag{5.183'}$$

where the functions $\tilde{f}_\gamma$ are components of vector functions belonging to the class $B_{E,c}$.

As in the case of neutral particles, it can be shown that the system of equations (5.183) considered within the class $\tilde{B}_A$ has a unique solution, coinciding with the corresponding components of the wave functions $\Psi_A(X, p_A)$. These components are equal to the sum of components (5.180').

If we single out the simplest terms $\psi_A\psi_A^{(c)}$, we get inhomogeneous equations of the type (4.45) for the components of the wave functions.

Similarly, one can determine the wave functions $\Psi_0(X, P)$. However, in this case we have to pay more attention to the analysis of the slowly decreasing terms of the asymptotics. Along with the distorted spherical waves we must single out other terms, describing the rescattering of particles. The detailed expressions for these terms were presented in Sections 1 – 4. We must take much care when stating boundary problems for systems containing particles with different signs of the charge. Here, according to (5.103'), the wave functions turn into infinity when the relative momentum k_α of such particles is zero. Therefore, when stating boundary problems for $\Psi_0(X, P)$, we must impose the restriction $k_\alpha \neq 0$. We will not consider and describe the concrete formulae any further. All results necessary for the differential formulation, have already be given in Sections 1 – 4.

The list of boundary problems for the wave functions would be incomplete, if we did not mention the differential analogue of equation (3.116), where the Coulomb potential is wholly included in the unperturbed Hamiltonian:

$$\left(-\Delta_X + \sum_\gamma \frac{n_\gamma}{|x_\gamma|} + v_\alpha(x_\alpha) - E\right)\Phi_\alpha(X) =$$
$$-v_\alpha(x_\alpha)\sum_{\beta\neq\alpha}\Phi_\beta(X). \tag{5.184}$$

Using results given above, one can show that if all particles in the system have the same sign of the charges $n_\alpha > 0$ ($\alpha = 1,2,3$), then these equations have a unique solution in the class $\tilde{B}_E$; the wave functions $\Psi_B(X, p_B)$ coincide with the sums of components over all α. Similarly, one can determine the wave functions $\Psi_0(X, P)$.

If there are particles with different sign of the charges, the asymptotics of the components Φ_α becomes even more complex than the asymptotics of the wave function, and the system (5.184) does not have the virtues mentioned above.

To complete this section, we will shortly describe the boundary problems for systems of N charged particles.

5.6.2 N charged particles

Let us assume that the interaction potentials $v_{a_{N-1}}(x_{a_{N-1}})$ of the pairs can be represented as a sum of the Coulomb and short-range parts

$$v_{a_{N-1}}(x_{a_{N-1}}) = \frac{n_{a_{N-1}}}{|x_{a_{N-1}}|} + v^{(s)}_{a_{N-1}}(x_{a_{N-1}}).$$

Let $\Omega_{a_{N-1}}(a,\nu)$ be the region in configuration space where the condition $|x_{a_{N-1}}| \leq a(1+|y_{a_{N-1}}|)^\nu$, $0<\nu<1/2$ is satisfied. Let us introduce a smooth finite function $\chi_{a_{N-1}}(X)$ equal to one in $\Omega_{a_{N-1}}(a,\nu)$ and to zero outside of $\Omega_{a_{N-1}}(a,\nu')$, $\nu<\nu'<1/2$.

To obtain the compact equations, with the help of the function $c_{a_{N-1}}$ let us split, as in (5.104) the two-particle potentials into sums of short- and long-range parts

$$v_{a_{N-1}}(x_{a_{N-1}}) = \hat{v}_{a_{N-1}}(X) + v^{(0)}_{a_{N-1}}(X),$$

where

$$\hat{v}_{a_{N-1}}(X) = \frac{n_{a_{N-1}}}{|x_{a_{N-1}}|}\chi_{a_{N-1}}(X) + v^{(s)}_{a_{N-1}}(x_{a_{N-1}});$$

$$v^{(0)}_{a_{N-1}}(X) = \frac{n_{a_{N-1}}}{|x_{a_{N-1}}|}\left(1 - \chi_{a_{N-1}}(X)\right).$$

In accordance with this splitting, let us represent the energy operator for N charged particles in the form

$$H = H_a + \sum_{a_{N-1}} \hat{V}_{a_{N-1}},$$

where the asymptotic Hamiltonian H_a is given by

$$H_a = H_0 + \sum_{a_{N-1}} V^{(0)}_{a_{N-1}}.$$

Next, let $R(z)$ be the resolvent of the Hamiltonian H. One can prove that the following representation holds

$$R(z) = R_0(z) + \sum_{A_2,B_2} \hat{R}_{A_2,B_2}(z)$$

with $R_a(z) = (Ha-z)^{-1}$ and where the components $\hat{R}_{A_2,B_2}(z)$ are determined by relations (4.87) and (4.88) in whose $R_0(z)$ is replaced by $R_a(z)$ and $V_{a_{N-1}}$ by $\hat{V}_{a_{N-1}}$. As in the case of neutral particles, it can be shown, that the components of the resolvent $\hat{R}_{A_2,B_2}(z)$ satisfy modified compact equations in the differential formulation. These equations can be obtained from (4.97) by replacing the kinetic energy operator H_0 by H_a, and $V_{a_{N-1}}$ by $\hat{V}_{a_{N-1}}$. The modified integral equations for $\hat{R}_{A_2,B_2}(z)$ are obtained from equation (3.102) with $k = 2$ by the replacement of $R_0(z)$ by $R_a(z)$ and $M^{(a_2)}_{A_3C_3}$ by the modified components $\hat{M}^{(a_2)}_{A_3C_3}$. The latter are given by equations (3.97) and (3.98) in which $R_0(z)$ is replaced by $R_a(z)$ and $V_{a_{N-1}}$ by $\hat{V}_{a_{N-1}}$.

Let us briefly describe the boundary conditions in the case of a system of charged particles. The components of the wave functions, corresponding to two-cluster channels, can be written as the sum (4.105), where the plane and spherical waves are distorted by Coulomb factors. The latter can be found by means of the eikonal approximations. As a result, the plane wave in (4.105) acquires the factor

$$\exp\left\{ i \sum_{\substack{\omega,\omega' \\ \omega\neq\omega'}} \frac{\sqrt{2\mu_{\omega\omega'}}q_\omega q_{\omega'}}{2|p_{\omega\omega'}|} \log\left(|p_{\omega\omega'}|\,|y_{\omega\omega'}| - (p_{\omega\omega'}, y_{\omega\omega'})\right)\right\}, \tag{5.185}$$

where q_ω is the total charge of particles in the subsystem ω, in the partition a_j, $y_{\omega\omega'}$ is the reduced relative coordinate of the subsystems ω and ω', $p_{\omega\omega'}$ is the momentum conjugated to it, and $\mu_{\omega\omega'}$ is the reduced mass of the subsystems ω and ω'. The spherical waves are distorted by the factor

$$\exp\left\{-i\sum_{\substack{\omega,\omega'\\ \omega\neq\omega'}}\frac{\sqrt{2\mu_{\omega\omega'}}q_\omega q_{\omega'}}{2\sqrt{E+\varkappa_A^2}}\frac{|y_A|}{|y_{\omega\omega'}|}\log 2\sqrt{E+\varkappa_A^2}|y_A|\right\}.$$

We left aside the questions concerning with the problem of description of the asymptotics in the singular regions where the phases (5.185) turn into infinity. In general case, this problem is still not solved. Let us only note that if we confine ourselves to two-cluster collisions, such asymptotics can be easily found. They have the form of a product of the eigenfunctions of the clusters and the Coulomb wave functions (5.1), corresponding to the effective interaction between the clusters. Such asymptotics were described in detail in the case of three particles.

CHAPTER 6

Mathematical Foundation of the Scattering Problem

This chapter is devoted to justification of the hypothesis about completeness of the wave operators. We will present the reasoning for the system consisting of two and three particles. First, we describe the methods used for the investigation of the continuous spectrum of the energy operator in the stationary formalism. In this part, the results of this chapter overlap with those of Chapter 2, where we have been working in the framework of the non-stationary approach. Together with these issues, we will proceed here with the study of the Coulomb problem: we will calculate the kernels of the scattering operator for charged particles and describe their properties. At the end of this chapter we will verify that the kernels of the wave operators, obtained by using the stationary formalism, coincide with kernels of the non-stationary wave operators (1.22) and (1.25) depending on the character of the pair interactions.

6.1 The System of Two Particles

We will begin with the investigation of a system of two particles. On this example, we will illustrate the basic methods for solving the above stated problems and then we will generalize them for the system of three particles.

6.1.1 Neutral particles

We describe the stationary approach to the proof of the assumption U of Chapter 1 concerning the structure of the continuous spectrum of the energy operator of a two-particle system. Let us, first of all, formulate this assumption for the analysed case. Decompose the energy operator h into the sum

$$h = h_d + h_c,$$

where h_d is a finite-dimensional operator

$$h_d f(x) = \sum_i -\varkappa_i^2 (f, \psi_i)\psi_i(x)$$

representing the invariant part of h in the subspace of discrete spectrum. The precise statement concerning the continuous spectrum gives the following proposition:

Proposition 6.1 *The operator h, acting in the subspace $\mathfrak{H}_c$ orthogonal to the subspace of the discrete spectrum, is unitary equivalent to the kinetic energy operator h_0, i.e., there exists an isometric operator u such, that the relation*

$$u^* u = I, \quad u u^* = I - P_d, \quad h u = u h_0, \tag{6.1}$$

is satisfied. Here P_d is the projector on the subspace of discrete spectrum. In terms of the operator u, one can take the wave operators defined by equation (3.22).

We will prove this statement by explicitly representing the kernels of the operator $u^{(\pm)}$ (3.22). Let us first check that the operators $u^{(\pm)}$ are isometric. We write the kernel of the product $u^* u$ in the form of the sum

$$(u^* u)(k, k') = \delta(k - k') + \delta T(k, k'),$$

where the kernel $\delta T(k, k')$ is a polynomial of second degree with respect to the two-body T-matrices:

$$\delta T(k, k') =$$
$$\frac{t(k, k', k^2 + i0)}{k'^2 - k^2 + i0} - \frac{t(k, k', k'^2 + i0)}{k'^2 - k^2 - i0} -$$

$$-\int dq \frac{t(k,q,k^2-i0)t(q,k',k'^2+i0)}{(q^2-k^2+i0)(q^2-k'^2-i0)}.$$

Note, that this kernel is equal to zero by virtue of the Hilbert identity for T-matrix (3.20), where we must put $z_1 = k^2 - i0$, $z^2 = k'^2 + i0$. This implies immediately $u^* u = I$.

To prove the interlacing property, it is sufficient to verify the equality $r(z)u = ur_0(z)$ and to use relation (2.6) afterwards. Let us show that this equality follows from the Hilbert identity. Indeed, the kernel of the product $r(z)u$ can be expressed by means of the T-matrices with help of equation (2.22) and (3.22). If we transform the quadratic term in T-matrix to a sum of two linear terms

$$(k^2-z)^{-1}\int dq \frac{t(k,q,z)t(q,k',k'^2+i0)}{(q^2-z)(q^2-k'^2-i0)} =$$

$$= (k^2-z)^{-1}\left(\frac{t(k,k',z)}{k'^2-z} - \frac{t(k,k',k'^2+i0)}{k'^2-z}\right),$$

we get after the reduction of similar terms the kernel of the operator $u^{(+)} r_0(z)$

$$\frac{t(k,k',k'^2+i0)}{k^2-k'^2-i0}(k'^2-z)^{-1}.$$

The case of $u^{(-)}$ can be handled similarly.

For the proof of the completeness of the wave operators, we will use the relation between the step of the resolvent on the continuous spectrum and the derivative of the spectral function:

$$\frac{\mathrm{d}E_\lambda}{\mathrm{d}\lambda} = \frac{1}{2\pi i}\lim_{\epsilon\downarrow 0}(r(\lambda+i\epsilon) - r(\lambda-i\epsilon)). \tag{6.2}$$

The right-hand side can be transformed with the help of the Hilbert identity

$$r(\lambda+i\epsilon) - r(\lambda-i\epsilon) = 2i\epsilon r(\lambda+i\epsilon)r(\lambda-i\epsilon)$$

or, in terms of the T-matrix,

$$2i\epsilon r(\lambda+i\epsilon)r(\lambda-i\epsilon) = (I - r_0(\lambda+i\epsilon)t(\lambda+i\epsilon))\times$$

$$\times 2i\epsilon r_0(\lambda+i\epsilon)r_0(\lambda-i\epsilon)(I - t(\lambda-i\epsilon)r_0(\lambda-i\epsilon)). \tag{6.3}$$

Further, note that for $\epsilon \downarrow 0$ the product of the free resolvents gives the d-function:

$$2i\epsilon((k^2 - \lambda)^2 + \epsilon^2) \to 2i\pi\delta(k^2 - \lambda). \tag{6.4}$$

The first factor on the right-hand side of (6.3) turns into the wave operator $u^{(+)}$ and the last one into its conjugate $u^{(+)*}$. As a result, we get the equation

$$\frac{\mathrm{d}\, E_\lambda}{\mathrm{d}\,\lambda}(k, k') = \int dq\delta(q^2 - \lambda)u^{(+)}(k, q)u^{(+)*}(q, k'). \tag{6.5}$$

Integrating and taking into account that

$$\int_0^\infty dE_\lambda = I - P_d,$$

we get the completeness relation $uu^* = I - P_d$.

Proposition 6.1 implies, in particular that any element f of the Hilbert space can be uniquely represented in the form $f = \sum_i c_i\psi_i + u^{(\pm)}f^{(\pm)}$, and for any bounded function φ

$$\varphi(h)f = \sum_i \varphi(-\varkappa_i^2)c_i\psi_i + u^{(\pm)}\varphi(h_0)f^{(\pm)};$$

the coefficients c_i and the functions $f^{(\pm)}$ are determined by the formulae

$$c_i = (f, \psi_i), \quad f^{(\pm)} = u^{(\pm)*}f$$

and

$$\|f\|^2 = \sum_i c_i^2 + \|f^{(\pm)}\|. \tag{6.6}$$

The formula above is an exact and simple representation of the expansion of the operator h in eigenfunctions and equation (6.6) is the Parceval equation.

Let us note that any operator of the form $u = u^{(\pm)}m$ with m being a unitary operator commuting with h_0, also has the above properties.

To conclude, let us present another way of calculating the kernel of the scattering matrix $s = u^{(-)*}u^{(+)}$. In the Hilbert identity, we take $z_1 = k^2 + 2i\epsilon$, $z_2 = k'^2 + i\epsilon$, so that $z_2 - z12 = k'^2 - k^2 - i\epsilon$. Passing to the limit $\epsilon \downarrow 0$ and taking into account equation (6.5), we immediately arrive at expression (3.22′) that we have obtained in Chapter 3 using the non-stationary definition (1.39).

6.1.2 Charged particles

Let us denote by $u^{(\pm)}$ the integral operators, given by the kernels

$$u^{(\pm)}(x,k) = (2\pi)-3/2\psi^{(\pm)}(x,k), \tag{6.7}$$

where $\psi^{(\pm)}(x,k)$ are the wave functions determined by relation (5.20). We will show that for a system of charged particles, Proposition 6.1 holds where as the operators $u^{(\pm)}$ we can take the Coulomb wave operators (6.7). We will act in the configuration space. Of course, this approach is suitable for neutral particles, as well.

First of all, let us note that the interlacing property is a compact notation of the Schrödinger equation (5.24′) in the operational form. To verify isometry of the wave operators, let us consider Schrödinger equations for the functions $\psi(x,k)$ and $\psi^*(x,k')$. We multiply each of these equations by the factor $\psi^8(x,k)$ and $\psi(x,k')$, respectively, and integrate over a sphere V_R of a large radius R. Taking the difference of the results, and applying Green formula (4.8), we have

$$\int_{V_R} \psi(x,k)\psi^*(x,k')dx =$$

$$= (k^2 - k'^2 - i0)^{-1} \int_{\partial V_R} \psi(x,k) \frac{\overleftrightarrow{\partial}}{\partial R} \psi^*(x,k')\, dS_R. \tag{6.8}$$

Here the symbol $\frac{\overleftrightarrow{\partial}}{\partial n}$ means the difference

$$u \frac{\overleftrightarrow{\partial}}{\partial n} v = u\frac{\partial v}{\partial n} - v\frac{\partial u}{\partial n}.$$

To find the limit of the right-hand side of (6.8) for $R \to \infty$, we multiply this relation by smooth finite functions $f(k)$ and $\overline{g(k')}$ and integrate over k and k'. Next, we calculate the asymptotics of the resulting integrals with the help of the following proposition.

Consider the integral

$$I^{(\pm)}(x,E) = \int d\hat{k}\psi^{(\pm)}(x,k)f(k), \quad E = k^2),$$

where $f(k)$ is a smooth function.

Proposition 6.2 *For $|x| \to \infty$, the asymptotics of the integral $I^{(\pm)}(x,E)$ has the form of a sum of incoming and outgoing distorted spherical waves:*

$$I^{(\pm)}(x,E) \sim -\frac{2\pi i}{\sqrt{E}} \frac{\exp\{-i\sqrt{E}|x| - iw_0(x)\}}{|x|} f(-\sqrt{E}\hat{x}) +$$

$$+\frac{2\pi i}{\sqrt{E}} F^{(\pm)}(\hat{x},E) \frac{\exp\{\pm i\sqrt{E}|x| \pm iw_0(x)\}}{|x|}, \tag{6.9}$$

where the amplitude $F^{(\pm)}$ is a smooth function.

We postpone the proof of this formula and will present it at the end of Section 2 with the help of the example of more general integrals.

According to equation (6.8), the problem of investigation of the asymptotics of the integrals obtained above reduces to the computation of the sum of expressions of the form

$$\int dudv \frac{\exp\{\pm iuR \mp ivR \pm iw_0(R) \mp iw_0(R)}{u - v - i0} \varphi(u,v),$$

$$u = k^2, \quad v = k'^2,$$

where φ are smooth, slowly oscillating functions. After changing the variable $u - v = t$, these integrals can be calculated with help of relations (2.35). As a result, we find that all terms except the integral with the exponential $\exp\{i(k'^2 - k^2)R\}$ corresponding to the first term in equation (6.9) vanish in the limit $R \to \infty$. The integral turns into the scalar product (f,g). This implies

$$\lim_{R\to\infty} \in dkdk' f(k)\overline{g(k')} \int_{V_R} \psi(x,k)\psi^*(x,k)dx = (f,g),$$

which is equivalent to the relation $(u^{(+)}f, u^{(+)}g) = (f,g)$. Similarly one can analyse the case of the operator $u^{(-)}$.

Now we are left with the proof of completeness of the wave functions. To do this, we use again the representation (6.2) and transform the difference of the Green functions by means of the relation

$$\lim_{\epsilon\downarrow 0}(r(x,x',\lambda+i\epsilon) - r(x,x',\lambda-i\epsilon)) =$$

$$= \lim_{R\to\infty} \int_{\partial V_R} r(x,y,\lambda - i0) \frac{\overleftrightarrow{\partial}}{\partial R} r(y,x',\lambda+i0)\, dS_R,$$

which can be obtained analogously to equation (6.8) for the wave functions. Computing the surface integral and making use of the asymptotic formulae (5.15), we get

$$\frac{d E_\lambda}{d \lambda}(x, k') = \frac{1}{(2\pi)^3} \int dk \delta(k^2 - \lambda) \psi(x, k) \psi^*(x', k),$$

which is analogous to (6.5). This immediately implies the completeness relation $uu^* = I - P_d$.

6.1.3 The scattering operator

The formula (6.9) used for the calculation of the operator u^*u proves to be helpful also for the determination of the kernel of the scattering operator. Let us recall that a similar expression (4.21) was encountered already in Chapter 4, where with the help of it, the alternative definition of the scattering operator was given. In the case of charged particles, we can act in exactly the same manner. Let us define the scattering operator as an integral operator defining the amplitude $F^{(\pm)}$ in equation (6.9) by the formula

$$F^{(\pm)}(\hat{k}, E) = \int dk' s(k, k') f(k'). \tag{6.10}$$

To describe this kernel, let us introduce a new distribution. Take the integral

$$I(z) = \int_0^1 t^{-z} f(t)\, dt. \tag{6.11}$$

Denote by $t^{-1-i\eta}$ the distribution defined with the help of the analytic continuation of this integral from the half plane $\operatorname{Re} z < 1$ into the point $z = 1 + i\eta$. Note, that for $z = 1 + i\eta$ the integral (6.11) admits the following regularization:

$$\int_0^1 t^{-1-i\eta} f(t)\, dt = \frac{i}{\eta} f(0) + \int_0^1 \frac{f(t) - f(0)}{t^{1+i\eta}} dt. \tag{6.12}$$

By $|\hat{p} - \hat{p}'|^{-2-2i\eta}$ we denote a function on the two-dimensional sphere $S^{(2)}$, which in the spherical coordinates $\cos\theta = (\hat{p}, \hat{p}')$, φ acts as the distribution $t^{-1-i\eta}$, $t = 1 - \cos\theta$

$$\int d\hat{p}' \frac{f(\hat{p}')}{|\hat{p} - \hat{p}'|^{2+2i\eta}} =$$

$$\int_{\pi/2}^{\pi} d\theta \int_0^{2\pi} d\varphi \frac{f(\theta,\varphi)}{(1-\cos\theta)^{1+i\eta}} + \int_0^{2\pi} d\varphi \int_0^1 dt \frac{f(t,\varphi)}{t^{1+i\eta}}. \tag{6.13}$$

In this notations the following statement holds.

Proposition 6.3 *The kernel $s(k,k')$ present in the representation (6.10) has the form*

$$s(k,k') = \frac{1}{\pi}\left(-\frac{2^{1+2i\eta}\eta e^{2i\arg\Gamma(1+i\eta)}}{|k|\,|\hat{k}-\hat{k}'|^{2+2i\eta}} + f_{cs}(\hat{k},k')\right)\delta(k^2-k'^2), \tag{6.14}$$

where f_{cs} is the short-range part of the scattering amplitude (5.22).

Thus, the kernel of the scattering amplitude is proportional to the scattering amplitude:

$$s(k,k') = i\pi^{-1}\delta(k^2 - k'^2 f(\hat{k},k'). \tag{6.15}$$

Here, the Coulomb part of the amplitude f_c should be understood as the generalized function defined above.

Let us note that the formula (6.9) can be rewritten in the notation similar to (4.5), i.e.,

$$\psi(x,k) \sim \frac{2\pi i}{\sqrt{E}}\left(\delta(-\hat{x},k)\frac{\exp\{-i\sqrt{E}|x| - iw_0(x)\}}{|x|} + \right.$$

$$\left. + s(|k|\hat{x},k)\frac{\exp\{i\sqrt{E}|x| + iw_0(x)\}}{|x|}\right). \tag{6.16}$$

In this form, the role of the scattering operator as the amplitude of the outgoing spherical wave can clearly be seen.

We will verify the representation (6.14) together with formula (6.9) at the end of Section 2.

Formula (6.14) for the kernel of the scattering operator can be obtained by direct evaluation of the expression $u^{(-)*}u^{(+)}$, as well. This can be done with help of the Green formula along the same scheme as in the case of the product u^*u. It is necessary to transform this expression to a surface integral over the sphere of radius R, and then calculate its limit for $R \to \infty$ using equation (6.9). We will not realize the alternative approach here. All necessary technical means for it have already be given above.

Thus, we see that the kernel of the scattering operator, unlike the case of neutral particles, does not contain the term one, and is proportional to the scattering amplitude. In this way, we proved a result heuristically established in Chapter 3. However, we must stress that according to the role of the δ-function like singularity, regularized formula (6.12), is accomplished by the singularity of the purely Coulomb scattering amplitude.

6.2 Continuous Spectrum of the Hamiltonian of the Three-Particle System

In this section, we will justify the hypothesis about the completeness of the wave operators for the system of three particles. We will consider here both neutral and charged particles.

Let us recall some notions concerning the three-particle problem, introduced in Chapter 2. The channels space $\hat{\mathfrak{H}}$ in this case is equal to the orthogonal sum

$$\hat{\mathfrak{H}} = \mathfrak{H}_0 \oplus \sum_{A \neq 0} \oplus \mathfrak{H}_A,$$

where the three-particle channel coincides with the Hilbert space $L_2(R^6)$ and the two-particle channels $\mathfrak{H}_A$, $A \neq 0$, with the Hilbert space $L_2(R^3)$.

The Hamiltonian of the channels $\hat{H}$ is an operator reducible in η

$$\hat{H} = \tilde{H}_0 \oplus \sum_{A \neq 0} \oplus \tilde{H}_A,$$

where the action of the energy operators in the channels is given by

$$\tilde{H}_0 f_0(P) = P^2 f_0(P),$$

$$\tilde{H}_A f_A(p_\alpha) = E_A(p_\alpha) f_A(p_\alpha), \quad E_A(p_\alpha) = p_\alpha^2 - \varkappa_A^2,$$

$$A = \{\alpha, t\}, \quad \alpha = 1, 2, 3; \quad i = 1, 2, \ldots, N.$$

Let P_d be the projector in $\mathfrak{H}$ on the eigensubspace and let $\mathfrak{H}_c$ be the orthogonal complement to $\mathfrak{H}_d$ in $\mathfrak{H}$, $\mathfrak{H} = \mathfrak{H}_c \oplus \mathfrak{H}_d$. Let us denote by H_c the invariant part of H in $\mathfrak{H}_c$.

In this section we will prove this statement:

Proposition 6.4 *The operators H_c and $\hat{H}$ are unitary equivalent, so that there exists an isometric operator U from $\hat{\mathfrak{H}}$ to $\mathfrak{H}$ such that the relations*

$$U^* U = \hat{I}, \quad UU^* = I - P_d, \quad HU = U\hat{H}. \tag{6.17}$$

hold.

6.2.1 System of three neutral particles

Let us begin the proof of this statement with the simpler case of neutral particles. We will show that as the operator U in Proposition 6.4, one can take the operator $\hat{U}^{(\pm)}$, whose action is described in terms of the wave operators $U_A^{(\pm)}$ by the equations

$$\hat{U}\hat{f} = \sum_A U_A f_A, \quad \hat{f} = \{f_A\}, \tag{6.18}$$

$$A = 0, \{\alpha, i\}, \quad \alpha = 1,2,3; \quad i = 1,2,\ldots,N_\alpha.$$

where the summation runs over all channels A, and f_A are elements of $\mathfrak{H}_A$.

To prove the isometry and the mixing property, in analogy with the case of two particles, we will make use of the Hilbert identity. Along with the Hilbert identity for the T-matrix $T(z_1) - T(z_2) = (z_2 - z_1)T(z_1)R_0(z_1)R_0(z_2)T(z_2)$, we will need also the Hilbert identity for the components $M_{\alpha\beta}(z)$

$$M_{\alpha\beta}(z_1) - M_{\alpha\beta}(z_2) =$$

$$(z_2 - z_1)\sum_{\gamma,\gamma'} M_{\alpha\gamma}(z_1)R_0(z_1)R_0(z_2)M_{\gamma'\beta}(z_2). \tag{6.19}$$

Calculation of the expression $U_A^* U_B$ for $A = B = 0$ can be performed with the help of the Hilbert identity for T-matrix. The corresponding formulae differ from the two-particle ones only by dimensions of the integration variables. Therefore, we omit the reasoning coinciding word for word with the one given above. We must stress that existence of many three-particle singularities does not reflect itself in the final result $U_0^* U_0 = I$, $HU_0 = U_0 H_0$. This can be explained by the fact that the singularities are defined as distributions continuously depending on the parameter $z = E + i\epsilon$ for all z up to the real axis.

Next, let us verify the analogous identities for the operators $U_A^{(\pm)}$, $A \neq 0$:

$$U_A^* U_B = \delta_{AB} I_A.$$

To this end, let us note first of all that the operators $T_{\gamma A}(z)$ determined by equations (3.65), satisfy the identities

$$\sum_{\alpha,\beta}\sum_{\gamma,\gamma'} T_{\gamma A}^*(z_1)R_0(z_1)R_0(z_2)T_{\gamma' B}(z_2) =$$

$$= (z_2 - z_1)^{-1} \left(\sum_{\beta} T^*_{\beta A}(z_1) R_0(z_2) L_B(\tilde{H}_B - z_2) - \right.$$

$$\left. - \sum_{\alpha} (\tilde{H}_A - z_2) L^*_A R_0(z_1) T_{\alpha B}(z_2) \right) . \tag{6.20}$$

Indeed, we get these expressions immediately if one multiplies Hilbert identity (6.19) for the components $M_{\alpha\beta}$ from the left by the operator $L^*_A R_0(z_1)$, from the right by the operator $R_0(z_2)L_B$, and use equations (3.65).

Putting $z_1 = p_\alpha^2 - \varkappa_A^2 - i\epsilon_1$ and $z_2 = {p'_\beta}^2 - \varkappa_B^2 + i\epsilon_2$, and taking the limits $\epsilon_1 \downarrow 0$, $\epsilon 2 \downarrow 0$, we obtain on the left side the kernel of the operator $U^*_A U_B$, acting from the channel $\mathfrak{H}_B$ into the channel $\mathfrak{H}_A$ and now we have to consider the right-hand side. If $A = B$ then there is a term having, for finite ϵ_1, ϵ_2, the form

$$\frac{\delta(p_\alpha - p'_\alpha)}{i(\epsilon_1 + \epsilon_2)} \left(i\epsilon_2 \int \frac{\varphi^*_A(k_\alpha)\psi_A(k_\alpha)}{k_\alpha^2 + \varkappa_A^2 - i\epsilon_2} dk_\alpha + \right.$$

$$\left. + i\epsilon_1 \int \frac{\psi^*_A(k_\alpha)\varphi_A(k_\alpha)}{k_\alpha^2 + \varkappa_A^2 + i\epsilon_1} dk_\alpha \right) .$$

Here the integral

$$I(\epsilon) = \int \frac{\varphi^*_A(k)\psi_A(k)}{k^2 + \varkappa_A^2 + i\epsilon} dk$$

for $\epsilon \downarrow 0$ turns into the normalization integral for the wave function $\psi_A(k)$, so that $I(0) = 1$. Therefore the whole expression gives in the limit the unit operator I_A with the kernel $\delta(p_\alpha - p'_\alpha)$. The remaining terms on the right-hand side have the form of a product of the singular denominators $(p_\beta^2 - \varkappa_B^2 - {p'_\alpha}^2 + \varkappa_A^2 - i\epsilon)^{-1}$ by the factors $i\epsilon_{1,2}$ tending to zero. These terms vanish in the limit $\epsilon_{1,2} \downarrow 0$. This implies the equality $U^*_A U_B = \delta_{AB} I_A$.

It remains to verify the relations

$$U^*_0 U_A = 0, \quad U^*_A U_0 = 0 \tag{6.21}$$

which express the mutual orthogonality of the domains of values of the operators U_0 and U_A, $A \neq 0$. To do this, we use again Hilbert identity and multiply it by the operator $R_0(z_2)L_B$ from the right:

$$\sum_{\gamma',\beta} T(z_1) R_0(z_1) R_0(z_2) T_{\gamma' B} =$$

$$= (z_1 - z_2)^{-1} (T(z_1) R_0(z_2) L_B(\tilde{H}_B - z_2) -$$

$$-\sum_{\alpha}(\tilde{H}_A - z_1)L_A^* R_0(z_1)T_{\alpha B}(z_2)). \tag{6.22}$$

We substitute $z_1 = P^2 - i\epsilon$, $z_2 = p_\beta'^2 - \varkappa_B^2 + i\epsilon$, so that $z_2 - z_1 = P^2 - p_\beta'^2 - \varkappa_B^2 - 2i\epsilon$ and take the limit $\epsilon \downarrow 0$. On the left-hand side we obtain the expression $U_0^* U_B$ and on the right-hand side, the sum of the kernels with singularities $\left(P^2 - p_\beta'^2 - \varkappa_B^2 - 2i\epsilon\right)^{-1}$ multiplied by the vanishing parameter ϵ. The lemma on singular integrals implies that the distribution on the right-hand side tends toward zero for $\epsilon \downarrow 0$ and we obtain equations (6.21).

Finally let us prove the interlacing property

$$R(z)U_A = U_A \tilde{R}_A(z),$$

where by $\tilde{R}_A(z)$ we denote the resolvent of the Hamiltonian of the channel $\tilde{H}_A$. In the operator equation (6.22) written in terms of the kernels, we substitute $z_1 = z$, $z_2 = -\varkappa_B^2 + p_\beta'^2 + i\epsilon$, so that $z_2 - z_1 = p_\beta'^2 - \varkappa_B^2 - z + i\epsilon$ and represent the denominator $(z_2 - z_1)^{-1}$ in the form

$$-\left(P^2 - p_\beta'^2 - \varkappa_B^2 - i\epsilon\right)^{-1} +$$

$$+(P^2 - z)\left(P^2 - p_\beta'^2 - \varkappa_B^2 - i\epsilon\right)^{-1}\left(p_\beta'^2 - \varkappa_B^2 - z + i\epsilon\right)^{-1}.$$

Multiplying the result by $R_0(z)$ and taking the limit $\epsilon \downarrow 0$, we get on the left-hand side

$$R(z)U_B - U_B \tilde{R}_B(z).$$

On the right side, for $\epsilon \neq 0$ we have a non-singular expression multiplied by ϵ. Thus in the limit $\epsilon \downarrow 0$ the right-hand side vanishes.

The next task is to show the completeness of the operators U_0 and U_A. To this end, we use the relation between the spectral function and the resolvent

$$\frac{\mathrm{d}\,E_\lambda}{\mathrm{d}\,\lambda} = \frac{1}{2\pi i}(R(\lambda + i\epsilon) - R(\lambda - i\epsilon)). \tag{6.2'}$$

By means of Hilbert identity, we write the difference of the resolvents in the form of a product

$$R(\lambda + i\epsilon) - R(\lambda - i\epsilon) =$$

$$= \left(I - R_0(\lambda + i\epsilon)\sum_{\alpha\beta} M_{\alpha\beta}(\lambda + i\epsilon)\right) 2i\epsilon R_0(\lambda + i\epsilon)\times$$

$$\times R_0(\lambda - i\epsilon)\left(I - \sum_{\alpha'\beta'} M_{\alpha'\beta'}(\lambda - i\epsilon)R_0(\lambda - i\epsilon)\right).$$

The expression on the right-hand side is a sum of singular kernels of two kinds. To the first belong the kernels containing products of complex conjugated denominators $\frac{2i\epsilon}{(P^2-\lambda^2)^2+\epsilon^2}$ or $\frac{2i\epsilon}{(p_\alpha^2 - \varkappa_A^2 - \lambda^2)^2 + \epsilon^2}$. In the limit $\epsilon \downarrow 0$ they give the d-function $\delta(P^2 - \lambda)$ or $\delta(E_A(p_A) - \lambda)$. The coefficients of these d-functions are expressed in terms of the products of the operators $U_0 U_0^*$ and $U_A U_A^*$.

To the second type belong products of all other singular denominators, defined for $\epsilon \downarrow 0$ in the sense of singular integrals of the type (3.7). Because these integrals are multiplied by factors tending to zero as $\epsilon \downarrow 0$, they vanished in the limit. As the result we get the equation

$$\frac{\mathrm{d}\,E_\lambda(P.P')}{\mathrm{d}\,\lambda} = \int dP'' U_0(P, P'')\overline{U_0(P', P'')}\delta(P''^2 - \lambda) +$$

$$+\sum_{A\neq 0}\int dp_A'' U_A(P, p_A'')\overline{U_0(P', p_A'')}\delta(E_A(p_A) - \lambda),$$

valid for λ not coinciding with singular points E_i. From this, as in the case of the two-particle system, the equation

$$\sum_A U_A U_A^* = I - P_d$$

follows.

The results obtained above yield the following proposition.

Proposition 6.5 *There exist operators U_0, U_A, $A \neq 0$, $A = \{\alpha, i\}$, $\alpha = 1, 2, 3, \ldots, N_\alpha$ acting from $\mathfrak{H}_0$ and $\mathfrak{H}_A$ into $\mathfrak{H}$, respectively and having the following properties:*

1. *Any function from the Hilbert space $\mathfrak{H}$ can be uniquely represented in the form*

$$f = f_d + \sum_A U_A f_A,$$

where $f_d \in \mathfrak{H}_d$ and $f_A \in \mathfrak{H}_A$, $A = 0, \{\alpha, i\}$.

2. For any $\varphi(t)$ bounded on the whole real axis, the relation

$$\varphi(H)f = \varphi(P_d H)f_d + \sum_A U_A \varphi(\tilde{H}_A) f_A.$$

holds

3. The function f_A is given by the equation

$$f_A = U_A^* f$$

4. The relations

$$\|f\|^2 = \|f_d\|^2 + \sum_A \|f_A\|^2.$$

hold.

Proposition 6.4 follows from this, if the operator $\hat{U}$ is defined by equation (6.18).

Let us finally consider the scattering operator $\hat{S} = \hat{U}^{(-)*}\hat{U}^{(+)}$. Proposition 6.2 implies that the operator $\hat{S}$ is unitary and commutes with any bounded function of the channel Hamiltonian $\hat{H}$. For the kernels of these operators, one can obtain explicit expressions in terms of the kernels $M_{\alpha\beta}$ by acting along the scheme explained on the example of the two-particle system. The corresponding formulae were given in Section 4 of Chapter 3. Finally, let us write the following relations for the matrix-kernels of the scattering operator, implied by the self-adjointness of the resolvent $R(z) = R^*(\bar{z})$ and similar identity for the kernels of the operators $M_{\alpha\beta}(z)$:

$$M_{\alpha\beta}(P, P', z) = \overline{M_{\beta\alpha}(P', P, \bar{z})}.$$

The kernels $G_{\alpha B}(P, p'_\beta, z)$ and $J_{B\alpha}(p'_\beta, P, z)$ are related by the equation

$$G_{\alpha B}(P, p'_\beta, z) = \overline{J_{B\alpha}(p'_\beta, P, \bar{z})}.$$

A similar relation is valid for the kernels $H_{AB}(p_\alpha, p'_\beta, z)$:

$$H_{AB}(p_\alpha, p'_\beta, z) = \overline{H_{BA}(p'_\beta, p_\alpha, \bar{z})}.$$

Thus, we have fully characterized the continuous spectrum of the operator H and justified the hypothesis about its completeness.

Let us stress that the method of proving Propositions 6.1 and 6.2 and formulae (3.71) presented in this section are based only on the stationary formulation of the scattering theory, and is therefore independent of the heuristic constructions carried out during derivation of the formulae for the N-particle system.

6.2.2 Charged particles

Let us prove relations (6.17) in the case of the three-particle system. We will again use the alternative reasoning in configuration space. We will show that as the operators U_A, determining the transformation $\hat{U}$, we can take the operators, defined by the kernels

$$U_0^{(\pm)}(X,P) = (2\pi)^{-3}\Psi_0^{(\pm)}(X,P),$$

$$U_A^{(\pm)}(X,P) = (2\pi)^{-3}\Psi_A^{(\pm)}(X,P), \quad A \neq 0.$$

First of all, let us note that as in the case of two charged particles, the interlacing properties $HU_A = U_A\tilde{H}_A$ reduce to the Schrödinger equations for the wave functions $\Psi_A(X,p_A)$ and, therefore, do not need to be further justified.

The central point in the proof of the isometry of the operators is taken by the following statement about the asymptotics of rapidly oscillating integrals of the type

$$I_A^{(\pm)}(X,E_A) = \int d\hat{p}_A \Psi_A^{(\pm)}(X,E_A) f_A(p_A),$$

where $f_A(p_A)$ are smooth functions.

For $|X| \to \infty$ the following asymptotic formulae hold

$$I_0^{(\pm)}(X,E) \sim \frac{e^{-\pi/4}(2\pi)^{5/2}}{E^{5/4}} Q_0^{(-)}(X,E) f_0(-\sqrt{E}\hat{X}) + \\ + \tilde{I}_0^{(\pm)}(X,E) \tag{6.23}$$

$$I_A^{(\pm)}(X,E_A) \sim \frac{2\pi i}{\sqrt{E_A}} \psi_A(x_A) Q_A^{(-)}(y_A,E_A) f_A(-|p_A|\hat{y}_A) + \\ + \tilde{I}_A^{(\pm)}(X,E_A), \quad A \neq 0 \tag{6.24}$$

where the functions $\tilde{I}_0^{(\pm)}$ and $\tilde{I}_A^{(\pm)}$ belong to the class $S_{E,c}$ for $E = P^2$ and $E = E_A(p_A)$, respectively.

The terms $\tilde{I}_0^{(\pm)}$ and $\tilde{I}_A^{(\pm)}$ will be described in more detail at the end of this section and there the proofs of these formulae will there be given, as well.

With the help of these formulae, we can demonstrate the isometry of the operators $U_0^{(\pm)}$ and $U_A^{(\pm)}$ and that their images are orthogonal, so that we have

$$U^*AU^B = \delta_{AB}I_A, \quad A, B = 0, \{\alpha, i\}. \tag{6.25}$$

Indeed, let us consider the bilinear form $I_{AB} = (U_A f_A, U_B g_B)$, where f_A, g_B are smooth finite functions. As in the two-particle system, this expression can be written as a limit

$$I_{AB} = \lim_{R\to\infty} \int dp_A dp'_B f(p_A)\overline{g(p'_B)} \int_{V_R} U_A(X, p_A)\times$$
$$\times\overline{U_B(X, p'_B)}dX$$

where the integration is over the ball $|X| \leq R$. Let us transform the inner integral with the help of Green formula to an integral over the sphere

$$(p_A^2 - {p'_B}^2 - i0)^{-1} \int_{\partial V_R} U_A(X, p_A) \frac{\overleftrightarrow{\partial}}{\partial R} \overline{U_B(X, p'_B)}dS_R.$$

We then integrate over the angular variables $\hat{p}_A$, $\hat{p}'_B$ and use equations (6.23) and (6.24). Integrating over variable $|p_A|$ and taking asymptotic formulae (2.35) into account, we obtain the relations $I_{AB} = \delta_{AB}(f_A, g_B)$, which imply (6.25).

Now, let us show that the wave functions Ψ_0 and Ψ_A generate a complete system of functions in the space corresponding to the continuous spectrum of H. To verify the completeness, it is sufficient to prove the following statement.

Let λ not be a singular point of equation (5.154). Then the spectral function E_λ is differentiable with respect to λ and the kernel of its derivative can be represented in the form

$$\frac{d E_\lambda(X.X')}{d\lambda} = \frac{1}{(2\pi)^6}\int \Psi_0(X, P)\Psi_0^*(X, P)\delta(P^2 - \lambda)+$$
$$+\frac{1}{(2\pi)^3}\sum_{A\neq 0}\int \Psi_A(X, p_A)\Psi_A^*(X, p_A)\delta(E_A(p_A) - \lambda)dp_A. \tag{6.26}$$

The proof of this expression can be given along the scheme used in the case of two charged particles.

Let us use the expression of the spectral function in terms of the resolvent (6.2′). We transform the difference of the kernels $R(X, X', \lambda \pm i0)$ with the help of Green formula to obtain:

$$\frac{\mathrm{d}\, E_\lambda(X.X')}{\mathrm{d}\,\lambda} =$$

$$= \lim_{R\to\infty} \int_{\partial V_R} R(X, Y, \lambda + i0) \,\frac{\overleftrightarrow{\partial}}{\partial R}\, R(Y, X', \lambda - i0) dS_R. \qquad (6.27)$$

Then, let us consider the bilinear form $\left(\frac{\mathrm{d}\,E_\lambda}{\mathrm{d}\,\lambda} f, g\right)$, where f and g are finite smooth functions. On the basis of formula (5.176) we can write the asymptotics of the integral on the right-hand side of equation (6.27) as a sum of terms generated by the products of all possible distorted spherical waves $Q_0^{(\pm)}$ and $Q_A^{(\mp)}$. With the help of relations (2.35), it can be show further that all terms generated by products of spherical waves with different labels $A \neq B$ vanish in the limit $R \to \infty$, a finite result giving only terms containing products of spherical waves of the same type . The amplitudes of these waves can be expressed through the wave functions by means of equations (5.178), (5.179). Hence one immediately obtains the representation (6.26).

As in the case of a system of neutral particles, the described results can be united into the Proposition 6.4.

6.2.3 The scattering operator

Let us further calculate the kernels of the scattering operator for a system of three charged particles. We will express these kernels through the scattering amplitudes which will be regarded as distributions. Therefore, it is convenient first to define these functions and to list the fundamental properties of the scattering amplitudes.

First of all, let us consider the amplitude corresponding to the functions $\Psi_A(X, p_A)$, $A \neq 0$. The amplitudes $F_{BA}(\hat{y}_B, p_A)$ of the distorted spherical waves Q_B, describing inelastic processes $B \neq A$, are smooth, bounded functions. If all particles have charges of the same sign, the amplitudes $F_{0A}(\hat{X}, p_A)$

of the distorted spherical waves in R^6 are smooth, as well. If in the system there are particles with charges of opposite signs, these amplitudes have singularities of the type $|x_\alpha|^{-1/2}|X|^{1/2}$ square integrable over the unit sphere $|X| = 1$. The amplitude $F_{AA}(\hat{y}_\alpha, p_\alpha)$, corresponding to absolutely elastic processes, has a non-integrable singularity for $(y_\alpha, p_\alpha) = 1$. This singularity can be explicitly written as

$$F_{AA}(\hat{p}', p) = \tilde{F}_{AA}(\hat{p}', p) + \frac{\tilde{f}_A(\hat{p}', p)}{|\hat{p} - \hat{p}'|^{2+2i\eta_A}}, \tag{6.28}$$

$$\eta_A = \frac{n_{\alpha\alpha}}{2|p_A|}.$$

where $\tilde{F}_{AA}$ and $\tilde{f}_A$ are bounded smooth functions. The second term in this formula coincides with the two-particle scattering amplitude for the operator $\tilde{h}_A^c$, whose leading singularity is purely Coulomb (5.5).

Bellow we will associate with the amplitude F_{AA} a distribution on the sphere $S^{(2)}$, which will be denoted by the same symbol F_{AA}. In local coordinates, the singular part (6.28) acts like a distribution $t^{-1-i\eta_A}$, defined by means of the analytic continuation of the integral (6.11), and coincides with the two-particle Coulomb amplitude (5.5).

Let us further consider the wave functions $\Psi_0(X, P)$.

The amplitudes of the three-dimensional spherical waves Q_B, describing the asymptotics of these wave functions, are smooth bounded functions. If the particles have charges of opposite signs, these amplitudes have singularities of the type $|ka|^{-1/2}$ square integrable over the unit sphere $S^{(2)}$. The amplitude of the six-dimensional spherical waves $F_{00}(\hat{X}, P)$ has an non-integrable singularity, reflected by the representation

$$F_{00} = F_c + \sum_\alpha F_\alpha + \sum_{\alpha \neq \beta} F_{\alpha\beta} + \tilde{F}_0. \tag{6.29}$$

Here $\tilde{F}_0$ is a smooth function bounded for $|k_\alpha| > 0$. If in the particles of the pair α attract each other, then for $|ka| = 0$ this function has a square integrable singularity on the unit sphere $|P| = 1$.

With the singular terms (6.29), we will associate distributions whose action will be described below. These distributions and their kernels will be denoted by the same symbols as the amplitudes.

The singularity of the term F_c has the form

$$F_c(\hat{P}, P') = \frac{A_c(\hat{P}, P')}{|\hat{P} - \hat{P}'|^{5+2i\eta_0}}, \quad \eta_0 = \sum_\alpha \frac{\eta_\alpha}{2|k_\alpha|}|P|, \tag{6.30}$$

where the bounded function A_c is given by formulae (5.121), (5.122). The distribution $|\hat{P} - \hat{P}'|^{-5-2i\eta_0}$ in spherical coordinates acts as $t^{-1-i\eta_0}$ (6.12).

By F_α we denote the distribution given by

$$F_\alpha(\hat{P}, P') = f_\alpha^{(s)}(\hat{k}_\alpha, k'_\alpha)\frac{A_\alpha(\hat{M}_\alpha, P')}{|p_\alpha - p'_\alpha|^{3+ia_\alpha}}. \tag{6.31}$$

Here A_α is a bounded function, and the action of the distribution $|p_\alpha - p'_\alpha|^{-3-ia_\alpha}$ in the spherical coordinates $u = |p - p'|$, $\widehat{p - p'}$, is defined by equation (6.12). The distribution $f_\alpha^{(s)}$ is defined by the equation

$$f_\alpha^{(s)}(\hat{k}_\alpha, k'_\alpha) = f_\alpha(\hat{k}_\alpha, k'_\alpha) -$$

$$-i \exp\left\{2i\arg\Gamma\left(1 + i\frac{n_\alpha}{2|k_\alpha|}\right)\right\}\delta(k_\alpha, k'_\alpha) \tag{6.32}$$

Here f_α is the singular amplitude for the pair of particles α, whose singularity is described by formula (5.5). The amplitude f_α acts as the distribution $t^{-1-i\nu}$, defined above. By $\delta(\hat{k}_\alpha, \hat{k}'_\alpha)$ we denote the d-function on the sphere $S^{(2)}$. Let us note that the distributions $f_\alpha^{(s)}$ and $|p_\alpha - p'_\alpha|^{-3-ia_\alpha}$ depend on different kinematic variables. Aside from the singular directions, F_α is an ordinary function, coinciding with the amplitude (5.134).

The distribution $F_{\alpha\beta}$ is given by the expression

$$F_{\alpha\beta}(\hat{P}, P') = f_\alpha^{(s)}(\hat{k}_\alpha, k_\alpha(p_\alpha, p'_\beta))f_\beta^{(s)}(k_\beta(p_\alpha, p'_\beta), k'_\beta)\times$$

$$\times\tilde{A}_{\alpha\beta}(\hat{P}, P')\left(k_\beta^2(p_\alpha, p'_\beta) - k'_\beta - i0\right)^{-1-ia_{\alpha\beta}}, \tag{6.33}$$

where the distributions $f_\alpha^{(s)}$, $f_\beta^{(s)}$ were defined above. The smooth function $\tilde{A}_{\alpha\beta}$ is described by equation (5.142), and the distribution $(k^2 - k'^2 - i0)^{-1-i\eta}$ is defined by means of the analytic continuation of the integral

$$\int_{-1}^{1} dt\varphi(t)(t - i\epsilon)^{-1-i\eta}$$

for $\epsilon \downarrow 0$.

Let us consider the integral operators given by the following kernels

$$S_{AB}(p_A, p'_B) = \frac{i}{\pi}\delta(E_A(p_A) - E_B(p'_B))F_{AB}(\hat{p}_A, p'_B),$$

$$S_{0B}(P, p'_B) = -i(2\pi)^{5/2}C_0^{-1}\delta(E - E_B(p'_B))F_{0B}(\hat{P}, p'_B), \qquad (6.34)$$

$$S_{A0}(p_A, P') = (2\pi^{-5/2}i\delta(E_A(p_A) - E_B(p'_B))F_{A0}(\hat{p}_A, P'), \quad P'^2 = E,$$

$$S_{00}(P, P') = 2\pi i C_0^{-1}\delta(P^2 - P'^2)F_{00}(\hat{P}, P'),$$

As above, we used the notation $C_0 = -e^{i\pi/4}(2\pi)^{5/2}\pi E^{3/4}$. By definition, the singular kernels F_{AA} and F_{00} are regarded as distributions defined above.

Let us consider the matrix kernels of the scattering operator $\hat{S}$ given by the equations

$$S_{AB} = U_A^{(-)*}U_B^{(+)}. \qquad (6.35)$$

The following statement holds.

The operator $\hat{S}$ is unitary and it commutes with any bounded function of the operator $\hat{H}$. Its matrix kernels S_{AB} are given by equations (6.34).

The first part of this statement can easily be proved with the help of Proposition 6.4 by the same reasoning as in the case of neutral particles. The representations (6.34) of the matrix kernels S_{AB} can be verified with the help of the following statement about asymptotics of the integrals (6.23), (6.24) which will be proved below.

The amplitudes of the spherical waves of the terms of the class $S_{B,c}$ in the formulae (6.23) *and* (6.24) *are expressed by means of the functions f_A and the operators S_{BA} by the relations*

$$f_{BA}(\hat{p}_B, E_B) = C_{BA}\int S_{BA}(p_B, p'_A)f_A(p'_A)dp'_A, \qquad (6.36)$$

where by C_{BA} we denoted the normalization coefficients

$$C_{BA} = -2\pi i(E_B + \varkappa_A^2)^{-1/2}. \quad A \neq 0, \quad B \neq 0;$$

$$C_{0A} = \frac{iC_0(E)}{\pi|p|^2(2\pi)^{3/2}}, \quad A \neq 0,$$

$$C_{B0} = -i(2\pi)^{5/2}E^{-2}, \quad A \neq 0;$$

$$C_{00} = i(\pi E^2)^{-1}C_0(E).$$

The formulae (6.23), (6.24), and (6.36) can be united into the following asymptotic representation, similar to (6.16)

$$\Psi_B(X, p_B) \sim C_{BB}\psi_B(x_B)\Theta^{(-)}(y_B, E_B)\delta(-y_B, p_B)+$$

$$+\sum_A C_{AB}\psi_A(x_A)\Theta^{(+)}(y_A, E_A)S_{AB}(|p_A|\hat{y}_A, p_B). \tag{6.36'}$$

To verify this, it is necessary to write the integral of the product of the wave functions $\Psi_A^{(-)*}\Psi_B^{(+)}$ by means of Green formula as the limit for large R of a surface integral over the sphere $S^{(2)}$. Taking the limit $R \to \infty$ and using equation (6.36), we get the results stated above.

As can be easily see from formula (6.34), unlike in the case of neutral particles, the diagonal elements of the scattering operator do not contain unit terms. All the matrix kernels are expressed only through the amplitudes of distorted spherical waves. However, the scattering amplitude has strong singularities which are at least as strong as δ-function singularities and are located at the same points as those of the kernels of the scattering operator for neutral particles. In particular, in the case of the operator S_{00} instead of the δ-function $\delta(P - P')$, the singularity (6.30) of the type $|P - P'|^{-5-2iq_0}$ appears, instead of the three-dimensional δ-function $\delta(p_\alpha - p'_\alpha)$ we have the singularity $|_\alpha - p'_\alpha|^{-3-ia_\alpha}$, and instead of the pole $(k_\alpha - k'_\alpha - i0)^{-1}$, the distorted Coulomb pole $(k_\alpha - k'_\alpha - i0)^{-1-ia_{\alpha\beta}}$.

It follows from the definition of the distributions $t^{-1-i\eta}$, generating singularities of the kernels of the scattering operators, that the integrals containing them can be represented in the regularized form (6.12). This formula can be written in the symbolic form

$$t^{-1-i\eta} = \frac{i}{\eta}\delta(t) + P(t^{-1-i\eta}), \tag{6.37}$$

where the principal value $P(t^{-1-i\eta})$ corresponds to the second term in (6.12). Comparing the definition of the distribution $f_\alpha^{(s)}$ with (6.37), we see that it corresponds to the principal value of the Coulomb amplitude

$$f_\alpha^{(s)} = P(f_\alpha(\hat{k}_\alpha, k'_\alpha)).$$

We can interpret this function as the true Coulomb scattering amplitude which remains after extraction from $f_\alpha(\hat{k}_\alpha, k'_\alpha)$ the δ-function localized at $\hat{k}_\alpha, \hat{k}'_\alpha) = 1$, distorted by the factor $i\exp\{2i\arg\Gamma(1+i\eta_\alpha)\}$.

The physical interpretation of the matrix elements remains unchanged in the case of charged particles. The effective cross-sections of the scattering processes $(B \to A)$ are proportional to $|F_{AB}|^2$. The presence of singularities in the kernel of the matrix elements F_{AB} reflect the fact that the total cross-sections of reactions with charged particles are infinite.

6.2.4 Coulomb rapidly oscillating integrals

To conclude this section, we will prove the previously formulated statement about asymptotics of integrals of the form I_A and I_0.

First of all, we will study the asymptotic behaviour of model integrals of the type

$$I_1 = \int d\hat{p} e^{i\sqrt{E}(\hat{p},x)} \Phi\left(i\nu + \frac{\mu}{2}, r_l + \mu, i\sqrt{E}|x|u\right) u^{\mu/2} g(\hat{p}) \tag{6.38}$$

where p and x are l-dimensional vectors, $u = 1 - (\hat{p}, \hat{x})$, $r_l = (l-1)/2$, and the integration is performed over the unit sphere $S^{(l-1)}$ in the l-dimensional space. The parameters n and m are real, and the function $g(\hat{p})$ is supposed to be smooth and vanishing for $u \geq 1$

$$g(\hat{p}) = 0, \quad \text{if } (\hat{x}, \hat{p}) \leq 0.$$

Proposition C *For $\sqrt{E}|x| \to \infty$, the following asymptotic formula holds*

$$I_l \sim \hat{C}_l A(\hat{x}, p)(\sqrt{E}|x|)^{\mu/2} \frac{\exp\{i\sqrt{E}|x| + i\nu \log 2\sqrt{E}|x|\}}{(\sqrt{E}|x|)^{r_n}} \tag{6.39}$$

where

$$C_l = \frac{\Gamma(r_l + \mu)}{\Gamma(i\nu + \mu/2)} e^{-\pi\nu - i\pi(r_l/2 + \mu/4)}.$$

The amplitude $A(\hat{x}, p)$ is given by the integral

$$A(\hat{x}, p) = \int d\hat{p} S_0(\hat{x}, p) g(\hat{p}), \quad p = \sqrt{E}\hat{p}, \tag{6.40}$$

where the singular kernel S_0*, defined by the equation*

$$S_0(\hat{p}', p) = (1 - (\hat{p}', \hat{p}))^{-r_l + i\nu},$$

acts in local coordinates $u = 1 - (\hat{p}', \hat{p})$ *on the sphere* $S^{(l-1)}$ *as the distribution* $t^{-1+\nu}$.

Let us prove this asymptotic formula. On the surface $S^{(l-1)}$ introduce spherical coordinates with respect to the $\hat{x}$ axis:

$$\hat{p} = \{\cos\theta, \hat{M}\sin\theta\}, \quad \cos\theta = 9\hat{x}, p), \quad , \hat{M} \in S^{(l-2)}.$$

The surface element $d\hat{p}$ is equal

$$d\hat{p} = \sin^{l-2}\theta d\theta \wedge d\hat{M} = (2u)^{r_l - 1} du \wedge d\hat{M},$$

where $d\hat{M}$ is the surface element on the sphere $S^{(l-1)}$. Represent further the integrand as the sum $g(\hat{p}) = g(\hat{x}) + (g(\hat{p}) - g(\hat{x}))$ and consider first the interval ΔI, corresponding to the difference $\Delta g(p) = g(\hat{p}) - g(\hat{x})$. Let us divide the integration domain $S^{(l-1)}$ into two parts $\Omega_{\hat{x}}$, $\tilde{\Omega}_{\hat{x}}$ where $\Omega_{\hat{x}}$ is a neighbourhoods of the point $\hat{p} = \hat{x}$ defined by

$$1 - \cos\theta \leq |x|^{-1/2+\epsilon}, \quad \epsilon > 0. \tag{6.41}$$

We will denote the corresponding integrals by $\Delta I_{\hat{x}}$ and $\Delta\tilde{I}_{\hat{x}}$. Since the difference Δg satisfies the estimate $|\Delta g(\hat{p})| \leq cu$ in the region $\Omega_{\hat{x}}$ and the measure of this region is of order $O(|x|^{-rn+\epsilon'})$, $1 > \epsilon' > 0$, the integral $\Delta I_{\hat{x}}$ can be written as a product of the factor $|x|^{-r_l}\exp\{i\sqrt{E}|x|\}$ by a function tending to zero as $|x|^{\epsilon_0}$, $\epsilon_0 > 0$. Therefore, this integral gives no contribution to the leading terms of the asymptotics.

In the integral $\Delta\tilde{I}_{\hat{x}}$, the function $\Phi(a, c, t)$ can be replaced by the leading asymptotic terms (5.3). The second term in (5.3) generates the distorted spherical wave (6.39), whose amplitude $\tilde{A}_{\hat{x}}$ is given by equation (6.40) with $g(\hat{p})$ replaced by $\Delta g(\hat{p})$. In this case, the integral $\int S_0(\hat{x}, p)\Delta g(\hat{p})d\hat{p}$ exists as an improper one. The asymptotic form of the integral, generated by the first term of (5.3), can be found by integrating by parts with respect to u. A non-trivial contribution comes only from the point $u = 1$, where $\Delta g = g(\hat{x})$.

We will denote this contribution by δI_1 but will not give an explicit formula for it, because eventually all contributions of the point $u = 1$ will cancel.

Next, let us consider the integral $I_{\hat{x}}$, corresponding to the function $g(\hat{x})$. In spherical coordinates it takes this form:

$$I_{\hat{x}} = g(\hat{x})e^{i\sqrt{E}|x|}\Omega_{l-2}\int_0^1 du u^{r_l-1+\frac{\mu}{2}}(2-u)^{r_l-1}\times$$

$$\times \exp\{-i\sqrt{E}u|x|\}\Phi(i\nu+\mu/2, r_l+\mu, i\sqrt{E}|x|u).$$

Here by Ω_{l-2} is denoted the surface area of the $(l-2)$-dimensional sphere:

$$\Omega_{l-2} = 2\pi^{\frac{l-1}{2}}\left(\Gamma\left(\frac{l-1}{2}\right)\right)^{-1}.$$

Before calculating the asymptotics of this integral, let us consider an additional auxiliary integral:

$$I_0 = g(\hat{x})\Omega_{l-2}2^{r_l+i\nu}\int_0^2 du u^{r_l-1+\frac{\mu}{2}}(2-u)^{-1-i\nu}\times$$

$$\times \exp\{i\sqrt{E}|x|(1-u)\}\Phi(i\nu+\mu/2, r_l+\mu, i\sqrt{E}|x|u).$$

Here the expression $(2-u)^{-1-i\nu}$ is understood as the distribution $t^{-1-i\nu}$, where $t = 1-u$. It is obvious that the singularity of the integrand in I_0 for $u = 0$ coincides with that of the original integrand in $I_{\hat{x}}$. However, unlike $I_{\hat{x}}$, the integral I_0 can be computed explicitly:

$$I_0 = g(\hat{x})\Omega_{l-2}2^{\frac{3r_l}{2}-i\nu-1}e^{-i\sqrt{E}|x|}\times$$

$$\times\frac{\Gamma(-i\nu)\Gamma\left(r_l+\frac{\mu}{2}\right)}{\Gamma\left(-i\nu+r_l+\frac{\mu}{2}\right)}\Phi\left(\frac{\mu}{2}, r_l+\mu, 2i\sqrt{E}|x|\right).$$

On the other side, this integral can be written as the sum of two integrals: $I_0 = I_{01}+I_{12}$, taken over the intervals $[0,1]$ and $[1,2]$. The integral I_{01} is of the type (6.38), whose asymptotics is known. In the second integral we can replace the confluent hypergeometric function $\Phi\left(\frac{\mu}{2}, r_l+\mu, 2i\sqrt{E}|x|\right)$ by its asymptotics (5.3). As a result, this integral takes the form of a sum of ordinary, rapidly oscillating integrals of the type (4.126). Their asymptotics are easy to calculate. Thus, for the integral I_{01}, having the same singularities as the original (6.38), we obtain the explicit asymptotic representation $I_{01} \sim I_0+I_{12}$. After reduction of similar terms, we get a sum of two terms. One has the form (4.39), where the amplitude $A(\hat{x},p)$ is given by the formula (6.40). As the function $g(\hat{p})$ we take in this case $(1-u/2)^{-i\nu-r_l}$. The second term describes the contribution of the point $u=1$. We do not write it explicitly here.

Finally, based on this result, let us find the asymptotics for $I_{\hat{x}}$. To this end, write $I_{\hat{x}}$ as the sum $I_{\hat{x}} = (I_{\hat{x}} - I_{01}) + I_{01}$. The integrand of the first term satisfies the estimate (6.41) and, therefore, its asymptotics can be investigated in the same manner as in the case of the integral $\Delta I_{\hat{u}}$. We obtain a sum of two terms. The first is the distorted spherical wave (6.39), whose amplitude has the form (6.40) where the function $g(\hat{p})$ must be replaced by the expression $g(\hat{p})((1-u/2)^{-i\nu-r_l}-1)$. The second is generated by the point $u=1$. The asymptotics of the integral I_{01} same form has the same form. We can verify that the sum of the terms corresponding to $u=1$, and those obtained above gives zero.

Thus, we investigated all the terms into which we have divided the original integral I_l. Gathering all these terms and reducing similar ones, we get the formula (6.39).

Let us proceed to the justification of equations (6.14) and (6.36).

Note that formula (6.14) follows immediately from Proposition C for $l=3$. As to the first term in (6.9), it describes the contribution of the point $u=2$, appearing after integration by parts with respect to u. A similar term appears in the case of neutral particles, as well (see equation (4.6)).

To conclude, let us prove the formulae (6.23) and (6.36). We will restrict ourselves to the more difficult integral I_0. The integrals I_A, $A \neq 0$, can be investigated in the same manner as the two-particle expressions (6.9).

It is clear that the contribution of the asymptotic terms $\tilde{\Phi}_0$, corresponding to the smooth parts of the scattering amplitude, can be represented in the form (6.36), so that it is sufficient to study the integrals with the functions L_c, Ψ_α, and $\Psi_{\alpha\beta}$. The asymptotics of such integrals can be found with the help of the saddle-point method. Here the plane eikonal has two critical points $\hat{p} = \pm\hat{X}$, where $(\hat{P},X) = \pm|X|$ and the single Z_α and the double $Z_{\alpha\beta}$ eikonals have one singular point, where $Z_\alpha = |X|$, $Z_{\alpha\beta} = |X|$.

The critical point of the plane eikonal $\hat{P} = -\hat{X}$ generates the first term in the asymptotic representation (6.23). The remaining critical points lie in the singular directions, where the functions L_c, Ψ_α and $\Psi_{\alpha\beta}$ can be described in terms of confluent hypergeometric functions. Therefore, we can apply Proposition C to find the asymptotics of the integrals in the neighbourhoods of the directions Ω_F, $\Omega_\alpha^{(0)}$, $\Omega_{\alpha\beta}^{(0)}$.

In the singular directions Ω_F and $\Omega_\alpha^{(0)}$, the asymptotics of the functions L_c and Ψ_α is explicitly expressed through the functions $\Phi(a,b,t)$ by the formulae (5.145′) and (5.128). Using relation (6.39), we get the terms in the asymptotic expression (6.23) corresponding to the amplitudes F_s and F_α.

To find the asymptotics of the integral containing the function $\Psi_{\alpha\beta}$, we can use the method developed during demonstration of Proposition C. To be able to use formula (6.41), we must express the function $\Psi(a,c,x)$ in (5.139) through the function $\Phi(a,c,x)$ with the help of the formula

$$\Psi(a,c,x) = \frac{\Gamma(1-b)}{\Gamma(a-b+1)}\Phi(a,b,x)+$$

$$+\frac{\Gamma(b-1)}{\Gamma(a)}x^{1-b}\Phi(a-b+1,2-b,x).$$

As a result, as in the case of integral (6.38), we get the asymptotic formula (6.39) where the amplitude of the spherical wave is expressed in terms of the distribution $(t-i0)^{-1-i\mu}$ by equation (6.40).

We must note that during integration of the amplitudes F_α and $F_{\alpha\beta}$ over the angular variables, we must put the two-particle amplitudes f_α for the pair α to zero in the direction of forward scattering $(\hat{k}_\alpha,\hat{x}_\alpha) = 1$. This is stipulated by the fact that in the asymptotic formulae (5.128) and (5.137), the given amplitudes are multiplied by cutoff functions, equal to zero for $(\hat{k}_\alpha,\hat{x}_\alpha) = 1$.

This concludes the investigation of the stationary scattering problem and proceed to the study the relation between the stationary and non-stationary definitions of wave functions.

6.3 Justification of the Non-Stationary Formulation of the Scattering Problem

In this section, we will show that the stationary wave operators coincide with the non-stationary ones, the latter being determined as limits of evolution operators. In this way, we will justify the formulation of the scattering problem.

6.3.1 The two-particle system

Let us show that the operator

$$u(t) = \exp\{iht\} \exp\{-ih_0 t\}$$

is strongly convergent and moreover

$$\lim_{t \to \mp\infty} u(t) = u^{(\pm)} \tag{6.42}$$

where the operators $u^{(\pm)}$ are given by equations (3.22).

To prove this, it is sufficient to show that the relation

$$\lim_{t \to \mp\infty} (e^{-iht} u^{(\pm)} f, e^{-ih_0 t} g) = (f, g) \tag{6.43}$$

holds with f and g being smooth, finite functions. Indeed, it follows from (6.1) that

$$\|(e^{iht} e^{-ih_0 t} - u^{(\pm)}) f\|^2 =$$

$$= 2(f, f) - 2\mathrm{Re}\,(u^{(\pm)} f, e^{iht} e^{-ih_0 t} f)$$

and the right-hand side vanishes for $t \to \pm\infty$, if the relation (6.43) holds.

Now, let us prove equation (6.43). Using (6.1), we transform the scalar product on the left-hand side of (6.43)

$$(e^{-iht} u^{(\pm)} f, e^{-ih_0 t} g) = (u^{(\pm)} e^{-ih_0 t} f, e^{-ih_0 t} g).$$

Using the expression of the kernel of $u^{(\pm)}$ in terms of the T-matrix (3.22), we find that this scalar product is equal to the sum of (f, g) and the term

$$\int dkdk' \frac{t(k,k',k'^2 \pm i0)}{k^2 - k'^2 \mp i0} e^{i(k^2-k'^2)t} f(k')\overline{g(k)}.$$

After integrating over the angular variables and the changing the variables $k^2 = u$, $k'^2 = v$, the integral takes the form

$$\int dudv \Phi(u,v) \frac{e^{i(u-v)t}}{u - v \mp i0}$$

and by virtue of (2.44), it vanishes in the limit $t \to \pm\infty$. Therefore, relation (6.43) is proved.

Now let us turn to a system of charged particles. We will prove that the operator

$$u(t) = \exp\{iht\} L_0 \exp\{-ip^2 t - iw_t\}$$

is strongly convergent and moreover

$$\lim_{t \to \mp\infty} u(t) = u^{(\mp)}$$

where the wave operators $u^{(\mp)}$ are defined by equations (6.7) and $w_t = \eta(p) \operatorname{sign} t \log 4p^2 t$.

As above, it is sufficient to prove

$$\lim_{t \to \mp\infty} (e^{-iht} u^{(\pm)} f, L_0 e^{-ipt - iw_t} g) = (f, g).$$

We will use the coordinate representation. For definitiveness, let us consider the case $t \to -\infty$.

Using the interlacing property, we change the order of the operators e^{-iht} and $u^{(+)}$, and write the scalar product as a limit $\epsilon \downarrow 0$ of the integral

$$I(t,\epsilon) = \frac{1}{(2\pi)^3} \int dx\, e^{-\epsilon|x|} \int dpdp' \psi(x,p) \times$$

$$\times \exp\{i(p',x) + it(p'^2 - p^2) + iw_t(p)\} f(p')\overline{g(p)}. \tag{6.44}$$

Note that the integral over the sphere V_R of the radius $R = o(t^\nu)$, $\nu < 1/3$, vanishes in the limit $|t| \to \infty$. Indeed, in this case, the parameter t is of higher order than $|x|$. Therefore, integrating first over x, the corresponding integral can be written in the form (1.28). The asymptotics of such an integral can be calculated by the saddle-point method. In the limit $|t| \to \infty$ we obtain zero.

Therefore, we can confine the integration over the exterior of a sphere of large radius $R_t = o(|t|^\nu)$. First integrate over the angular variables $\hat{p}$ and $\hat{p}'$. Since $|x|$ is large, we can use formula (6.9) to calculate the asymptotics of the integral for $\hat{p}$. The asymptotics for $\hat{p}'$ can be obtained using the similar equation (4.6). As a result, we get the following asymptotic equation:

$$I(t,\epsilon) \sim \int_0^\infty ds\, s \int_0^\infty ds'\, s' \int_{|x|>R_t} dx e^{-\epsilon|x|} \times$$

$$\times \exp\{it(s'^2 - s^2) - iw_t(s)\} \times$$

$$\times \left(\overline{g(s'\hat{x})} \frac{e^{-is'|x|}}{|x|} - \overline{g(-s'\hat{x})} \frac{e^{is|x|}}{|x|} \right) \times$$

$$\times \left(f(-s\hat{x}) \frac{e^{-is|x| - iw_0(s)}}{|x|} + F(\hat{x}) \frac{e^{is|x| + iw_0(s)}}{|x|} \right) \qquad (6.44')$$

Thus, we can write $I(t,e)$ as the sum of four rapidly oscillating integrals, containing the exponent $\exp\{\mp i|x|s' \pm i|x|s\}$. Let us now integrate over the radial variable $|x|$. We complement the integration interval $[R_t, \infty)$ to the whole semi-axis $[0,\infty)$ by adding terms vanishing in the limit $|t| \to \infty$. The arising integral can be calculated explicitly. As the result, we get the following functions, determining the non-trivial singular and rapidly oscillating parts of the integral

$$\frac{\exp\{it(s'^2 - s^2) - iw_t(s)\}}{(\pm s' \mp s + i0)^{1+i\eta}} e^{\frac{\pi\eta}{2}} \Gamma(1 + i\eta). \qquad (6.44'')$$

Finally, let us integrate over s and use the asymptotic formula (2.44) from Chapter 2. Only the term corresponding to the denominator $(s' - s + i0)^{-1}$ will have a non-zero limit. This limit equals the scalar product (f,g), what was to be established.

Thus, we have completely investigated the system of two particles and can proceed to the study of the three-particle wave operators.

6.3.2 The three-particle system

The basic result that we will prove in the case of the system of three charged particles is the following statement.

The operators

$$U_A(t) = e^{iHt} L_A e^{-i\tilde{H}t}, \quad A = 0, \{\alpha, i\}, \tag{6.45}$$

are strongly convergent for $t \to \mp\infty$, *and*

$$\lim_{t \to \mp\infty} U_A(t) = U_A^{(\pm)}, \quad A = 0, \{\alpha, i\}, \tag{6.46}$$

where the operators $U_A^{(\pm)}$ *are defined by equations* (3.63) – (3.63").

These relations are analogous to equations (6.42) for the two-particle system. But, there are evident differences in the results, corresponding to the evolution operators e^{iht} and e^{iHt}. Comparing equations (6.42) and (6.46) we can see that the operators $u(t)$ and $U_0(t)$ are constructed similarly. An important property of the asymptotics of the operator $u(t)$, i.e., of the operators $u^{(+)}$ and $u^{(-)}$, is the coincidence of their domains of values. The operators $U_0^{(\pm)}$ do not possess this property. As it follows from Proposition 6.4, only the orthogonal sums of the domains of values of the operators $U_0^{(\pm)}$ and $U_0^{(\pm)}$, $A \neq 0$ coincide.

This difference is, of course, caused by the above mentioned distinction between the systems of two and three particles. In terms of the scattering theory, the former is a one channel problem, and the latter a multichannel one. The results proved in this chapter allow to give exact mathematical meaning to the concepts of channel, wave operators and scattering operator for the multichannel system as developed in Chapter 2 on the example of a system of three particles.

As in the case of the system of two particles, the basis of the proof of equation (6.46) is the following statement

1. *Let $f(P)$ and $g(P)$ be finite smooth functions vanishing in the neighbourhood of the singular surface $P^2 = E_i$. Then*

$$\lim_{t\to\mp\infty}(e^{-iHt}U_0 f, L_0 e^{-i\tilde{H}_0 t} g) = (f, g). \tag{6.47}$$

2. *Let f_A and g_A be finite smooth functions, and let $E_A(p_A) = E_i$ in the neighbourhood of singular surfaces. Then*

$$\lim_{t\to\mp\infty}(e^{-iHt}U_A f_A, L_0 e^{-i\tilde{H}_A t} g_A) = (f_A, g_A). \tag{6.48}$$

Equations (6.46) follows immediately from these relations, because

$$\|(U_0(t) - U_0^{(\pm)} f\|^2 =$$

$$= 2\left(\|f\|^2 - \mathrm{Re}\,(e^{-iHt}U_0^{(\pm)} f, L_0 e^{-i\tilde{H}_0 t} f)\right),$$

$$\|(U_A(t) - U_A^{(\pm)} f_A\|^2 =$$

$$= 2\left(\|f_A\|^2 - \mathrm{Re}\,(e^{-iHt}U_A^{(\pm)} f_A, L_A e^{-i\tilde{H}_A t} f_A)\right),$$

and the right hand sides vanish in the limit $t \to \mp\infty$ because of equations (6.47) and (6.48).

The proof of equations (6.47) and (6.48) can be obtained along the same scheme as in the two-particle system. It is sufficient to verify that the expression

$$I_A(t) = \left((U_A^{(\pm)} - L_A)e^{-i\tilde{H}_A t} f_A, e^{-i\tilde{H}_A t} f_A\right), \quad A = 0, \{\alpha, t\}.$$

vanishes in the limit $t \to \mp\infty$. To this end, one must substitute the expressions for the kernels of the wave operators in terms of the components of the T-matrices and change to spherical integration variables. Calculating the asymptotics of the obtained integrals with respect to the radial variables by means of equation (2.44), we get the desired equality

$$\lim_{t\to\mp\infty} I_A(t) = 0.$$

6.3.3 Charged particles

Now, let us assume that the particles are charged. In this case, the generalized wave operators are defined in the non-stationary approach by means of the relations

$$U_A^{(\pm)} = \lim_{t\to\mp\infty} e^{-iHt} L_A e^{-i\tilde{H}_A t - iW_t^{(A)}}, \tag{6.49}$$

where the Coulomb phase operators are given by the formulae

$$W_t^{(0)} = \sum_\alpha \frac{n_\alpha}{2|k_\alpha|} \operatorname{sign} t \log 4P^2|t|, \quad A = 0,$$

$$W_t^{(A)} = \frac{n_{\alpha\alpha}}{2|p_\alpha|} \log 4|p_\alpha|^2|t|, \quad A \neq 0 \tag{6.50}$$

We will prove that, as in the case of a system of neutral particles, the strong limits (6.49) coincide with the stationary wave operators.

This statement can be also proved along the same scheme as the similar proposition in the case of neutral particles. It is sufficient to verify the following proposition.

Let f_0 and g_0 be smooth finite functions, vanishing in the neighbourhood of the singular surfaces $P^2 = E_i$. Then the equation

$$\lim_{t\to\mp\infty} \left(e^{-iHt} U_0^{(\pm)} f, L_0 e^{-i\tilde{H}_0 t - iW_t^{(0)}} g\right) = (f, g) \tag{6.51}$$

holds.

Let, moreover, f_A, g_A be finite smooth functions vanishing in the neighbourhood of the singular surfaces $E_A(p_A) = E_i$. Then the relations

$$\lim_{t\to\mp\infty} \left(e^{-iHt} U_A^{(\pm)} f_A, L_A e^{-i\tilde{H}_A t - iW_t^{(A)}} g_A\right) = (f_A, g_A) \tag{6.52}$$

hold.

These relations can be verified in exactly the same way as in the case of a two-particle system. Let us outline this scheme on the example of (6.51).

With the help of the relation

$$\exp\{iHt\} U_0^{(\pm)} = U_0^{(\pm)} \exp\{i\tilde{H}_0 t\}$$

which follows from Proposition 6.4, the scalar product on the left hand side of equation (6.51) can be written as

$$\left(U_0^{(\pm)} e^{-i\tilde{H}_0 t} f, L_0 e^{-i\tilde{H}_0 t - iW_t^{(0)}} g\right). \tag{6.53}$$

Let us introduce the integral similar to (6.44):

$$I(t,\epsilon) = \frac{1}{(2\pi)^6} \int dX\, e^{-\epsilon|X|} \int dP dP' \Psi_0(x,p) \times$$

$$\times \exp\{i(P',X) + it(P'^2 - P^2) + iW_t^{(0)}(p)\} f(P') \overline{g(P)} \qquad (6.54)$$

The following reasoning repeats the argument employed in the two-particle system. First, in equation (6.54), we can restrict the integration domain to the exterior of the sphere $|X| \geq R_t$. Then we must integrate over the angular variables $\hat{P}$ and $\hat{P}'$ calculate the asymptotics of these integrals with the help of formula (6.23) and by similar relation for plane waves $\exp\{i(X,P')\}$. We get a formula similar to (6.44′), but different in one point, namely instead of the function $F(\hat{x})\frac{e^{is|x|+iw_0(s)}}{|x|}$ the sum over all channel $\sum_A F_{A^0}(\hat{y}_A)\psi_A(x_A)Q_A^{(+)}(y_A, E_A)$, $E_A = s^2 - \varkappa_A^2$ appears. But in the limit $t \to -\infty$ only the terms containing a product of six-dimensional spherical waves remains. Here, the singular and rapidly oscillating functions, appearing after integration over $|X|$, are again given by (6.44″), where the new Coulomb parameter $\eta_0 = \sum_\alpha \frac{n_\alpha}{2s \cos\omega_\alpha}$, $\cos\omega_\alpha = \frac{|x_\alpha|}{|X|}$ should is introduced. The limit for $t \to -\infty$ of the so obtained expression can be found with the help of equation (2.44). We get the required relation

$$\lim_{t\to-\infty} \lim_{\epsilon\downarrow 0} I_0(t,\epsilon) = (f,g).$$

Thus, we concluded the study of the problem having principal importance from the point of view of foundations of the scattering theory. Another circle of important problems is related with applications of the methods developed above to the calculation of physically interesting variables as the amplitudes of various scattering problems in the three-particle reactions. An approach to the solution of such problems will be considered in Chapter 7.

CHAPTER 7

Some Applications

In this chapter we will describe methods of calculation of wave function based on stationary approaches to scattering theory developed in previous chapters. We have no intention to present an overall review of methods for solving the Schrödinger and compact equations. The selection of subjects is based on our evaluation of the effectiveness of various approaches and on our own research in this field.

In what follows we will assume that the interaction potential is spherically symmetric: $v(x) = v(r)$, $r = |x|$. Practically, all low energy scattering problems may be reduced to this case. Numerical solutions of few-body scattering problems with such interactions are based on decomposition of the wave function in angular bases which are selected according to various considerations. However, all such bases are essentially related to the standard spherical functions used for solving the two-body equation. Therefore, for completeness, we first introduce the traditional and well known two-body partial wave expansion and then we generalize the results to the three-particle case.

7.1 Partial Waves in Two-Body Systems

In this section, we separate the angular variables in the Schrödinger equation for wave function and in integral equations for T-matrix. In this way we reduce the problem of finding the scattering amplitude to solving ordinary differential or one-dimensional integral equations.

7.1.1 Schrödinger equation

The solution of the Schrödinger equation with the spherically symmetric potential can be expanded into a series in Legendre polynomials

$$\psi(x,k) = \frac{1}{4\pi qr}\sum_{l=0}^{\infty}(2l+1)i^l P_l(\cos\theta)\psi_l(r,q). \tag{7.1}$$

Here θ denotes the angle between the vectors k and x and ψ_l is a solution of the differential equation ($q = |k|$)

$$\left(-\frac{\mathrm{d}^2}{\mathrm{d}r^2} + \frac{l(l+1)}{r^2} + v(r)\right)\psi_l(r) = q^2\psi_l(r), \tag{7.2}$$

called the partial (radial) Schrödinger equation The asymptotics of the wave functions consists of in-coming and out-going waves

$$\psi_l \sim \frac{1}{2}\left(\exp\left\{-iqr + i\frac{\pi}{2}(l+1)\right\} - \right.$$

$$\left. - \exp\left\{-i\frac{\pi}{2}(l+1) + 2i\delta_l\right\}\exp\{iqr\}\right). \tag{7.3}$$

At the origin $r = 0$, the wave function ψ_l vanishes

$$\psi_l(0) = 0.$$

The quantity Δ_l is called the partial scattering phase and the function $s_l = e^{2i\delta_l}$ the partial S-matrix. The plane wave $e^{i(k,x)}$ can be expanded as in (7.1) in the form of the analytic series

$$e^{i(k,x)} = \frac{1}{4\pi kr}\sum_{l=0}^{\infty}(2l+1)i^l P_l(\cos\theta)j_l(rk)$$

where j_l are the spherical Bessel functions

$$j_l(rq) = \sqrt{\frac{\pi qr}{2}}J_{l+1/2}(qr)$$

asymptotically equal to one half of the sum of in-coming and out-going waves

$$j_l(rq) \sim \frac{1}{2}\left(\exp\left\{-iqr + i\frac{\pi}{2}(l+1)\right\} - \right.$$

$$\left. - \exp\left\{iqr - i\frac{\pi}{2}(l+1)\right\}\right). \tag{7.4}$$

Comparing the asymptotic formulae (7.3) and (7.4) with the asymptotics of the wave function (7.4) we obtain the following expression for the scattering amplitude

$$f(\hat{x},k)=\frac{1}{2i|k|}\sum_{l=0}^{\infty}(2l+1)\left(e^{2i\delta_l}-1\right)P_l(\cos\theta) \tag{7.5}$$

Finally, by expanding both sides of equation (4.18) in spherical harmonics, we obtain the integral representation of partial S-matrix

$$s_l=1-\int_0^{\infty}dr\, j_l((qr)v(r)\psi_l(r,q). \tag{7.6}$$

The important characteristic of a scattering process is the value of the scattering amplitude $f(\hat{x},k)$ at zero energy called scattering length. This term originates from the fact that in the case of scattering on a solid sphere, the amplitude $f(\hat{x},0)$ equals the radius of the sphere. In the limit $q\to 0$, the spherical Bessel functions $j_l(qr)$ tend to zero for fixed r as and the solutions clearly behave in the same way. Hence, it follows from the integral representation (7.6) that the partial S-matrix tends to one in the limit $q\to 0$

$$s_l\sim 1+c_lq^{2l+1} \tag{7.7}$$

Therefore, the scattering length for short-range potential c_0 equals the value of the partial amplitude at $l=0$ and zero energy

$$f(\hat{x},0)=\lim_{q\to 0}(e^{2i\delta_0(q)}-1)q^{-1}=c_0.$$

7.1.2 Charged particles

In the case of Coulomb particles the solution of the partial Schrödinger equation (7.2) can be expressed in terms of analytic functions. The regular solution vanishing at the origin reads

$$\psi_{l,c}(r,q)=\frac{(2qr)^{l+1}}{2}e^{iqr}f_{l,c}(r,q), \tag{7.8'}$$

where the function $f_{l,c}$ satisfies the confluent hypergeometric equation

$$rf_{l,c}''+2(l+1+iqr)f_{l,c}'+2q(i(l+1)-\eta)f_{l,c}=0,$$

$$\eta = \frac{n}{2q} \tag{7.8}$$

and is normalized as

$$f_{l,c}(r,q) =$$

$$= e^{-\frac{\pi\eta}{2}} \frac{\Gamma(l+1+i\eta)}{2l+1} \Phi(i\eta+l+1, 2l+2, -2iqr).$$

For $qr \to \infty$, the partial wave function ψ_l asymptotically equals the sum of the distorted in-coming and out-going waves

$$\psi_{l,c}(r,q) \sim \exp\left\{-iqr + i\frac{\pi}{2}(l+1) + i\eta \log 2qr\right\} -$$

$$-s_{l,c} \exp\left\{iqr - i\frac{\pi}{2}(l+1) - i\eta \log 2qr\right\}.$$

where $s_{l,c}$ is the Coulomb partial scattering matrix

$$s_{l,c} = e^{2i\delta_{l,c}}$$

and the Coulomb phase shifts are given by the equality

$$\delta_{l,c} = \arg \Gamma(1+l+i\eta).$$

The Coulomb scattering amplitude is expressed in terms of the phase shifts as follows

$$f_c(\hat{x}, k) = \frac{1}{2iq} \sum_{l=0}^{\infty} (2l+1) e^{2i\delta_{l,c}} P_l(\cos\theta). \tag{7.9}$$

For $\theta \neq 0$ this series is conditionally convergent. In the forward direction, this series diverges and the sum should be understood in the sense of a distribution as was already discussed in Chapter 5, Section 6. If, in addition to the Coulomb part, the potential has also a short-range term

$$v(r) = \frac{n}{r} + v_s(r),$$

then the partial equation has a solution of the form

$$\psi_l(r,q) \sim \psi_{l,c}(r,q) + \varphi_l(r,q),$$

where the asymptotics of the function φ_l contains only outgoing waves

$$\varphi_l(r,q) \sim s_{l,cs} \exp\left\{iqr - i\frac{\pi}{2}(l+1) - i\eta \log 2qr\right\}.$$

The sum of amplitudes of the outgoing Coulomb and short-range waves

$$s_l = s_{l,c} + s_{l,cs}$$

is called the partial S-matrix. It is convenient to express this function in the exponential form as

$$s_l = \exp\{2i(\delta_{l,c} + \delta_l^{cs})\},$$

where $\delta_{l,c}$ is the Coulomb phase shift and δ_l^{cs} is phase shift corresponding to the short-range part of the interaction. The last term of the full amplitude, i.e., of the sum of the Coulomb and short-range amplitudes (5.22) can be expressed as

$$f_{cs}(\hat{x}, q) = \frac{1}{2i|q|} \sum_{l=0}^{\infty} (2l+1) e^{2i\delta_{l,c}} \left(e^{2i\delta_l^{cs}} - 1 \right) P_l(\cos\theta) \qquad (7.10)$$

which, is contrary to (7.9), is absolutely convergent. It should be noted that the quantity $e^{2i\delta_l^{cs}}$ does not coincide with the partial amplitude of a short-range potential $v_s(r)$ in absence of the Coulomb interaction. The term "short-range" part of the amplitude has a conditional meaning.

7.1.3 Low energy behaviour of Coulomb amplitudes

At low energies the partial Coulomb S-matrices do not tend to any definite limit. This property, in particular, causes slow convergence of the series (7.9). Indeed, the behaviour of the Coulomb S-matrix for $q \to 0$ is determined by the asymptotics of Γ-function at large values of the argument and is described by the formula

$$s_{l,c} \sim \exp\left\{ i\omega(q) - i\frac{\pi}{4} + i\frac{\pi}{2}\epsilon(\eta) \right\},$$

where $\omega(q) = 2\eta(\log|\eta| - 1)$ and $\epsilon(h)$ is the Heaviside function: $\epsilon(\eta) = 1$ for $\eta > 0$ and $\epsilon(\eta) = 0$ for $\eta < 0$. The behaviour of the functions $s_{l,cs}$ can be studied by means of integral equations of perturbation theory (5.20). These functions behave as

$$s_{l,cs} \sim \chi_c(q)\tilde{s}_l^{cs}$$

where the function $\tilde{s}_l^{cs}$ has a finite limit for $q \to 0$ and the singular factor χ_c equals

$$\chi_c = \exp\{i\omega(q) - 2\pi\eta\epsilon(\eta)\}.$$

The limit of $\tilde{s}_l^{cs}$ for $q \to 0$ is, in general, different from zero.

If the short-range part of the interaction decreases rapidly, the short-range part of the amplitude which is given by rapidly convergent series (7.9) behaves analogously:

$$f_{cs}(\hat{x}, k) \sim \frac{1}{|k|}\chi_c \tilde{f}_s(\hat{x}, k). \tag{7.11}$$

It is natural to call the function $\tilde{f}_s$ which has a finite limit for $|k| \to 0$ the Coulomb scattering length. This function coincides with the scattering length defined in (7.7) if the Coulomb potential is absent.

If the short-range part of the potential contains multipole terms (5.25), then the low energy behaviour of the amplitude $f_{cs}(\hat{x}, k)$ can no longer be described by means of the simple formula (7.11). In this case, the series (7.10) does not converge absolutely and generally speaking, the behaviour of its general term does not reflect the behaviour of the sum. In particular, it can be proved that for potentials $v_s(x)$ which for $|x| \to \infty$ decreases as $|x|^{-2}$ i.e., $v_s(x) \sim \frac{\mu(\hat{x})}{|x|^2} + O(e^{-\epsilon|x|})$, the short-range part of the scattering amplitude f_{cs} can be represented in the form of a sum

$$f_{cs}(\hat{x}, k) = f_d(\hat{x}, k) + \chi_c \tilde{f}_s(\hat{x}, k)|k|^{-1},$$

where, apart from the singular term $\chi_c \tilde{f}_s$ caused by the action of the potential at short-range, there is the additional singularity f_d corresponding to the dipole potential. The function $\tilde{f}_s$ has a finite limit for $|k| \to 0$ and the leading term in its expansion in powers of the energy, can be regarded as the scattering length for the potential $v_s(x)$.

Without proof we present formulae for the low energy behaviour of the function f_d.

Let $\cos\theta = (\hat{k}, \hat{k}')$ and let $\mathcal{D}_\epsilon$ be the region on the sphere $|k'| = 1$ where $\cos\frac{\theta}{2} \geq \epsilon > 0$. In this region, the following representation of the function f_d holds:

$$f_d(\hat{k}', k) = (\sin\theta)^{-1}\sigma_d(|k|, \theta)b_\pm(\hat{k}', \hat{k})(1 + O(|k|)), \tag{7.12}$$

where the term σ_d which becomes singular for $|k| \to 0$ is defined as

$$\sigma_d(|k|, \theta) = \frac{-\exp\{i\arg\chi_c)\}}{2|k|}\left(\sin\frac{\theta}{2}\right)^{-2i\eta}$$

and the functions $b_\pm$ are defined in terms of the angular part of the dipole potential

$$b_\pm(\hat{k}', \hat{k}) = \int_{l_\pm} d\hat{x}\, \mu(\hat{x}) \mp \theta \frac{\mu(\mp\hat{x}) + \mu(\pm\hat{k}')}{2}.$$

Here the sign plus (minus) corresponds to the Coulomb attraction (repulsion) and l_+ (l_-) is a unit semicircle in the plane of the vectors k and k' for which $(\widehat{k - k'}, \hat{x}) \leq 0$ $((\widehat{k - k'}, \hat{x}) > 0)$.

For backward scattering, where $\hat{k}' = -\hat{k}$, the function f_d at zero energy is more singular than for any other scattering angles:

$$f_d(-\hat{k}, k) = e^{\pm i \frac{\pi}{4}} \sqrt{\pi|\eta|}\sigma_d(|k|, \pi)\hat{b}_\pm(\hat{k}); \tag{7.13}$$

here the function $\hat{b}_\pm$ is of the form

$$\hat{b}_\pm(\hat{x}) = \frac{1}{2\pi} \int_{S_\pm} d\hat{x}\, \mu(\hat{x})(1 - (\hat{k}, \hat{x})^2)^{-1/2} \mp \frac{\pi\mu(\mp\hat{k})}{2},$$

where S_+ (S_-) is unit semisphere on which $(\hat{x}, \hat{k}) > 0$ $((\hat{x}, \hat{k}) < 0)$.

Intermediate form of the asymptotics of the function f_d for nearly backward scattering is described by the following representation

$$f_d(\hat{k}', k) =$$

$$= (\sin\theta)^{-1}\sigma_d(|k|, \theta)C_\pm(\hat{k}', \hat{k})e^{\pm i\frac{\pi}{4}}\Phi_\pm\left(\frac{\pi^2}{4}\eta\cos^2\frac{\theta}{2}\right), \tag{7.14}$$

where $C_\pm(\hat{k}', \hat{k}) = b_\pm(\hat{k}', \hat{k})$ for $\hat{k}' \neq -hatk$ $C_\pm(-\hat{k}, \hat{k}) = 2\hat{b}_\pm(\hat{k})$, Φ_+ is the Fresnel integral $2\int_{\sqrt{t}}^{\infty} e^{i\tau^2} d\tau$ and $\Phi_-(t) = \Phi_+^*(t)$. There exists a close relation between low energy and angular singularities described in Chapter 5 Section 1. Namely, the asymptotics of the angular singularity (5.26) for $|k| \to 0$ away from the residue of the pure Coulomb part of f_c coincides in the leading order with the angular singularity of the energetic singularity of f_d:

$$f_d(\hat{k}', k)|_{\hat{k}' \to \hat{k}} \sim (f(\hat{k}', k) - f_c(\hat{k}', k))|_{k^2 \to 0}.$$

At the same time, the dipole part of the wave function $\psi(x, k)$ defined by equation (5.25) takes into account not only all angular singularities of the scattering amplitude but also the leading term of the low energy singularity resulting from the dipole interaction.

Let us briefly discuss some problems arising when the scattering amplitude is computed in terms of partial wave functions. It is not a difficult task to solve numerically the partial wave equations at the given value of momentum l. For instance, one can use finite difference approximation of differential operators. At the final stage, the problem reduces to inversion of a sparse matrix having non-zero matrix elements only on the diagonal and on the super/sub-diagonals.

However, the partial wave expansion method is efficient only if the partial wave series converges rapidly. According to equation (7.7), the phase shifts for rapidly decreasing potentials tend to zero for $q \to 0$:

$$\delta_l(q) \sim q^{2l+1} \tag{7.15}$$

and therefore, at low energies only few partial waves are needed to obtain very acurrate scattering amplitude.

In contrast to the case of neutral particle, the partial wave series for charged particles converges slowly. It is necessary therefore to write the scattering amplitude as the sum

$$f = f_c + f_{cs}^{(N)}, \tag{7.16}$$

where the Coulomb amplitude is known explicitly and the term $f_{cs}^{(N)}$ is represented by a finite number of terms of the series (7.10) N. The partial amplitudes $s_{l,cs}$ are obtained by solving the ordinary differential equations (7.8). It can be shown that these amplitudes decrease rapidly to zero at large l and hence only a small number of terms yields a high accuracy in spite of the fact that at $q \to 0$, the partial amplitudes have no additional decreasing terms of the type (7.15). If the potential has dipole parts, it is reasonable to isolate explicitly the singular terms f_d in equation (7.16) produced by the dipole interaction.

7.1.4 Partial T-matrix

Let us now turn to the calculation of the amplitudes by solving integral equations in the momentum space. In contrast to differential equations, the integral equations (3.4) are not very convenient for calculations with local potentials. Their kernels are singular and the matrices obtained by their discretization are not sparse. Therefore, to approximate local potentials, separable potentials are usually employed i.e., integral operators with kernels

$$v(k,k') = \sum_{j=1}^{N} \lambda_j \varphi_j(k)\overline{\varphi_j(k')}. \tag{7.17}$$

The perturbation equation (3.4) reduces in this case to equation with degenerate kernel and its solution can be obtained by inversion of an $N \times N$ matrix. In the particular case of the first rank separable potential, the T-matrix kernel simply reads

$$t(k,k',z) = \frac{\varphi_j(k)\overline{\varphi_j(k')}}{\Delta(z)}, \tag{7.18}$$

where

$$\Delta(z) = \lambda^{-1} + \int dk \frac{|\varphi(k)|^2}{k^2 - z}. \tag{7.18'}$$

Let us note that the function $\Delta(k^2 + i0)$ is monotone. Hence, it cannot have more than one root and consequently, the energy operator with separable potential of rank one cannot possess more than one bound state. The corresponding form-factor in this case equals the function $\varphi(k)$ and the normalized eigenfunction $\psi(k)$ is

$$\psi(k) = \frac{\varphi(k)}{k^2 + \varkappa^2},$$

where $-\varkappa^2$ is the bound state energy.

In the general case (7.17), the T-matrix equals the sum

$$t(k,k',z) = \sum_{i,j} \lambda_j \varphi_i(k)\overline{\varphi_j(k')}(\Delta^{-1})_{ij}. \tag{7.19}$$

where Δ_{ij} denotes an $N \times N$ matrix defined as

$$\Delta_{ij} = \frac{1}{\lambda}\delta_{ij} + \int dk \frac{\varphi_j(k)\overline{\varphi_i(k')}}{k^2 - z}.$$

Similarly to the wave function, the T-matrix kernel function can be also expanded in Legendre series

$$t(k,k',z) = \frac{1}{2\pi^2} \sum_{l=0}^{\infty} (2l+1) t_l(q,q',z) P_l(\cos\theta). \tag{7.20}$$

$$\cos\theta = (\hat{k}, \hat{k}'), \quad q = |k|, \quad q' = |k'|,$$

where the on-shell partial T-matrices coincide with the partial S-matrices

$$t_l(q, q, q^2 + i0) = \frac{1}{2iq}\,(e^{2i\delta_l} - 1)$$

and the function $t_l(q, q', z)$ satisfies the integral equation

$$t_l(q, q', z) = v_l(q, q') - \int_0^\infty \frac{v_l(q, q'')}{q''^2 - z} t_l(q'', q', z)\, dq'' \tag{7.21}$$

which can be solved explicitly by replacing the potential v_l by its separable approximation (7.17). For instance, in the case of a spherically symmetric potential of rank one

$$v_l(q, q') = \lambda\varphi(q)\varphi^*(q')$$

the kernel of the T-matrix is given by equation (7.18) in which the integral over k is replaced by one-dimensional integral over q on the interval $[0, \infty)$.

If particles are charged, i.e., if the potential equals to the sum of Coulomb and separable potentials, the T-matrix takes the form

$$t(z) = t_c(z) + t_{cs}(z), \tag{7.22}$$

where the Coulomb T-matrix is given analytically (3.121) and the additional term t_{cs} resulting from the separable part of interaction is represented in an integral form. In the case of separable rank one potential, then

$$t_{cs}(k, k', z) = \frac{g_{cs}(k)\overline{g_{cs}(k')}}{\Delta_{cs}(z)}, \tag{7.22'}$$

where the quantity $\Delta_{cs}(z)$ in determined analogously to (7.18') as

$$\Delta_{cs}(z) = \lambda^{-1} + \int dk dk'\, r_c(k, k', z)\varphi(k')\varphi^*(k),$$

and g_{cs} is a modified Coulomb form-factor

$$g_{cs}(k) = \varphi(k) + \int \frac{t_c(k, q, z)\varphi(q)}{q^2 - z} dq. \tag{7.22''}$$

In the general case, the kernel t_{cs} is determined by equation of the type (7.19) with functions φ_i replaced by the Coulomb modified form-factors $g_{cs}^{(i)}$ which are expressed in terms of φ_i by formulae (7.22'').

In many cases, for example when the functions $\varphi_i(k)$ are rational functions, it is possible to calculate all required integrals explicitly. However, other transformations of the formulae (7.22″) will not be discussed here.

Therefore, the problem of calculating the scattering amplitude for local potentials can be reduced in this approach to the problem of solving a set of independent systems of algebraic equations of finite rank.

We will not discuss methods for approximating local potentials by separable ones. We only stress that the efficiency of such method depends on the number of terms in the potential expansion

$$v(k-k') \sim \sum_i \lambda_i \varphi_i(k) \overline{\varphi_i(k')}$$

determining the accuracy of the approximation as well as the number of partial waves required in expansion (7.20). For example, in nuclear physics the convergence rate of such expansion is fast.

7.2 Partial Equations for Components

In this section, we will consider three-body systems. After isolating angular variables in equations for components of wave functions, we describe methods of solving partial equations. Having in mind applications of such methods in nuclear physics, we will discuss systems of identical particles. In this case, in order not to complicate the notation, we will assume that the particles have spin zero. This means that the wave functions must be symmetric with respect to the interchange of particles. All constructions can be however applied in the standard way to the case of particles of arbitrary spin and isospin.

Let us note that from the numerical point of view, the equations for components are more convenient than the Schrödinger equation. This is related to the fact mentioned in Section 5 Chapter 5 that the boundary conditions for the component $\Phi_\alpha(x_\alpha, y_\alpha)$ are expressed in terms of a fixed pair of Jacobi coordinates $\{x_\alpha, y_\alpha\}$, whereas to describe the asymptotics of the wave functions $\Psi(X)$ in the regions Ω_α, Ω_β, $\alpha \neq \beta$ one has to use all pairs of Jacobi coordinates. This makes it almost impossible to prescribe correctly the boundary conditions for numerical solution of the scattering problem.

7.2.1 Bispherical basis

Let us recall that the wave function components are solutions of the following system of differential equations

$$(-\Delta + v_\alpha(x_\alpha) - E)\,\Phi_\alpha(X) = -v_\alpha(x_\alpha)\sum_{\alpha\neq\beta}\Phi_\beta(X), \tag{7.23}$$

and the wave function equals their sum

$$\Psi(X) = \sum_\alpha \Phi_\alpha(X).$$

We will discuss only scattering processes of one particle on a bound state of two others $(2 \to 2)$, $(2 \to 3)$. In this case, solutions of the system (7.23) are in the class of functions $B_{E_A}(\hat{p}_A)$ described in Chapter 4, Section 3.

Identity of the particles leads to a simple relation between components Φ_α Let us denote by $P^{\pm}$ the operators of cyclic permutation of particles defined as

$$P^+(123) = (312), \quad P^-(123) = (231).$$

Then $\Phi_2 = P^+\Phi_1$, $\Phi_3 = P^+\Phi_1$ and, consequently, the decomposition of the wave function into components takes the form

$$\Psi(X) = (I + P^+ + P^-)\Phi(X), \tag{7.24}$$

where $\Phi \equiv \Phi_1$. This representation ensures symmetry of the wave function with respect to interchanges of any of two particles.

After selecting out the initial state χ_A, $\Phi = \chi_A + \tilde{\Phi}$, the system of equations (7.23) reduces to a single inhomogeneous equation for the function $\tilde{\Phi}$

$$(-\Delta + v(x) - E)\,\tilde{\Phi} = -v(x)(P^+ + P^-)(\chi_A + \tilde{\Phi}). \tag{7.25}$$

The next stage is angular analysis of this equation.

Let $\vec{l}$ and $\vec{\lambda}$ denote operators of the orbital angular momenta of the pair of particles (23) and the particle 1 with respect to the centre of mass of the particles (23),respectively:

$$\vec{l} = -i[x_1, \nabla_{x_1}], \quad \vec{\lambda} = -i[y_1, \nabla_{y_1}].$$

Their eigenfunctions are spherical functions $Y_l^{m_l}(\hat{x})$ and $Y_\lambda^{m_\lambda}(\hat{y})$.

The transformation to the representation of the total angular momentum $\vec{L} = \vec{l} + \vec{\lambda}$ is carried out by the momentum summation rule

$$Y_{\lambda l L}(\hat{x}, \hat{y}) = \sum_{m_l + m_\lambda = M} < l m_l \lambda m_\lambda | LM > Y_l^{m_l}(\hat{x}) Y_\lambda^{m_\lambda}(\hat{y}), \tag{7.26}$$

where $< l m_l \lambda m_\lambda | LM >$ are Clebsch-Gordan coefficients. The functions Y_{aL}, where a denotes the multiple index $a = \{l, \lambda\}$, form a basis in the Hilbert space of functions on a four-dimensional torus $S^{(2)} \times S^{(2)}$ which is known as the bispherical basis.

Let the initial state of a pair of particles be described by an eigenfunction $\psi_l(r)$ with the orbital momentum l. Then the cluster plane wave can be written as

$$\chi(X) = \frac{4\pi}{|p|} \sum_{\lambda, L} i^\lambda Y_\lambda^{m_\lambda} \mathcal{Y}_{aL}(\hat{x}, \hat{y}) \chi_{aL}, \tag{7.27}$$

where summation runs over λ and L for fixed l and

$$\chi_{aL} = j_\lambda(|p|,|y|)\psi_l(|x|) \tag{7.27'}$$

is the partial cluster wave. Analogous expansion holds for the component $\Phi(x,y)$:

$$\Phi(x,y) = \sum_{a,L} \frac{\Phi_{aL}(|x|,|y|)}{|x|\,|y|}\mathcal{Y}_{aL}(\hat{x},\hat{y}). \tag{7.28}$$

In order to carry out the angular analysis of equations (7.25), we express matrix elements of the operators $P^\pm$ in the basis $\{\mathcal{Y}_{aL}\}$. The operators $P^\pm$, when applied to function of coordinates, change values of the arguments::

$$P^+\frac{\Phi_{aL}(|x|,|y|)}{|x|\,|y|}\mathcal{Y}_{aL}(\hat{x},\hat{y}) = \frac{\Phi_{aL}(|x'|,|y'|)}{|x'|\,|y'|}\mathcal{Y}_{aL}(\hat{x}',\hat{y}')$$

where the variables x' and y' are related to the original ones as follows

$$x' = -\frac{1}{2}x + \frac{\sqrt{3}}{2}y, \quad y' = -\frac{1}{2}y - \frac{\sqrt{3}}{2}x. \tag{7.29}$$

The coefficients $-\frac{1}{2}$ and $\frac{\sqrt{3}}{2}$ represent the coefficients $c_{\alpha\beta}$ and $s_{\alpha\beta}$ in the case of equal masses.

Expanding the right hand side of the above expression for P^+ into angular functions $\mathcal{Y}_{aL}(\hat{x},\hat{y})$, we obtain the representation

$$P^+\frac{\Phi_{aL}(|x|,|y|)}{|x|\,|y|}\mathcal{Y}_{aL}(\hat{x},\hat{y}) =$$

$$= \sum_{a'} \frac{1}{2}\frac{\mathcal{Y}_{aL}(\hat{x},\hat{y})}{|x|\,|y|}\int_{-1}^{1} du\, h^L_{a'a}(u)\Phi_{aL}(|x'|,|y'|), \tag{7.30}$$

where $u = (\hat{x},\hat{y})$ and now the problem consists of calculating the weight functions $h^L_{a'a}$.

According to equation (7.29), the radial part of $\Phi_{aL}(|x'|,|y'|)$ depends on directions of the vectors x and y only through the scalar product (x,y) and therefore expanding in Legendre polynomials we get

$$\frac{\Phi_{aL}(|x'|,|y'|)}{|x|^{l+1}|y|^{\lambda+1}} = \sum_{k=0}^{\infty}(-1)^k 4\pi\sqrt{2k+1}\mathcal{Y}_{kk0}(\hat{x},\hat{y})\times$$

$$\times\frac{1}{2}\int_{-1}^{1} du P_k(u)\frac{\Phi_{aL}(|x'|,|y'|)}{|x|^{l+1}|y|^{\lambda+1}}.$$

Making use of the simple relation

$$|\alpha x + \beta y| Y_l(\widehat{\alpha x + \beta y}) =$$

$$= \sum_{l_1+l_2=l} (\alpha|x|)^{l_1} (\beta|y|)^{l_2} \left(\frac{4\pi(2l+1)!}{(2l_1+1)!(2l_2+1)!} \right)^{1/2} \mathcal{Y}_{l_1 l_2 l}(\hat{x}, haty)$$

we get the following expression

$$\mathcal{Y}_{aL}(\hat{x}', haty')|x'|^{l}|y'|^{\lambda} =$$

$$= \sum_{\substack{\lambda_1+\lambda_2=\lambda \\ l_1+l_2=l}} D_{\lambda_1\lambda_2} \tilde{D}_{l_1 l_2} \sum_{\lambda' l'} [(2\lambda+1)(2l+1)(2\lambda'+1)(2l'+1)]^{1/2} \times$$

$$\times \begin{pmatrix} \lambda_1 & \lambda_2 & \lambda \\ l_1 & l_2 & l \\ \lambda' & l' & L \end{pmatrix} |y|^{\lambda_1+l_1} |x|^{\lambda_2+l_2} [\mathcal{Y}_{\lambda_1 l_1 \lambda'}(\hat{y}, haty) \otimes \mathcal{Y}_{\lambda_2 l_2 l'}(\hat{x}, hatx)]_L,$$

where the coefficients are of the form

$$D_{\lambda_1\lambda_2} \tilde{D}_{l_1 l_2} = 4\pi \left(\frac{(2\lambda+1)!(2l+1)!}{(2\lambda_1+1)!(2\lambda_2+1)!(2l_1+1)!(2l_2+1)!} \right)^{1/2} \times$$

$$\times \frac{(-1)^{\lambda+l_2} (\sqrt{3})^{\lambda_2+l_1}}{2^{\lambda+l}}. \tag{7.31}$$

Finally, uniting the obtained angular function $\mathcal{Y}_{kk0}(\hat{x}, haty)$ with the function $\mathcal{Y}_{\lambda'' l'' L}(\hat{x}, haty)$, we obtain the operator $P+$ in the form

$$P+ \frac{\Phi_{aL}(|x|,|y|)}{|x|\,|y|} \mathcal{Y}_{aL}(\hat{x}, haty) = \frac{(-1)^{l+\lambda+L}}{4\pi} \sum_{k=0}^{\infty} (-1)^k \sqrt{2k+1} \times$$

$$\times \frac{1}{2} \int_1^{-1} P_k(u) \frac{\Phi_{aL}(|x'|,|y'|)}{|x|^{l+1}|y|^{\lambda+1}} du \sum_{\substack{\lambda_1+\lambda_2=\lambda \\ l_1+l_2=l}} D_{\lambda_1\lambda_2} \tilde{D}_{l_1 l_2} \times$$

$$\times \sum_{\lambda' l'} [(2\lambda+1)(2l+1)(2\lambda_1+1)(2l_1+1)(2\lambda_2+1)(2l_2+1)]^{1/2} \times$$

$$\times |y|^{\lambda_1+l_1} |x|^{\lambda_2+l_2} \begin{pmatrix} \lambda_1 & \lambda_2 & \lambda \\ l_1 & l_2 & l \\ \lambda' & l' & L \end{pmatrix} < \lambda_1 0 l_1 0 | \lambda' 0 >< \lambda_2 0 l_2 0 | l' 0 > \times$$

$$\times \sum_{\lambda'' l''} \sqrt{(2\lambda'+1)(2l'+1)(2k+1)} \times$$

$$\times < k0\lambda'0|\lambda''0 >< k0l'0|l''0 > \begin{pmatrix} l'' & \lambda'' & L \\ \lambda' & l' & k \end{pmatrix} \mathcal{Y}_{\lambda''l''L}(\hat{x}, haty).$$

If we compare this expression with equation (7.30) which defines the kernel $h^L_{a'a}(|x|,|y|,u)$, we find

$$h^L_{a'a} = \frac{|x|\,|y|}{|x'|\,|y'|}(-1)^{l+L}\frac{(2\lambda+1)(2l+1)}{2^{\lambda+l}}\times$$

$$\times[(2\lambda)!(2l)!(2\lambda''+1)(2l''+1)]^{1/2}\times$$

$$\times\sum_{k=0}^{k_{max}}(-1)^k(2k+1)P_k(u)\sum_{\substack{\lambda_1+\lambda_2=\lambda\\ l_1+l_2=l}}\frac{|y|^{\lambda_1+l_1}|x|^{\lambda_2+l_2}}{|y'|^{\lambda}|y'|^{l}}\times$$

$$\times(-1)^{l_2}(\sqrt{3})^{\lambda_2+l_1}[(2\lambda_1)!(2\lambda_2)!(2l_1)!(2l_2)!]^{-1/2}\times$$

$$\times\sum_{\lambda'l'}(2\lambda'+1)(2l'+1)\begin{pmatrix} \lambda_1 & l_1 & \lambda' \\ 0 & 0 & 0 \end{pmatrix}\begin{pmatrix} \lambda_2 & l_2 & l' \\ 0 & 0 & 0 \end{pmatrix}\times$$

$$\times\begin{pmatrix} k & \lambda' & \lambda'' \\ 0 & 0 & 0 \end{pmatrix}\begin{pmatrix} k & l' & l'' \\ 0 & 0 & 0 \end{pmatrix}\begin{pmatrix} l'' & \lambda'' & L \\ \lambda' & l' & k \end{pmatrix}\begin{pmatrix} \lambda_1 & \lambda_2 & \lambda \\ l_1 & l_2 & l \\ \lambda' & l' & L \end{pmatrix}, \quad (7.32)$$

$$k_{max} = \frac{1}{2}(\lambda'' + l'' + \lambda + l).$$

The action of the operator P^- is determined by equation (7.30) in which the kernel $h^L_{a'a}$ should be replaced by $(-1)^{\lambda+\lambda'}h^L_{a'a}$

It should be pointed out that the potential $v(x)$ is diagonal in the basis $\{\mathcal{Y}_{aL}\}$:

$$< a'L'|v|aL >= \delta_{L'L}\delta_{\lambda'\lambda}\delta_{l'l}v_l(r).$$

We introduce the following notation for the centrifugal potential

$$v^{(orb)}_{aL} = \frac{l(l+1)}{|x|^2} + \frac{\lambda(\lambda+1)}{|y|^2}.$$

Comparing coefficients of identical basis elements, we obtain an infinite system of integro-differential equations for radial components of wave functions

$$\left(-\frac{\partial^2}{\partial x^2} - \frac{\partial^2}{\partial y^2} + v^{(orb)}_{aL} - E\right)\tilde{\Phi}(|x|,|y|) =$$

$$= -v_l(|x|)\left(\tilde{\Phi}_{aL}(|x|,|y|) + \frac{1}{2}\sum_{a''}\int_{-1}^{1}(1+(-1)^{\lambda+\lambda'})\times\right.$$

$$\times\, h^{L}_{aa''}\left(\tilde{\Phi}_{a''L}(|x'|,|y'|)\chi_{a''L}(|x'|,|y'|)\right)du\bigg). \tag{7.33}$$

The radial components $\tilde{\Phi}_{aL}$ defined in the first quadrant $x>0$, $y>0$, vanish at the coordinate axes

$$\tilde{\Phi}_{aL}|_{x=0}=\tilde{\Phi}_{aL}|_{y=0}=0.$$

by virtue of equation (7.28).

The system of equations (7.33) must be supplemented by asymptotic boundary conditions. To this end, we expand the asymptotic representations (4.25) - (4.28) for components from the class $B_{E_A}(\hat{p}_A)$ in bispherical basis and we take into account the fact that by virtue of isotropy of space, the amplitudes F_{BA} and F_{0A} do not depend on projection of angular momentum. As a result we obtain the asymptotic equalities

$$\tilde{\Phi}_{aL}(|x|,|y|)\sim f^{L}_{aa_0}\chi^{(+)}_{aL}(|x|,|y|)+A^{L}_{aa_0}(\theta)\frac{e^{i\sqrt{E}\rho}}{\rho^{1/2}} \tag{7.34}$$

where $\rho^2=x^2+y^2$, $\theta=\mathrm{arccot}|y|/|x|)$. $\chi^{(+)}_{aL}$ denotes the outgoing cluster wave defined by (7.27) with spherical Bessel function $j_\lambda(|q|\,|y|)$ replaced by spherical Hankel function

$$h_\lambda(|q|\,|y|)=\left(\frac{\pi|q|\,|y|}{2}\right)^{1/2}H^{(1)}_{\lambda+1/2}(|q|\,|y|).$$

The coefficients $f^{L}_{aa_0}$ are called the partial amplitudes and the quantities $S^{L}_{aa_0}=\delta_{aa_0}+2if^{L}_{aa_0}$ the partial S-matrices. They describe processes of elastic scattering and internal rearrangement. According to (7.34) and (7.28), the physical amplitudes $F(\hat{y},p_A)$ are expressed as

$$F(\hat{y},p_A)=\frac{4\pi}{|p_A|}\sum_{a,a_0,L}f^{L}_{aa_0}\mathcal{Z}^{*}_{a_0L}(\hat{p}_A)\mathcal{Z}_{aL}(\hat{y}), \tag{7.34'}$$

where

$$\mathcal{Z}_{aL}(\hat{y})=i^{-\lambda}<lm_l\lambda m_\lambda|LM>Y^{m_\lambda}_{\lambda}(\hat{y}).$$

The function $A_{aL}(\theta)$ so called partial breakup amplitudes determine component $F_1(\hat{X},p_A)$ of the full amplitude as follows

$$F_1(\hat{X},p_A)\frac{E_A}{|p_A|}\sum_{a,a_0,L}\frac{A^{L}_{aa_0}(\theta)}{\sin\theta\cos\theta}\mathcal{Y}_{aL}(\hat{x},\hat{y})\mathcal{Z}^{*}_{a_0L}(\hat{p}_A). \tag{7.35}$$

According to (7.24), the breakup amplitude equals the sum of components

$$F_0(\hat{X}, P) = (1 + P^+ + P^-)F_1(\hat{X}, P). \qquad (7.35')$$

It is also possible to represent this amplitude in the form of the series (7.35) if one makes use of equation (7.30). In this case the functions A_{aL} on the right hand side of (7.35) should be replaced by

$$\mathcal{A}^L_{aa_0}(\theta) = A^L_{aa_0}(\theta) + \int_{-1}^{1} du \sum_{a'} H^L_{aa'}(\theta, u) A^L_{a'a_0}(\theta')$$

where $\theta' = \arctan \frac{y'}{x'}$.

Let us point out that the amplitudes $f^L_{aa_0}$ and $\mathcal{A}^L_{aa_0}$ are related to each other by the unitarity condition

$$1 = \sum_{\alpha} |S^L_{aa_0}|^2 + \frac{4E_A(p_A)}{|p_A|} \int_0^{\frac{\pi}{2}} d\theta \sum_{\alpha} \mathcal{A}^{L*}_{aa_0}(\theta) A^L_{aa_0}(\theta).$$

In this way, the original differential equation (7.25) in six-dimensional space has been reduced to the infinite set of integro-differential equations (7.33) where the differential operator acts on two-dimensional variables and the integration is performed along arcs centred at the origin of the coordinates. For numerical calculation such system must be truncated.

This means that the original task with a potential $v(r)$ has to be replaced by a model problem with interactions in the form of a matrix integral operator acting on angular variables. Starting from certain value of the two-particle angular momentum l, the matrix elements $v_{ll'}$ vanish. Efficiency of this approach must of course be tested in real calculations. It should be pointed out, however, that the number of partial waves in which the potential acts should equal the number of terms giving non-trivial contribution to expansion (7.5) and this number is small at low energies .

As a result, for each value of the total angular momentum L we obtain a system of equations whose rank equals the number of partial waves in which the potential $v_{ll'}$ effectively acts. There exist well developed numerical methods to solve such systems. Here we describe one such method based on finite difference approximation of differential and integral operators.

7.2.2 Numerical solution of the scattering problem

For simplicity, we will assume, as it is usually done in nuclear physics that the particles interact only in the s-state ($l = 0$). In this case the only non-zero matrix elements are $< 0\lambda L|v|0\lambda L >= v_0(x)$. This means that in expansion (7.28) we should only keep terms $a = \{\lambda, l\}$ with $l = 0$:

$$\Phi(|x|, |y|) = \sum_L \frac{\Phi_L(|x|, |y|)}{|x|\,|y|} \mathcal{Y}_{L_0L}(\hat{x}, \hat{y}). \tag{7.36}$$

Assume further that the two-body subsystems possess one bound state $\psi(r) = r^{-1}\varphi(r)$ the energy $-\varkappa^2$. In this case, the system (7.33) reduces to the set of independent integro-differential equations

$$\left(\frac{\partial^2}{\partial x^2} + \frac{\partial^2}{\partial y^2} - v(x) - \frac{L(L+1)}{y^2} - E\right) \Phi_L(x, y) =$$

$$= v(x) \int_{-1}^{1} du\, h^L(x, y, u)(\Phi_L(x', y') + \chi_L(x', y')); \tag{7.37}$$

$$x, y > 0.$$

The geometric function $h^L \equiv h^L_{0L,0L}$ calculated by means of equation (7.32) can be written as

$$h^L = \frac{xy}{x'y'} \frac{4}{\sqrt{3}} \left(\frac{-1}{2\sin\theta'}\right)^L \times$$

$$\times \sum_{k=0}^{L} \frac{L!}{k!(l-k)!} P_k(u)(\sqrt{3}\cos\theta)^k(\sin\theta)^{L-k}, \tag{7.38}$$

where

$$u = -\frac{\cos 2\theta + 2\cos 2\theta'}{\sqrt{3}\sin\theta}$$

$$x' = \frac{1}{2}(x^2 - 2\sqrt{3}xyu + 3y^2)^{1/2}$$

$$y' = \frac{1}{2}(3x^2 + 2\sqrt{3}xyt + y^2)^{1/2}.$$

The unknown functions $\Phi_L(x, y)$ must satisfy the boundary conditions

$$\Phi_{aL}(x, y)|_{x=0} = \Phi_{aL}(x, y)|_{y=0} = 0 \tag{7.39}$$

and moreover the following asymptotic conditions

$$\Phi_L(x, y) \sim \lambda_L \psi(x) \exp\left\{iqy + i\pi\frac{L+1}{2}\right\} +$$

$$+A_L\left(\frac{x}{y}\right)\frac{\exp\{i\sqrt{E}\rho\}}{\rho^{1/2}}. \tag{7.40}$$

The coefficient λ_L are usually written as $\lambda_L = \frac{1}{2i}\left(\eta_L e^{2i\delta_L} - 1)\right]$ where η_L are called the absorption coefficients and δ_L the phase shifts. Below the breakup threshold, the absorption coefficients are equal to one, above the threshold they are smaller than one. Note that in this case elastic scattering, the amplitude is expressed in terms of partial amplitudes in analogy to equation (7.5):

$$F(\hat{y}, p_A) = \frac{1}{2i|p_A|}\sum_L (2L+1)\lambda_L P_L(\cos\theta), \quad \chi os\theta = (\hat{y}, \hat{p}_A). \tag{7.41}$$

To calculation $\Phi_L(x, y)$, it is convenient to apply the finite-difference approximation to the equation (7.37) in polar coordinates $\rho = (x^2 + y^2)^{1/2}$, $\theta = \arctan\frac{x}{y}$. This choice allows for the most natural expression of the integral on the right hand side of equation (7.37). Let the grid for finite-difference approximation have N_θ mesh points on the arc $\rho =$const and $N_\rho + 1$ mesh points on the ray $\theta =$const. (see Figure 20). Let χ, $\chi = \{\chi_1, \ldots, \chi_{N_m}\}$ denotes the vector of values of Φ_L at the mesh points

$$\chi \in R^{N_m}, \quad N_m = (N_\rho + 1)N_\theta. \tag{7.42'}$$

After passing to finite differences, equation (7.37) can be written in the matrix form

$$(B - E)\chi = \tilde{\chi}_0. \tag{7.42}$$

Here B is a linear operator operating in the finite-dimensional vector space R^{N_m} corresponding to the Laplacian and to the integral (7.37). $\tilde{\chi}_0$ denotes the vector of values of the inhomogeneous term in (7.37) evaluated at the grid. To solve equation (7.42) means to find χ for $\tilde{\chi}_0$ given.

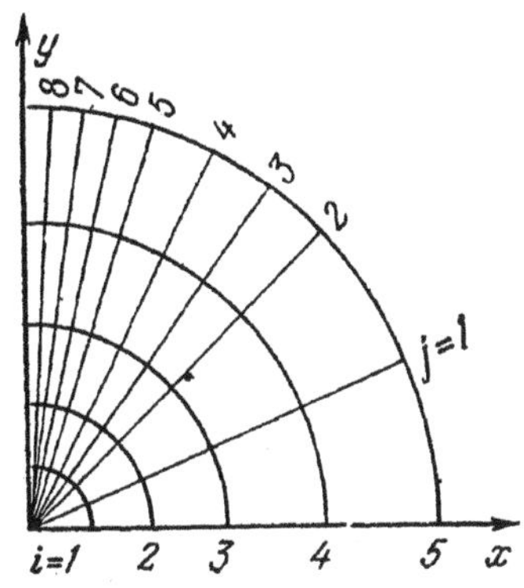

Fig. 20

It is convenient to label the mesh points for the description of χ and χ_k in the following way. Let $\tilde{\chi}_j$ be the vector from the space $R^{N_{\theta j}}$ describing values of the function Φ_L on the j-th ray $\rho = \rho_j$. The components of this vector χ_k^j are defined as

$$\chi_k^j = \Phi_L(\rho_j \cos\theta_k, \rho_j \sin\theta_k),$$

$$j = 1, 2, \ldots, N_{\theta_j}, \quad 0 \le \theta_j \le \pi/2.$$

The components are labelled in order of increasing angle θ. The space R^{N_m} can be written in the form of the orthogal sum

$$R^{N_m} = \sum_{j=1}^{N_\theta} \bigoplus R^{N_{\theta j}}, \quad \chi = \{\hat{\chi}^1, \hat{\chi}^2, \ldots, \hat{\chi}^{N_\theta}\}. \tag{7.43}$$

In this representation, the finite-difference equation can be reduced to the form

$$L_j\hat{\chi}^{j-1} + M_j\hat{\chi}^j + R_j\hat{\chi}^{j+1} = I_j, \tag{7.44}$$

where the operators L_j, M_j,and R_j map from the subspaces $R^{N_{\theta j-1}}$, $R^{N_{\theta j}}$, and $R^{N_{\theta j+1}}$ into $R^{N_{\theta j}}$, respectively. These operators are represented by real square matrices of rank N_{θ_j}. The matrices L_j R_j are diagonal. They are generated by the radial part of the Laplacian

$$\frac{\partial^2}{\partial\rho^2} + \frac{1}{\rho}\frac{\partial}{\partial\rho}, \tag{7.45}$$

which, after applying the finite-difference approximation, couples values of the functions $\Phi_L(x,y)$ at the neighbouring rays to each other. The matrix M_j of rank N_{θ_j} corresponds to the contribution of the angular part of the Laplacian $\rho^{-2}\frac{\partial^2}{\partial\theta^2}$ and to integration along the arc $\rho = \rho_j$. The presence of the integral operator on the right hand side of (7.37) makes this matrix non-diagonal. I_j denotes the vector from $R^{N_{\theta j}}$ corresponding to the inhomogeneous term $\tilde{\chi}_0$:

$$\tilde{\chi}_0 = \{I_1, I_2, \ldots, I_{N_\rho}\}.$$

The matrix B in equation (7.42) in thus a sparse matrix and in the representation (7.43) it has a band structure (see Figure 21).

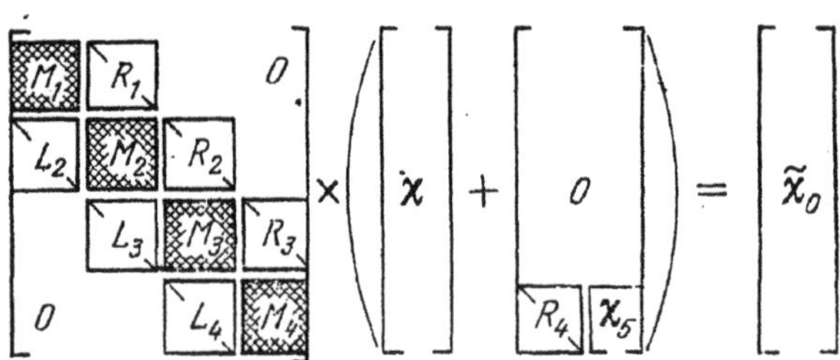

Fig. 21

There are $(N_\rho + 1)N_\theta$ unknowns in equations (7.44) to determine which we have $N_\rho N_\theta$ relations. For equation (7.44) to have a unique solution corresponding to the physical wave function, the boundary condition (7.40) must be supplemented.

Neglecting terms of the order $O(\rho^{-5/2})$ in equation (7.42) we obtain

$$\hat{\chi}^{N_\rho} = \lambda_L \hat{\Psi}^{N_\rho} + \frac{\exp\{i\sqrt{E}\rho_{N_\rho}}{\rho^{1/2}_{N_\rho}} \hat{A}_{N_\rho}$$

(7.46)

$$\hat{\chi}^{N_{\rho+1}} = \lambda_L \hat{\Psi}^{N_{\rho+1}} \frac{\exp\{i\sqrt{E}\rho_{N_{\rho+}}}{\rho^{1/2}_{N_{\rho+}}} \hat{A}_{N_\rho},$$

where $\hat{\Psi}^{N_\rho}$ and $\hat{A}_{N_\rho}$ are vectors obtained from values of the function $\chi_L(x,y)$ and the amplitude $A_L\left(\binom{x}{y}\right)$ evaluated at the grid points. After eliminating the unknown vector $\hat{A}_{N_\rho}$, we obtain a relation between the vectors $\hat{\chi}^{N_\rho}$ and $\hat{\chi}^{N_{\rho+1}}$ which yields the deficient condition for determination of the vector χ.

$$\hat{\chi}^{N_{\rho+1}} = C_{N_\rho}\hat{\chi}^{N_\rho} + \lambda_L : \hat{\Phi}_{N_\rho}. \tag{7.47}$$

Here

$$C_{N_\rho} = \left(\frac{\rho_{N_\rho}}{\rho_{N_{\rho+1}}}\right)^{1/2} \exp\left\{i\sqrt{E}\left(\rho N_{\rho+} - \rho_{N_\rho}\right)\right\},$$

$$\hat{\Phi}_{N_\rho} = C_{N_\rho}\hat{\Psi}_{N_\rho} - \hat{\Psi}_{N_{\rho+1}}.$$

Equation (7.44) supplemented by the boundary condition (7.47) takes the form

$$\hat{M}_\chi = \tilde{\chi}_0 + \lambda_L\hat{\Phi}, \tag{7.48}$$

schematically represented in Figure 22.

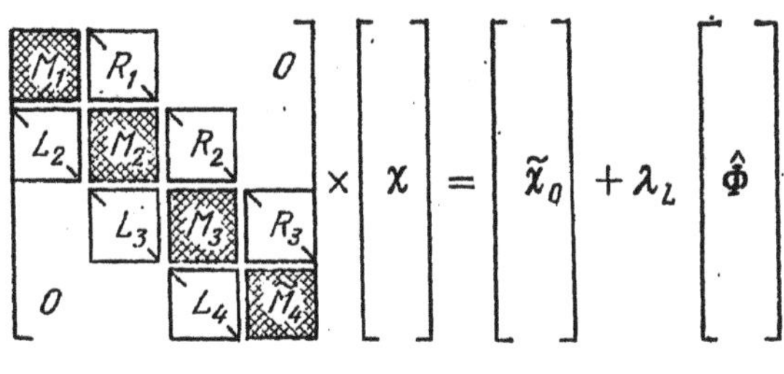

Fig. 22

All the vectors are given in representation (7.43) and the vector $\hat{\Phi}$ has only one non-zero components $\hat{\Phi}_{N_\rho}$ in the space R^{N_ρ}. In figure 22, $\tilde{M}_{N_\rho}$ denotes the linear operator in R^{N_ρ} defined as

$$\tilde{M}_{N_\rho} = M_{N_\rho} + C_{N_\rho}R_{N_\rho}.$$

Note, that on the right hand side of (7.44) an unknown quantity λ_L appears which represents the partial amplitude of absolutely elastic scattering. By virtue of linearity, a solution or (7.48) can be written as the sum

$$\chi = A + \lambda_L B, \tag{7.49}$$

where the vectors A and B are determined by equations

$$\hat{M}A = \tilde{\chi}_0, \quad \hat{M}B = \hat{\Phi} \tag{7.50}$$

with known inhomogeneous terms.

Solving equations (7.50) is equivalent to inverting a matrix of the order $N_\rho N_\theta$. In general, inversion of large matrices is a difficult task. However, as long as the matrix $\hat{M}$ under consideration is sparse and has a band structure, it can be efficiently inverted for example by the Gauss algorithm. This matrix can be reduced to a triangular form by subsequent elimination of the unknowns. Then, with the help of the inverse procedure, the vectors A and B can be found and the solution χ can be written in the form (7.49).

The unknown coefficients λ_L can be determined, for instance, by means of the asymptotic formula (7.40).

Compare now equations (7.46) and (7.49) in the region Ω_1, where $x \gg y$. At large values of ρ_{N_ρ}, the second term in (7.46) is much larger than the first one (at energies below the threshold it is exponentially small, whereas at $E > 0$ it is of order of $O(\rho^{-1/2})$). If the second term is neglected, λ_L is given by the following relation

$$\lambda_L \hat{\Phi}_i^{N_\rho} = A_i^{N_\rho} + \lambda_L B_i^{N_\rho},$$

from which it follows that

$$\lambda = A_i^{N_\rho} (\hat{\Phi}_i^{N_\rho} - B_i^{N_\rho})^{-1}.$$

Here $\hat{\Phi}_i^{N_\rho}$ denotes components of vectors $\hat{\Phi}^{N_\rho}$ corresponding to small values of the angles θ_i. For example, one can set $i = N_\theta - 1$. Once λ_L in known, by means of equality (7.49) it is possible to find the vector $\hat{\chi}^{N_\rho}$ corresponding to the values of the function Φ_L on the outer arc $\rho = \rho_{N_\rho}$. The breakup amplitude A_L is then specified by the equation

$$\hat{A}_{N_\rho} = \left(\hat{\chi}^{N_\rho} - \lambda_L \hat{\Phi}^{N_\rho}\right) \rho_{N_\rho}^{1/2} \exp\left\{-i\sqrt{E}\rho^{N_\rho}\right\}.$$

Thus, to determine the physically interesting quantities λ_L and A_L it is not necessary to compute the function Φ_L at the "inner" points, they are determined by values of this function at the boundary.

One must emphasize that in order to increase the accuracy and efficiency of the calculation, the grid points must be spaced non-uniformly. The grid size must be finer near the axis $x = 0$ and more widely spaced at $x \to \infty$ (see Figure 20). Such distribution of the mesh points is necessary to describe accurately the intricate behaviour of the wave function in the region Ω_1. In the region Ω_{10}, where $x \to \infty$, the wave function takes the asymptotic form (7.40) and depends smoothly on the angular variable. The grid size in this region may be chosen to be more widely spaced. In this connection it is of importance to fulfil the conditions of self-consistency which mean that the error due to the introduction of the "cut-off radius" in the vicinity of the boundary $\rho = \rho_{N_\rho}$ should be smaller than the error of discretisation. This condition is satisfied when

$$|\Delta h|^2 \gg (\rho_{N_\rho})^{-3/2}$$

where Δh denotes the minimal grid size in vicinity of the boundary.

7.2.3 Charged particles

The modified equations for components (5.183) can be used for calculation of scattering amplitudes in the case of charged particles.

The partial wave analysis is carried out exactly as in the case of neutral particles. As the result, a homogeneous infinite set of integro-differential equations of the type (7.33) is obtained. The left hand side of the equations must be supplemented by matrix elements of long range forces for $\beta \neq \alpha$:

$$\int d\hat{x}_\alpha d\hat{y}_\alpha \mathcal{Y}_{aL}(\hat{x}_\alpha, \hat{y}_\alpha) \mathcal{Y}^*_{a'L'}(\hat{x}_\alpha, \hat{y}_\alpha) \sum_{\beta \neq \alpha} \frac{n_\beta}{|x_\beta|}(1 - \chi_\beta) =$$

$$= \delta_{LL'} v^{(0)}_{aLa'L'}, \tag{7.51}$$

and on the right hand side, the potential $v_{ll'}$ must be replaced by the short-range term

$$\hat{v}_{ll'} \delta_{ll'} \delta_{\lambda\lambda'} =$$

$$\int d\hat{x}_\alpha d\hat{y}_\alpha \mathcal{Y}^*_{aL}(\hat{x}_\alpha, \hat{y}_\alpha)\mathcal{Y}_{a'L'}(\hat{x}_\alpha, \hat{y}_\alpha)\left(\frac{n_\alpha}{|x_\alpha|}\chi_\alpha + v^{(s)}(x_\alpha)\right). \qquad (7.51')$$

Note that the long-range interaction potential (7.51) is, contrary to the short-range parts, non-diagonal in the basis $\{\mathcal{Y}_{aL}\}$.

For numerical calculation, the infinite system of equations must be truncated and one should be considered only systems of finite rank. To this end, all matrix elements of short- and long–range potentials should be set to zero starting with some $l = l_0$. This means that $\hat{v}_{ll'} = 0$ and $v^{(0)}_{aLa'L'} = 0$ for $l, l' > l_0$.

As in the case of neutral particles, the cut-off parameter l_0 depends on the number of terms in the partial wave expansion (7.20) needed for required accuracy. The cutting-off procedure of the matrix elements $v^{(0)}_{aLa'L'}$ must be checked independently by comparing the results for various values of l_0. The results should stabilize and depend weakly on l_0. As the quantitative condition of admissibility of the cut-off procedure one uses the smallness of the elements $v^{(0)}_{aLa'L'}$ in the region where the wave function is significantly different from zero. This can always be achieved by taking the cut-off parameter of the Coulomb potential sufficiently large. Asymptotic conditions which define the physical solution can be obtained either by construction of the eikonal approximations to the partial equation or by the direct expansion of the asymptotic formulae (5.183) in the basis $\mathcal{Y}_{aL}$. We will not describe these conditions in the general case and will discuss only the simple case of particles interacting in the s-state.

In number of cases, for example when the particles have charges o the same sign and the Coulomb interaction is much weaker then the additional one, the computation may be done by using the simplest variant of the modified equations (5.184) in which Coulomb interaction is incorporated in the "unperturbed" Hamiltonian.

We will describe the formulation of the boundary problem for the case of three identical particles interacting in s-state. This means that all matrix elements of the potentials vanish for $l, l' > 0$. Only the simplest modified equation (5.184) will be considered. In this case, we have a set of independent equations (7.37), to the left hand side of whose, the Coulomb matrix elements must be added. In accordance with assumptions concerning the character of the interactions, it is sufficient to calculate only matrix elements $v^{(0)}_L =$

$(\mathcal{Y}_{L_0L}, (v_2^{(c)} + v_3^{(c)})\mathcal{Y}_{L_0L})$. Since the Coulomb potential depends only on the angle between vectors x and y, the following representation holds for these matrix elements

$$v_L^{(0)} = 2n \int_{-1}^{1} du((x^2 - 2\sqrt{3}xyu + 4y^2)^{-1/2} +$$

$$+(x^2 + 2\sqrt{3}xyu + 4y^2)^{-1/2}) = \frac{2n\mu(\theta)}{\rho}.$$

where

$$\mu(\theta) = \begin{cases} \frac{2}{\sqrt{3}\sin\theta}, & \theta \geq \frac{\pi}{6} \\ \frac{2}{\cos\theta}, & \theta < \frac{\pi}{6}. \end{cases}$$

As a result, we obtain the equation

$$\left(\frac{\partial^2}{\partial x^2} + \frac{\partial^2}{\partial y^2} - \frac{L(L+1)}{y^2} - \frac{2n\mu(\theta)}{\rho} + E\right)\Phi_L(x,y) =$$

$$= v_s(x)\left(\Phi_L(x,y) + \int_{-1}^{1} du h^L(\theta,u)\Phi_L(x',y')\right). \tag{7.52}$$

According to (5.69), the initial state of the system L_A is a product of an eigenfunction of the operator h_1 and a wavefunction of the Hamiltonian $h_1^c = -\Delta_y + v_1^c$ where the potential

$$v_1^c(y) =$$

$$= \int dx|\psi(x)|^2 \left(\left|-\frac{1}{2}x + \frac{\sqrt{3}}{2}\right|^{-1} + \left|\frac{1}{2}x + \frac{\sqrt{3}}{2}\right|^{-1}\right)$$

describes scattering of one particle on a pair of bound particles which generates the effective long range potential $v_1^c(y)$. This potential is purely Coulomb for $y \to \infty$: $v_1^c(y) \sim \frac{2n}{\sqrt{3}y}$. After separating angular variables, we obtain radial components of the initial state $\chi_l = \psi(x)\tilde{\psi}_L(qy)$. The function $\tilde{\psi}_L(qy)$ is given by equation (7.8′) in which the Coulomb parameter n is replaced by the effective charge $2n/\sqrt{3}$.

For $U_L = \Phi_L - \chi_L$, the following inhomogeneous equation holds

$$\left(\frac{\partial^2}{\partial x^2} + \frac{\partial^2}{\partial y^2} - v(x) - \frac{L(L+1)}{y^2} - \frac{2n\mu(\theta)}{\rho} + E\right)U_L(x,y) =$$

$$= \left(\frac{2n}{\sqrt{y}} - \frac{2n}{x}\right)\chi_L\epsilon\left(\frac{\pi}{6} - \theta\right) -$$

$$-v(x)\int_{-1}^{1} duh^L(\theta,u)\,(\chi_L(x',y') + U_L(x',y')) \qquad (7.53)$$

where $\epsilon(t)$ is the unit step function. The unknown functions satisfy zero boundary conditions (7.39) on the coordinate axes and the following conditions for $\rho \to \infty$

$$U_L \sim \lambda_L \psi(x) \exp\left\{iqy + iW_1 - \frac{i\pi(l+1)}{2}\right\} +$$

$$A^L(\theta)\frac{\exp\{i\sqrt{E}\rho + iW_0\}}{\rho^{1/2}}. \qquad (7.54)$$

The Coulomb phases W_1 and W_0 are given by the eikonal equations (5.50) which, in this case, read

$$W_1 = -\frac{n}{\sqrt{3}}\log 2qy, \quad W_0 = -\frac{n}{2\sqrt{E}}\mu(\theta)\log 2\sqrt{E}\rho.$$

The coefficients λ_L and A^L determine elastic scattering and breakup amplitudes, respectively. According to (5.21) and (5.22), the scattering amplitude F is written in the form of a sum of smooth and singular terms

$$F(\hat{y}_1, p_A) = f_c(\hat{y}_1, p_A) + f_{cs}(\hat{y}_1, p_A).$$

Here f_c denotes Coulomb scattering amplitude corresponding to the potential $\frac{2n}{\sqrt{3}y}$ and the smooth part f_{cs} is expressed, in analogy to (7.10), in terms ofthe partial amplitudes λ_L

$$f_{cs} = \frac{1}{2iq}\sum_{L=0}^{\infty}(2L+1)\lambda_L P_L(\cos\theta),$$

$$\cos\theta = (\hat{y}_1, \hat{p}_A). \qquad (7.55)$$

The partial phase shifts $\delta_L = \delta_L^{cs} + \arg\Gamma\left(L+1+i\frac{n}{\sqrt{3}q}\right)$ and the absorption coefficients η_L are given at the same time by

$$\lambda_L = \frac{1}{2i}\left(\eta_L e^{2i\delta_L^{cs}} - 1\right)\exp\left(2i\arg\Gamma\left(L+1+i\frac{n}{\sqrt{3}q}\right)\right),$$

so that δ_L^{cs} corresponds to the additional phase shift due to the presence of short-range potential v_s

The relation between breakup amplitude and the coefficients A^L is the same as in the case of neutral particles, i.e., (7.35) and (7.35')

Thus, we obtained integro-differential equations (7.53) for the radial functions and described their boundary conditions. For numerical solution of these equations the same methods as in the neutral particles case may be employed.

Concluding this section, we should note that the homogeneous equations (7.37) or (7.53) treated in the space of square integrable functions possess non-trivial solutions at the points $-E_i$ equal to the eigenvalues of the energy operator. Approximate values of eigenvalues can be obtained at any value of the total momentum L with the help of a finite rank system which is obtained by truncating the potential matrix elements. For their determination, one can, for instance, make use of the finite-difference approximation of the equations supposing that the wave function components are equal to zero at sufficiently large distances from the origin. The eigenvalues $-E_i$ correspond in this case to zeros of the Fredholm determinant of the system like (7.42) which are located below the threshold of continuous spectrum of the operator H.

Efficiency of this approach of determination of eigenvalues of the energy operator has been verified in many cases. One of the reasons is the fact that the structure of components of the wave function is usually simpler than that of the wave function itself. This can be seen for example, on the asymptotic behaviour of the eigenfunctions described at the end of Chapter 4. Consequently, the components are more easily approximable that the wave functions themselves.

7.3 Integral Equations for Separable Potentials

This section will be devoted to description of methods for numerical solution of compact equations in momentum space. We will consider integral equations for kernels $T_{\alpha B}$ which, by virtue of equation (3.63), define wave operators U_B.

7.3.1 Compact equations in bispherical basis

The set of equations (3.28) for the kernels $T_{\alpha B}$ reduces, in the case of identical particles discussed in previous section, to one equation

$$K_1 = Q_1 - T_1(E + i0)R_0(E + i0)(P^+ + P^-)K_1, \tag{7.56}$$

where K_1 denotes an operator with the kernel $T_{1B}(P, p'_B)$ acting from the space $\mathfrak{H}_1$ in $\mathfrak{H}$.

First we perform the partial wave analysis of this equation. By analogy with (7.28), we expand the kernel K_1 in the bispherical basis

$$K_1(P, p'_1) = \sum_{a,L} \frac{K_{aL}(|k|, |p|)}{|k|\,|p|} \mathcal{Y}_{aL}(\hat{k}, \hat{p}). \tag{7.57}$$

The kernel of the T-matrix $T_1(E + i0)$ is given in this basis by the series

$$T(P, P', E + i0) = \sum_{a,L,a',L'} \delta_{ll'}\delta_{\lambda\lambda'}\mathcal{Y}_{aL}(\hat{k}, \hat{p})\mathcal{Y}_{a'L'}(\hat{k}', \hat{p}')\times$$

$$\times \frac{1}{|k|\,|k'|\,|p|\,|p'|} t_{aLa'L'}(|k|, |k'|, E - p^2 + i0)\delta(|p| - |p'|),$$

where $t_{aLa'L'}$ denotes partial T-matrices (7.20). For partial components K_{aL} we obtain an infinite set of two-dimensional integral equations

$$K_{aL}(k, p) =$$

$$= \varphi_l(k)\delta(p - p')\delta_{aL,a_0L_0} - \sum_{a',L'} \int_0^\infty dk'dp'\delta(p - p')\times$$

$$\times \left(\Omega^{(+)}_{aL,a'L'}(k,p,k',p')K_{a'L'}(k'_+,p'_+) + \right.$$

$$\left. + \Omega^{(-)}_{aL,a'L'}(k,p,k',p')K_{a'L'}(k'_-,p'_-)\right), \tag{7.58}$$

where $\varphi_l(k)$ is the form-factor corresponding to the eigenfunction $\psi_l(k)$. Here $\Omega^{(\pm)}_{aL,a'L'}$ denote the singular kernels which can be expressed in terms of the partial T-matrices and the geometric coefficients $h^L_{aa'}$ by the equalities

$$\Omega^{(\pm)}_{aL,a'L'}(k,p,k',p') = \frac{t_{aL,a'L'}(k,k',E-p^2+i0)}{k'^2+p'^2-E-i0} h^L_{aa'}(\theta), \quad \theta = \operatorname{arccot}\frac{p'}{k'}.$$

Here, in accordance with definitions (3.32) of the Chapter 3, we introduced the notation

$$k'_\pm = -\frac{1}{2}k' \pm \frac{\sqrt{3}}{2}p', \quad p'_\pm = -\frac{1}{2}p' \mp \frac{\sqrt{3}}{2}k'.$$

If we truncate this set, as it was done in the previous section, we obtain a set of equations of finite rank for each value of the total angular momentum L. However, even in the simplest case of potentials acting only in the S-state, the numerical solution of this set which reduces to single equation represents a very complicated problem. We will not discuss the problems of how to calculate the pair T-matrices by means of singular integral equations (7.21) and related difficulties; we will just indicate the following circumstances.

The integral equations are usually solved with the use of a finite difference approximation of the integrals. If we take a not very dense grid, say 30×30, then even in the simplest case, the solutions of equation (7.58) reduces to the inversion of a complex 900×900 matrix which is not sparse. Such a problem reaches the present day limits of computational technology. Several approximate methods have been proposed to avoid this difficulty. We present only one such method consisting in replacing the potentials $v(r)$ by their separable approximations.

7.3.2 Separable potentials

We will consider in detail only the simplest case of separable potential of rank one and we start with reduction of the original six-dimensional integral equation (7.56). The equation for partial components will be described later.

Assume that the two-particle Hamiltonian with separable potential has one spherically symmetric bound state with the form-factor $\varphi(k)$.

As we saw in Section 1, the T-matrix is in this case also given by the degenerate kernel (7.18) and, therefore, the solution of equation (7.56) takes form of the product

$$K_1(P, p_1') = \varphi(k_1)Q(p_1, p_1'), \tag{7.59}$$

where the kernel $Q(p_1, p_1')$ satisfies the three-dimensional integral equation

$$Q(p_1, p_1') = \delta(p - p') + \int \frac{\Omega(p, q, E)Q(q, p')dq}{\Delta(E - q^2)}. \tag{7.60}$$

The function $\Delta(E)$ is given by equation (7.18) and the kernel $\Omega(p, q, E)$ reads

$$\Omega(p, q, E) = \frac{\varphi(k_1(p, q))[\varphi(k_2(p, q)) + \varphi(k_3(p, q))]}{k_1^2(p, q) + p^2 - E - i0},$$

where, as above, the variables k_α with arguments denote the Jacobi momenta given as linear functions of p_α and p'_β by equation (3.32), i.e.,

$$k_1(p, q) = -\sqrt{3}p - 2q, \quad k_2(p, q) = \sqrt{3}q - 2p,$$

$$k_3(p, q) = -2p - \sqrt{3}q.$$

Note that the function $\Delta^{-1}(E - q^2)$ has a pole at the point $q^2 = E - \varkappa^2$ and can be written as

$$\Delta^{-1}(E - q62) = \frac{\tilde{v}(q)}{q^2 - E - \varkappa^2 - i0},$$

where $\tilde{v}(q)$ is a smooth function. Thus, the obtained equation can be interpreted as a generalized equation of perturbation theory of the type (3.4) in which the kernel $\Omega(p, q, E)\tilde{v}(q) \equiv \tilde{\Omega}(p, q, E)$ plays a role of effective potential.

It should be mentioned that at positive energies, i.e., above the breakup to three free particles threshold, this potential is singular. However, as we have seen in Chapter 3, its singularities $(k_1^2(p, q) + p^2 - E - i0)^{-1}$ correspond to the secondary singularities of compact equations (3.29) and disappear in the course of integration. By analogy with formulae given in Chapter 3, solution of equation (7.60) on the energy shell coincides, up to multiplicative factor, with the scattering amplitude of elastic scattering

$$F(\hat{p}, p') = -2\pi 62Q(|p'|\hat{p}, p'), \quad E = p'^2 - \varkappa^2.$$

By comparing the breakup amplitude $F_0(\hat{p}, p')$ defined as residuum of the resolvent at the pole $p^2 - z)^{-1}$, with equations (4.32′) and (7.59), we find that its component A_1 reads

$$A_1(\hat{P}, p') = -e^{i\pi/4}\left(\frac{\pi}{2}\right)^{1/2} E^{3/4}\varphi(k)Q(p, p')|_{p^2=E^2+\varkappa^2}.$$

The total amplitude is given by the symmetrised expression

$$F_0 = (I + P^+ + P^-)A_1.$$

Therefore, in the case of separable potentials of rank one, the problem of computing the scattering amplitudes reduces to solving the three dimensional integral equation (7.60). If the potential has rank N, we obtain a system of N equations similar to (7.60).

Let us separate the angular variables $\hat{p}$ in equation (7.60) as in the two-particle problem. Since now the effective potential $\tilde{\Omega}(p, p', E)$ is not symmetric we obtain, unlike equation (7.21), an infinite set of coupled one dimensional integral equations. These equations follow from (7.58) by replacing kernels of partial T-matrices by equations (7.19) for separable potentials. Truncating the infinite set, we obtain a finite set of equations for each value of the total angular momentum L. In particular, if the potential acts only in the S-state, the set reduces to single equation. Let us write down, for example, the equation for L=0

$$Q(p, p') =$$
$$= \delta(p - p') - \int \Omega_0(p, q, E)\Delta_0^{-1}(E - q^2)Q(q, p')\, dq.$$

The kernel $\Omega_0(p, q, E)$ defined as

$$\Omega_0(|p|, |q|, E) = \int d\hat{q}\Omega(p, q, p')\left(k_1^2(p, q) + p^2 - E\right)^{-1}$$

has for $E > 0$ logarithmic singularities that can be evaluated explicitly by integrating by parts with respect to $\xi = (\hat{p}\hat{q})$. These singularieties have the following form

$$\log((\sqrt{3}|p| \pm 2|q|)^2 + p^2 - E).$$

On the energy shell, the function $Q(p,p',E)$ equals the partial wave amplitude of elastic scattering

$$Q(p,p',E)|_{p^2-\varkappa^2=p'^2-\varkappa^2=E} = (2i\sqrt{E})^{-1}\left(E^{2i\delta_l}-1\right), \quad p,p'>0,$$

and the total amplitude is given by equation (7.41). A relation analogous to equation (7.35) exists between the partial components of the breakup amplitude and the total breakup amplitude:

$$A_1 = \varphi(k)Q(p,p',E)|_{p^2+k^2=p'^2-\varkappa^2=E}.$$

The problem of calculating the scattering amplitudes has therefore been reduced to solutving one dimensional integral equations or systems of such equations. This method has been widely used in physics. There exist numerous ways of converting equations with singular kernels to equations with smooth kernels, however, we will not discuss them here. Such problems are solved by using the standard approaches and rich literature on their solution exists.

7.3.3 Superposition of Coulomb and a separable potentials

If the particles are charged and the short-range potentials are separable, then the set of equations (7.56) can also be written in the form of three-dimensional integral equations (7.60). The kernel $\Omega(p,q,E)$ is then expressed in terms of two-particle T-matrices corresponding to the sum of Coulomb and separable interaction. To illustrate problems arising in calculations based on that equations, we will consider, as an example, a system in which only two particles (23) are charged. The separable potentials are of rank one and are assumed to be identical for each pair. We want to find a solution of (7.60) antisymmetric with respect to interchange of the particles. Such solution can have the form of the product (7.59) with function $Q(p,p')$ satisfying equation (7.60) with

$$\Omega(p,p',E) = \frac{\varphi(k_1(p,p'))}{k_1^2(p,p')+p^2-E-i0}\varphi(k_2(p',p)) -$$

$$-\int dq \frac{\varphi(k_1(p,q))\varphi(k_1(p',q))}{(k_1^2(p,q)+p^2-E-i0)(k_1^2(p',q)+p'^2-E-i0)}\times$$

$$\times[t(k_3(q,p), k_3(q,p'), E - q^2 + i0) - t(k_3(q,p), k_3(-q,-p'), E - q^2 + i0)].$$

The total three-particle T-matrix and the amplitude of elastic scattering are expressed in terms of the kernel Q. However, here, contrary to the case of neutral particles, the kernel $Q(p,p')$ must be renormalized in accordance with the procedure described in Chapter 4.

We recall that the most singular term in the kernel $\Omega(p,p',E)$ coincides at negative energies with the pure Coulomb potential describing interaction of one free particle with the bound pair. Therefore, by replacing

$$t(p,p',E) \to \frac{n}{\sqrt{3}\pi^2}(|p-p'|^{-2} + \mu^2),$$

the first T-matrix $t(k,k',E)$ ceases to depend on the integration variable q, whereas the second exchange T-matrix $t(k-k',E)$ remains to be dependent and its singularity smooths after integration.

As a result, we find that the kernel $\Omega(p,p',E)$ for $\mu \to 0$ has a strong Coulomb singularity

$$\Omega_{sing}(p,p',E) = \frac{n}{\sqrt{3}\pi^2}|p-p'|^{-2}(p'^2 - E - \varkappa^2 - i0)^{-1},$$

which coincides with the kernel of the two-particle perturbation theory equation for the Coulomb T-matrix. Carrying out the renormalization $Q \to e^{-i\eta \log \mu} Q$ described in Section 4 of Chapter 4, we obtain the elastic scattering amplitude.

Note, that within the framework of our model, we can describe a system consisting of one neutron and two protons. The scattering length and the amplitude of elastic scattering at energies below the breakup threshold can be calculated on the basis of the renormalized equations. These equations become non-compact above the breakup threshold as a result of circumstances discussed in detail in Chapter 3.

7.4 Cluster Integrals

In this section we will consider a problem which nicely illustrates the use of the methods of quantum scattering theory in other fields of theoretical physics. Namely, this is the calculation of cluster integrals and closely related virial coefficients in quantum statistical physics.

7.4.1 Formulation of the problem

Let us start with definition of cluster integrals and virial coefficients. Expansions of pressure in powers of density v^{-1}

$$\frac{pv}{kT} = \sum_{N=0}^{\infty} A_{N+1}(T) v^{-N}, \tag{7.61}$$

the so called virial expansions, have been used systematically in the theory of non-ideal gases. The temperature dependent coefficients $A_N(T)$ are called the virial coefficients.

Cluster integrals are defined as coefficients $B_N(T)$ in the expansion of pressure in powers of activity ζ:

$$\frac{p}{kT} = \sum_{N=1}^{\infty} B_N(T) \zeta^N.$$

The virial coefficients are related to the cluster integrals by the relation

$$A_{N+1} =$$

$$= -\frac{N}{N+1} \sum_{n} (-1)^{\sum_i n_i - 1} \frac{(N - 1 + \sum_i n_i)}{N!} \prod_i \frac{((i+1)B_{i+1})^{n_1}}{n_i!}.$$

The summation runs over all partitions a of an N-particle system. n_i denotes the number of subsystems of the partition a consisting of i particles, so that $\sum_i i n_i = N$. The first two relations are

$$A_2 = -B_2,$$

$$A_3 = -2B_3 + 4B_2^2.$$

In quantum statistical physics the cluster integrals are expressed directly through energy operator $H^{(N)}$ of an N-particle system and through energy operators $H_{a_k}^{(N)}$ of the partitions. For example, in the case of Boltzman statistics, we have

$$B_N = \frac{1}{N!\lambda^3}\mathrm{Sp}\sum_a(-1)^{\sum_i n_i-1}\left(\sum_i n_i - 1\right)!\times$$

$$\times\left(e^{-\beta H_a^{(N)}} - e^{-\beta H_0^{(N)}}\right), \tag{7.62}$$

where the summation runs over all partitions a_k $(k = 1, 2, \ldots, N-1)$ of an N-particle system. The parameter β is proportional to the inverse temperature, $\beta = \frac{1}{kT}$, and λ denotes the so-called thermal wave length. Similar relations hold also for Bose–Einstein and Fermi–Dirac statistics. We will consider the calculation of cluster integrals in Boltzmann statistic only. The generalization to other types of statistics is a standard procedure and will not be treated here. All necessary formulae can be found in statistical physics textbooks.

The definition of the cluster integrals contains trace of the connected part of the operators $e^{-\beta H^{(N)}}$. Let us recall that the term "connected part" was introduced in Chapter 2 in connection with the classification of singularities of the resolvent. The cluster integrals are apparently closely related to the spectral properties of many-particle Hamiltonians. They can be expressed, for example, in terms of wave functions by using equations (1.36). The statement that these characteristics reduce to the scattering operators is nontrivial, however, and we devote this section to its justification.

More precisely, the problem is to express explicitly the cluster integrals in terms of the scattering operators. We will obtain formulae for the third and fourth cluster integrals. For other N, $N \geq 4$, such formulae are unknown. It is possible, however, to accept a hypothesis that the cluster integral B_N even for $N \geq 4$ are explicitly expressible, as the third integral is, in terms of the scattering operator of N-particle system and the scattering operators of its subsystems.

Let us proceed to the calculation of the cluster integrals. We start with the simplest transformations of equation (7.62). Using the relation between the resolvent and the function of a self-adjoint operator (2.6) we can write the cluster integral B_N in the form of a sum

$$B_N = \frac{1}{\lambda^3 N!}\left(\sum_{i=1}^{n_0} e^{-\beta\varkappa_i^2} + \int_{-\varkappa^2}^{\infty} e^{-\beta E}\Omega_+^{(N)}(E)\,dE\right), \qquad (7.63)$$

where the first term describes contribution from the discrete spectrum of the operator $H^{(N)}$ and the second term the contribution from its continuous components. The function $\Omega_+(E)$ is the trace of the jump of the connected part of the resolvent $[R^{(N)}(z)]_c$ on the continuous spectrum. This function can be written, with the help of the fact that the resolvent is self-adjoint $R^{(N)}(\bar{z}) = R^{(N)}(z)$, as the limit

$$\Omega_+^{(N)}(E) = \lim_{\epsilon\downarrow 0} \operatorname{Sp}\operatorname{Im}[R^{(N)}(E+i\epsilon)]_c, \qquad (7.64)$$

where the notation of the imaginary part of the operator A

$$2i\operatorname{Im} A = A - A^*$$

has been introduced.

Hence, the problem now is to express the function $\Omega_+^{(N)}(E)$ in terms of scattering operators of an N-particle system and its subsystems. The corresponding formulae are known in mathematical literature as the trace formulae. Let us look first how this problem is solved in the case of two-particle system.

7.4.2 The second cluster integral

The second cluster integral is defined as

$$B_2 = \frac{1}{2\lambda^3}\operatorname{Sp}\left(e^{-\beta h} - e^{-\beta h_0}\right) \qquad (7.65)$$

where for two-particle energy operators we used the notation introduced in Section 1 of Chapter 3. Accordingly, equation (7.64) takes the form

$$\omega_+(E) = \lim_{\epsilon\downarrow 0}\omega(E+i\epsilon),$$

where

$$2i\omega(z) = \operatorname{Sp}\operatorname{Im}(r(z) - r_0(z)).$$

First we derive a number of auxiliary identities. Consider the unitary operator

$$\tilde{s}(z) = \frac{h-\bar{z}}{h-z}\frac{h_0-z}{h_0-\bar{z}}, \qquad z = E + i\epsilon. \tag{7.66}$$

We will transform this operator as follows. First we use the obvious relations

$$\frac{h-\bar{z}}{h-z} = I + 2i\epsilon r(z), \qquad \frac{h_0-z}{h_0-\bar{z}} = I - 2i\epsilon r_0(z).$$

Then we express the resolvent in terms of the T-matrix:

$$I + 2i\epsilon r(z) = I + 2i\epsilon r_0(z) - 2i\epsilon r_0(z)t(z)r_0(z)$$

and simplify the obtained expression with the help of the identity

$$2i\epsilon r_0(z)r_0(\bar{z}) = r_0(z) - r_0(\bar{z}). \tag{7.67}$$

Arranging similar terms we obtain

$$\tilde{s}(z) = I - 2i\epsilon r_0(z)t(z)r_0(z). \tag{7.68}$$

Differentiating the expression

$$\mathrm{Sp}\,\log\frac{h-\bar{z}}{h-z}\frac{h_0-z}{h_0-\bar{z}}, \qquad z = E + i\epsilon$$

with respect to E, we find that

$$2i\omega(z) = \frac{\partial}{\partial E}\mathrm{Sp}\,\log\tilde{s}(z).$$

On the other hand, the right hand side of this equation can be written, by virtue of unitarity of the operator $\tilde{s}(z)$, in the form of the trace

$$2i\omega(z) = \mathrm{Sp}\,\tilde{s}^*(z)\frac{\partial\tilde{s}(z)}{\partial E}. \tag{7.69}$$

Now, we can make use of representation (7.68) of the operator $\tilde{s}(z)$. This relation will be called the fundamental preparatory formula.

Now we rearrange the right hand side of equation (7.69) in such a way that the limit $\omega_+(E)$ could be expressed directly in terms of scattering operator.

We introduce some new notation symbols. It is convenient to use the scattering matrix defined in Section 6 of Chapter 2 instead of the scattering operator. In the two-particle case, the latter is the integral operator $s(E)$

acting in Hilbert space of functions $L_2(S^{(2)})$ square integrable on the two dimensional unit sphere. The operator $s(E)$ is unitary for all energies E. The kernel of the scattering matrix $s(\hat{p}, \hat{p}', E)$ can be expressed on energy shell in terms of the T-matrix as

$$s(\hat{p}, \hat{p}', E) = \delta(\hat{p}, \hat{p}') - \pi i \sqrt{E} t(\sqrt{E}\hat{p}, \sqrt{E}\hat{p}', E + i0). \qquad (7.70)$$

$\frac{\mathrm{d}\, s(E)}{\mathrm{d}\, E}$ denotes the derivative of S-matrix with respect to the parameter E. It is defined by the kernel (7.70) differentiated with respect to E at fixed $\hat{p}$ and $\hat{p}'$.

Let us denote by $\nabla_0 t(z)$ the operator defined by the kernel

$$\nabla_0 t(z) = |p|^{-1} \left((\frac{\partial}{\partial p^2} + \frac{\partial}{\partial p'^2} + \frac{\partial}{\partial E} \right) |p| t(p, p', E + i\epsilon),$$

in which the differentiations is carried out with respect to the radial variables p^2 and p'^2, while the angular variables $\hat{p}$ and $\hat{p}'$ are kept fixed. Note that on energy shell $p^2 = p'^2 = E$, the kernel of the operator $\nabla_0 t(E + i0)$ coincides with the kernel of the operator $(i\pi\sqrt{E})^{-1} \frac{\mathrm{d}\, s(E)}{\mathrm{d}\, E}$.

The reconstructed preparatory formula mentioned above is of the following form

$$2i\omega(z) = k_0(t(z)), \qquad (7.71)$$

with the operator function $k_0(t(z))$ defined as

$$k_0(t(z)) = -2i\epsilon r_0 r_0^* \nabla_0 t - (2i\epsilon)^2 r_0 r_0^* t^* r_0 r_0^* \nabla_0 t. \qquad (7.72)$$

The variable $z = E + i\epsilon$ on the right hand side of this equation has been omitted and the identity $r_0(\bar{z}) = r^*(z)$ was used.

From here the desired trace formula follows immediately. Indeed, passing to the limit $\epsilon \downarrow 0$ in equation (7.71) and taking into account the relation

$$2i\epsilon r_0(E + i0) r_0^*(E + i0) \to 2\pi i \delta(p^2 - E), \qquad (7.73)$$

we express $\omega_+(E)$ through the T-matrix on energy shell

$$2i\omega_+(E) =$$

$$-2\pi i \int dp \delta(p^2 - E) E^{-1/2} \frac{\partial}{\partial E} E^{1/2} t(\sqrt{E}\hat{p}, \sqrt{E}\hat{p}, E + i0) +$$

$$+2\pi i \int dp dp' \delta(p^2 - E)\delta(p'^2 - E)\overline{t(p,p',E+i0)}\times$$

$$\times E^{-1/2}\frac{\partial}{\partial E}E^{1/2}t(\sqrt{E}\hat{p},\sqrt{E}\hat{p}',E+i0). \tag{7.74}$$

This relation can be rewritten in terms of the S-matrix

$$\omega_+(E) = \frac{1}{2i}\mathrm{Sp}\, s^*(E)\frac{\mathrm{d}\, s(E)}{\mathrm{d}\, E}. \tag{7.75}$$

The trace in this case should be understood as a trace of a integral operator in $L_2(S^{(2)})$, i.e.,

$$\mathrm{Sp}\, A = \int d\hat{p} A(\hat{p},\hat{p}).$$

It remains to prove equation (7.71). Let us note first of all that the fundamental preparatory formula (7.69) can be rewritten as

$$2i\omega(z) =$$

$$= -2i\epsilon\frac{\partial}{\partial E}\mathrm{Sp}\, r_0 r_0^* t - (2i\epsilon)^2 \mathrm{Sp}\, r_0 r_0^* t^* \frac{\partial}{\partial E} r_0 r_0^* t. \tag{7.76}$$

This follows from the identities

$$\mathrm{Sp}\, r_0 t^* r_0^* \frac{\partial}{\partial E} r_0 t r_0^* - \mathrm{Sp}\, r_0 r_0^* t^* \frac{\partial}{\partial E} r_0 r_0^* t =$$

$$= \mathrm{Sp}\, r_0 t^* r_0 r_0^* t r_0^{*2} - \mathrm{Sp}\, r_0 r_0 t^* r_0^* r_0^{*2} t = 0,$$

and from the commutativity of the operator t and t^*:

$$t r_0 r_0^* t^* = t^* r_0 r_0^* t. \tag{7.77}$$

Let us examine in detail the first term in (7.76)

$$2i\epsilon\frac{\partial}{\partial E}\mathrm{Sp}\, r_0 r_0^* t = \frac{\partial}{\partial E}\int dp \frac{-2i\epsilon}{(p^2-E)^2+\epsilon^2} t(pp, E+i\epsilon).$$

. We introduce the spherical integration variables $|p|$, $\hat{p}$ so that $dp = p^2 d|p| \wedge d\hat{p}$. Differentiating the first term, we turn to the differentiation with respect to p^2:

$$\frac{\partial}{\partial E}\frac{1}{(p^2-E)+\epsilon^2} = -\frac{\partial}{\partial p^2}\frac{1}{(p^2-E)^2+\epsilon^2}$$

and then integrate by parts with respect to p^2.As a result of this transformation, we obtain the first term in equation (7.74). The similar transformation of the second term in (7.76) yields

$$-(2i\epsilon)^2 \mathrm{Sp}\, r_0 r_0^* t^* r_0 r_0^* \nabla_0 t + (2i\epsilon)^2 \mathrm{Sp}\, r_0 r_0^* t^* r_0 r_0^* \partial_2 t -$$

$$-(2i\epsilon)^2 \mathrm{Sp}\, r_0 r_0^* (\partial_2 t^*) r_0 r_0^* t.$$

The symbol ∂_2 denotes the differentiation of the kernel $t(p, p', z)$ with respect to the "energy" variable p^2 in the second argument p'^2:

$$\partial_2 t(p, p', z) = \frac{\partial}{\partial p'^2} t(p, p', z);$$

$$\partial_2 t^*(p, p', z) = \frac{\partial}{\partial p'^2} \overline{t(p, p', z)}. \tag{7.78}$$

The terms containing the derivative ∂_2 can be written in the form

$$(2i\epsilon)^2 \mathrm{Sp}\, r_0 r_0^* \partial_2 t^* r_0 r_0^* t - (2i\epsilon)^2 \mathrm{Sp}\, (\partial_2 r_0 r_0^* t^*) r_0 r_0^*$$

and this sum is equal to zero by virtue of equation (7.77). As a result we obtain representation (7.74).

The trace formula (7.75) represents the main result of our considerations. Substituting equation (7.75) in the integral (7.63) we express explicitly the second cluster integral through the scattering matrix. At this point we conclude our discussion of the second cluster integral.

7.4.3 Preparatory formulae for the N-th cluster integral

Consider the trace

$$\Omega^{(N)}(z) = \mathrm{Sp}\, \mathrm{Im}\, [R^{(N)}(z)]_c.$$

For z mowing towards the real axis, this quantity has a finite limit (7.64) which was used in formula (7.63).

Next, consider the unitary operators

$$\tilde{S}()z) = \frac{H - \bar{z}}{H - z} \frac{H_0 - z}{H_0 - \bar{z}}, \tag{7.79}$$

$$\tilde{S}_{a_k}(z) = \frac{H_{a_k} - \bar{z}}{H_{a_k} - z}\frac{H_0 - z}{H_0 - \bar{z}}.$$

Here and in what follows, we will omit the superscript N in the operators. As in the two-particle case, it is easy to verify validity of the representations

$$\tilde{S}(z) = I - 2\pi i R_0(z) T(z) R_0^*(z),$$

$$\tilde{S}_{a_k}(z) = I - 2\pi i R_0(z) T_{a_k}(z) R_0^*(z). \tag{7.80}$$

Let us denote by $[\log \tilde{S}(z)]_c$ the connected part of the operator $\log \tilde{S}(z)$. It is defined by equation (7.62) in which the operators $\left(e^{-\beta H^{(N)}} - e^{-\beta H_0^{(N)}}\right)$ and $\left(e^{-\beta H_a^{(N)}} - e^{-\beta H_0^{(N)}}\right)$ should be replaced by the operators $\log \tilde{S}(z)$ and $\log \tilde{S}_{a_k}(z)$, respectively. Taking the operators $\tilde{S}(z)$ and $\tilde{S}_{a_k}(z)$ in the form (7.76) and differentiating each term of trace of the connected part $[\log \tilde{S}(z)]_c$ with respect to E, we obtain

$$2i\Omega(z) = \frac{\partial}{\partial E}\mathrm{Sp}\,[\log \tilde{S}(z)]_c.$$

Taking finally into account unitarity of the operators $\tilde{S}(z)$ and $\tilde{S}_{a_k}(z)$, we find the following fundamental preparatory formula

$$2i\Omega(z) = \mathrm{Sp}\left[\tilde{S}^*(z)\frac{\partial \tilde{S}(z)}{\partial E}\right]_c, \tag{7.81}$$

where $z = E + i\epsilon$ and

$$\left[\tilde{S}^*(z)\frac{\partial \tilde{S}(z)}{\partial E}\right]_c = \sum_a (-1)^{\sum_i n_i - 1}\left(\sum_i n_i - 1\right)!\tilde{S}^*(z)\frac{\partial \tilde{S}(z)}{\partial E}. \tag{7.81'}$$

Before proceeding to discussion of this formula, we make the assumption that the energy operator H of the N-particle system as well as the energy operators H_{a_k} of the subsystems possess no bound states. Such assumption does not change the nature of the problem; all reasoning can easily be generalized to the general case. It however allows us to avoid the necessity of introducing numerous cumbersome notations.

A further transformation of the fundamental preparatory formula can be carried out along the same scheme which was used in the two-particle case.

Representation (7.80) is to be used for all terms (7.81). To calculate the trace of the obtained expression, it is necessary to integrate by parts with respect to P^2 in order to transfer the derivative from the kernel $\frac{\partial}{\partial E} R_0 R_0^*$ to the kernels T and T_{a_k}. As a result, we obtain the preparatory formula

$$2i\Omega(z) = \mathrm{Sp}\,[K_0(T)]_c, \tag{7.82}$$

where K_0 denotes expressions analogous to (7.72), to wit

$$K_0(T) = -2i\epsilon R_0 R_0^* \nabla_0 T - (2i\epsilon)^2 R_0 R_0^* T^* R_0 R_0^* \nabla_0 T.$$

Here, $\nabla_0 Q$ denotes integral operators defined by the kernel

$$(\nabla Q)(P, P', z) =$$
$$= |P|^{-3N+5}\left(\frac{\partial}{\partial P^2} + \frac{\partial}{\partial P'^2} + \frac{\partial}{\partial E}\right)|P|^{3N+5} Q(P, P', E + i\epsilon). \tag{7.83}$$

The next step consists in performing the limit $\epsilon \downarrow 0$ in equation (7.82). Proceeding formally by analogy with the two-body case, we should replace $i\epsilon R_0 R_0^*$ in the limit $\epsilon \downarrow 0$ by δ-functions $2\pi i(P^2 - E)$ and the kernels T and ∇_0 by scattering matrices and their derivatives. As a result, a simple formula would follow

$$\Omega_+(E) = \frac{1}{2i}\mathrm{Sp}\left(S^*(E)\frac{\mathrm{d}\, S(E)}{\mathrm{d}\, E}\right)_c, \tag{7.84}$$

where $S(E)$ is the N-particle scattering matrix and $\left(S^* \frac{\mathrm{d}\, S}{\mathrm{d}\, E}\right)_c$ denotes the connected part of the operator $\left(S^* \frac{\mathrm{d}\, S}{\mathrm{d}\, E}\right)$ defined according to (7.81′).

However, such formal generalization of equation (7.75) to the many-particle case is incorrect. The point is that the equation (7.84) has no immediate meaning. Whereas its left hand side is defined correctly by equation (7.64), the trace appearing on the right hand side of (7.84) contains infinite terms which do not mutually cancel. The appearance of such singularities is a consequence of the fact that kernel of the operator $S(E) - I$, which up to a multiplicative factor coincides with the kernel of the T-matrix on energy shell, contains in many-particle case strong singularities on the diagonal. As we have seen in Chapter 3, these singularities correspond to the processes of multiple rescattering. We recall that in the two-particle case the operator $S(E) - I$ is given by a smooth kernel.

One may naturally expect that the indefinite quantity Sp $(S^* drSE)_c$ can be somehow regularized. Then, equation (7.84) becomes meaningful and correct. Even this statement is however not true. The point is that in the case of the model operators (3.74), discussed in Section 4 of Chapter 3, the trace Sp $(S^* drSE)_c$ is well defined and even equals zero, whereas the trace of the equation (7.64) clearly differs from zero. This we will see at the end of this section.

Hence, to calculate $\Omega_+(E)$, we must consider in details all non-singular terms and express them in terms of the S-matrices of subsystem. We are going to show how this can be done in the three-particle case. Attempts to realize such a program in the general N-body case were so far unsuccessful.

7.4.4 Trace formula for three-particle systems

Now we turn to considering in details the passage to the limit $\epsilon \downarrow 0$ in the case of three particles. For this purpose, it is convenient to use the abbreviated notations introduced in Chapter 3. To start with, we write down relations, obtained in the general case, for the three-particle problem.

The third cluster integral is defined as

$$B_3 = \frac{1}{3!\lambda^3}\mathrm{Sp}\left(e^{-\beta H} - e^{-\beta H_0} - \sum_\alpha \left(e^{-\beta H_\alpha} - e^{-\beta H_0}\right)\right).$$

It can be expressed in terms of the limiting value of the function

$$\Omega(z) = \frac{1}{2i}\mathrm{Sp}\,\mathrm{Im}\,[R(z)]_c$$

on the real axis $\Omega_+(E) = \lim_{\epsilon\downarrow 0}\Omega(E+i\epsilon)$ with the help of equation (7.63). The connected part of the resolvent now reads

$$[R(z)]_c = R(z) - R_0(z) - \sum_\alpha (R_\alpha(z) - R_0(z)).$$

It is noteworthy that the operator $[R(z)]_c$, contrary to the operator $\mathrm{Im}\,[R(z)]_c$, has no trace because of slow decrease of the kernel. The fundamental preparatory formula takes in this notation the form

$$2i\Omega(z) = \mathrm{Sp}\left(\tilde{S}^*(z)\frac{\partial \tilde{S}(z)}{\partial E} - \sum_\alpha \tilde{S}^*_\alpha(z)\frac{\partial \tilde{S}_\alpha(z)}{\partial E}\right), \tag{7.85}$$

where the operators $\tilde{S}(z)$ and $\tilde{S}_\alpha(z)$ are expressed in terms of the T-matrix by equation (7.80). Finally, the operator $[K_0(T)]_c$ in the preparatory formula (7.82) is given as

$$[K_0(T)]_c = K_0(T) - \sum_\alpha K_0(T_\alpha),$$

with $K_0(T)$ and $K_0(T_\alpha)$ defined by (7.83) with $N = 3$.

In what follows, we will assume that the scattering is absolutely elastic, i.e., that the two-body subsystems have no bound states. In such a case, the scattering matrix $S(E)$ for three-particle systems represent an integral operator acting in the space of square integrable functions on the unit sphere in the six-dimensional space $L_2(S^{(5)})$. This operator is defined by the singular kernel

$$S(\hat{P}, \hat{P}', E) = \delta(\hat{P}, \hat{P}') - \pi i E^2 T(E^{1/2}\hat{P}, E^{1/2}\hat{P}', E + i0). \qquad (7.86)$$

Here, the energy E can attain only non-negative values $E \geq 0$. It is clear that this kernel contains all singularities of the T-matrix kernel $T(P, P', z)$. The kernel of the scattering matrix of the operator H_α $(\alpha = 1, 2, 3)$ is analogously expressed in terms of the two-particle T-matrices on energy shell as

$$S_\alpha(\hat{P}, \hat{P}', E) = \delta(\hat{P}, \hat{P}') - \pi i E^2 \delta(p_\alpha - p'_\alpha) \times$$

$$\times t_\alpha\left(\sqrt{E - p_\alpha^2}\hat{k}_\alpha, \sqrt{E - p_\alpha^2}\hat{k}'_\alpha, E - p_\alpha^2 + i0\right). \qquad (7.86')$$

It is convenient to write the scattering matrices $S(E)$ and $S_\alpha(E)$ in the form

$$S(E) = I - 2\pi i\hat{T}(E), \quad S_\alpha(E) = I - 2\pi i\hat{T}_\alpha(E), \qquad (7.87)$$

with the unit operator separated. Expressions for kernels of the operators $\hat{T}(E)$ and $\hat{T}_\alpha(E)$ in terms of on energy shell T-matrices follow from the simple comparison of equations (7.86), (7.86′), and (7.87).

Let us now turn to transformation of the preparatory formula (7.82). The difficulties in performing the limiting process in this equation are related to three-particle singularities (3.31) of the kernels $M_{\alpha\beta}$ present in the definition of the kernel $T(P, P, z)$, equation (3.24). If such singularities were absent, the kernel of the operator Sp $(S^* dr S E)_c$ would have no singularities on the diagonal and we could have used the limiting relation

$$i\epsilon R_0 R_0^* \to 2\pi i \delta(P^2 - E), \qquad (7.88)$$

which is valid for smooth bounded functions. As we have already stated, equation (7.84) would result. However, since the kernels $M_{\alpha\beta}$ are singular, this relation cannot be used in general. The result of the formal limiting procedure has no meaning and requires generalizations.

Let us now discuss in more detail which singularities appear on the right hand side of the equation (7.82). The strongest three-particle singularities of the kernel $M_{\alpha\beta}$ and consequently of the T-matrix kernel $T(P, P', z)$ result from singularities of the kernels T_α and $Q^{(0)}_{\alpha\beta}$, i.e., of kernels of the free terms and the first iterations of compact equations (3.28). We isolate them by writing

$$T(z) = T_0(z) + \bar{T}(z)$$

where

$$T_0(z) = \sum_\alpha T_\alpha + \sum_{\alpha\beta}' Q^{(0)}_{\alpha\beta}. \tag{7.89}$$

By $S_0(E)$ we denote the singular part of the scattering matrix which results from replacing the kernel $T(P, P', z)$ by $T_0(P, P', z)$ in equation (7.86). Later we will discuss contribution of this part to the trace formula in details. On the other hand, the singularities of the kernel of the remaining part of the T-matrix $\bar{T}(z)$, which correspond to the second and third iteration of equation (3.28), are weak and cannot affect the passage to the limit in equation (7.82). This can be verified by using the explicit formulae (3.40) and (3.43) for such singularities. Hence, if the singular terms (7.89) are isolated beforehand, the limiting transition $\epsilon \downarrow 0$ in remaining terms (7.82) can be carried out. As a result, we obtain the following trace formula

$$\Omega_+(E) = \frac{1}{2i}\text{Sp }(S^* dr SE - S_0^* dr S_0 E) + \frac{1}{2i}\Delta(E), \tag{7.90}$$

where

$$\Delta(E) = \lim_{\epsilon\downarrow 0} \text{Sp}\left(K_0(T_0) - \sum_\alpha K_0(T_\alpha)\right). \tag{7.91}$$

This formula, however, is not fully satisfactory because although $(S_0^* dr S_0 E)$ and $\Delta(E)$ are expressed in terms of the scattering characteristics, these characteristics do not reduce to scattering matrices. More precisely, they contain the two-particle T-matrices consisting of two, three or four terms. Also the fact that the operator $(S_0^* dr S_0 E)$ contains, besides the true singular terms, also regular terms having the trace should be counted as a drawback of this formula.

Let us note further that regularization of trace of the operator Sp $(S^* drSE)_c$ can be carried out with the help of two-particle scattering matrices. Indeed, as it was shown in Section 3 of Chapter 3, the polar singularities of the kernels $Q^{(0)}_{\alpha\beta}(P, P', z)$ on energy shell are contained in the kernel $T_{\alpha\beta}(P, P')$, equation (3.79). Therefore, in equation (7.90), we can use instead of the operator $S_0(E)$ the operator $\tilde{S}_0(E)$ which is expressed explicitly in terms of two-particle scattering matrices. This operator is defined by equations (7.89) and (7.86) in which the kernel $Q^{(0)}_{\alpha\beta}(P, P', z)$ must be replaced by $T_{\alpha\beta}(P, P')$. The formula regularized in this way will read

$$\Omega_+ = \frac{1}{2i}\mathrm{Sp}\left(S^* drSE - \tilde{S}_0^* dr\tilde{S}_0 E\right) + \frac{1}{2i}\tilde{A}_0(E) \tag{7.92}$$

where the operator $\tilde{A}_0(E)$ is defined as

$$\tilde{A}_0(E) = \mathrm{Sp}\left(\tilde{S}_0^* dr\tilde{S}_0 E - S_0^* drS_0 E\right) + \Delta(E). \tag{7.93}$$

It is not a priori clear that the operator $\tilde{A}_0(E)$ can be expressed in terms of two-particle scattering matrices. This is however true and this statement represents the fundamental result of this section. An explicit expression of this operator will be given in slightly modified form below.

The product $\tilde{S}_0^* dr\tilde{S}_0 E$ of singular operators contains also regular terms having a trace. The aforementioned modification consists in retaining only the "true" singular terms in the operator $\tilde{S}_0^* dr\tilde{S}_0 E$. The other terms for which the trace operation is meaningful are united with the quantity Sp $\tilde{A}_0(E)$. Finally, we obtain the trace formula

$$\Omega_+(E) = \frac{1}{2i}\mathrm{Sp}\left((S^* drSE)_c - A(E)\right) + \frac{1}{2i}\mathrm{Sp}\,\tilde{A}_0(E) \tag{7.94}$$

in which the operators $A(E)$ and $\tilde{A}(E)$ are explicitly given in terms of two-particle scattering matrices. This formula is the main result of the present section. It allows to express the third cluster integral in terms of scattering matrix.

Prior to the derivation of this formula, we describe the operators $A(E)$ and $\tilde{A}(E)$. We start with the operators $A(E)$ which regularizes the trace

(7.94). For this purpose it is convenient to use a new operator $\hat{T}_{\alpha\beta}$ which includes singularities of the kernel $Q^{(0)}_{\alpha\beta}(P, P', z)$ and which, contrary to $T_{\alpha\beta}$, contains a symmetric combination of arguments of the kernels t_α and t_β on energy shell. Namely, we set

$$\hat{T}_{\alpha\beta} = \frac{1}{2}(T^{(+)}_{\alpha\beta} + T^{(-)}_{\alpha\beta}),$$

where $T^{(\pm)}_{\alpha\beta}$ are integral operators defined by the kernels

$$T^{(+)}_{\alpha\beta}(\hat{P}, \hat{P}', E) = \\ = \frac{2E^2|s_{\alpha\beta}|^{-3}}{(E_\alpha E_{\beta\alpha})^{1/2}} \frac{\hat{t}_\alpha(\hat{k}_\alpha, \hat{k}'_{\alpha\beta}, E_\alpha)\hat{t}_\beta(\hat{k}_{\beta\alpha}, \hat{k}'_\beta, E_{\beta\alpha})}{E'_{\alpha\beta} - E_\alpha - i0}, \tag{7.95}$$

$$T^{(-)}_{\alpha\beta}(\hat{P}, \hat{P}', E) = \\ = \frac{-2E^2|s_{\alpha\beta}|^{-3}}{(E'_{\alpha\beta} E'_\beta)^{1/2}} \frac{\hat{t}_\alpha(\hat{k}_\alpha, \hat{k}'_{\alpha\beta}, E'_{\alpha\beta})\hat{t}_\beta(\hat{k}_{\beta\alpha}, \hat{k}'_\beta, E'_\beta)}{E_{\beta\alpha} - E'_\beta - i0}, \tag{7.95'}$$

where $k'_{\alpha\beta} = k_\alpha(p_\alpha, p'_\beta)$, $E'_{\alpha\beta} = {k'_{\alpha\beta}}^2$, $k_{\beta\alpha} = k_\beta(p_\alpha, p'_\alpha)$, $E_{\beta\alpha} = k^2_{\beta\alpha}$, and $k^2_\alpha + p^2_\alpha = {k'_\beta}^2 + {p'_\beta}^2 = E$. By $\hat{t}_\alpha(\hat{k}, \hat{k}', E)$ we denote kernel of the operator $\hat{t}_\alpha$ related to the pair S-matrix as follows

$$s_\alpha = I - 2\pi i \hat{t}_\alpha.$$

In this notation, the operator $A(E)$ is given by the equality

$$A(E) = 2\pi i B(E) + C(E), \tag{7.96}$$

where

$$B(E) = \sum_{\alpha\neq\beta} \left[\hat{T}^*_{\alpha\beta} \frac{\mathrm{d}}{\mathrm{d}E} S_\alpha S_\beta - (S^*_\alpha + S^*_\beta - I)\frac{\mathrm{d}}{\mathrm{d}E}\hat{T}_{\alpha\beta}\right],$$

$$C(E) = \sum_{\alpha\neq\beta} (S^*_\alpha - I)\frac{\mathrm{d}\,S_\beta}{\mathrm{d}\,E}. \tag{7.96'}$$

Description of the operator $\tilde{A}(E)$ requires introduction of new notation symbols. This operator contains, in particular, special symbols of differentiation ∂_α. This differentiation is defined by its action on kernels and is applied to the sums of products of the kernels $\hat{t}_\alpha$ and of complex conjugate kernels and to integrals containing corresponding products. The operation ∂_α acts on a product as a differentiation such that each term of the type $\hat{t}_\alpha(\hat{k}_\alpha, \hat{k}'_\alpha, E_\alpha)$ is differentiated with respect to E_α. On other terms, even though they depend on E_α, this differentiation acts like on constants. When applied to integrals the operation ∂_α is assumed to act on integrands following the same rule. The composition $\partial_\alpha \frac{\mathrm{d}}{\mathrm{d}E}$ should be understood as

$$\partial_\alpha \frac{\mathrm{d}}{\mathrm{d}E} = \frac{\mathrm{d}}{\mathrm{d}E}\partial_\alpha + \frac{1}{E}\partial_\alpha.$$

Finally, ∂ denotes here the sum $\sum_\alpha \partial_\alpha$.

Further, let an operator Q be given by a singular kernel $Q_0\varphi$ where Q_0 is a smooth kernel and φ is a function of either of two types:

$$\varphi_1 = (E'_{\alpha\beta} - E_\alpha \mp i0)^{-1}, \quad \varphi_2 = \delta(E'_{\alpha\beta} - E_\alpha).$$

In such a case $[Q]_r$ should be understood as a regularized operator defined by the kernel Q_0.

In this notation the following formula holds

$$\tilde{A}(E) = \frac{1}{4\pi i}\partial[C]_r + i\pi\left(\frac{1}{E}\frac{\mathrm{d}}{\mathrm{d}E}E - \frac{1}{2}\partial\right)B_R + i\pi\tilde{B}_R, \tag{7.97}$$

where B_R and $\tilde{B}_R$ denote particular regularizations of the operator $B(E)$, to wit

$$B_R = \sum_{\alpha\neq\beta}\left[\hat{T}_{\alpha\beta}\right]_r^* \frac{\mathrm{d}}{\mathrm{d}E}S_\alpha S_\beta - (S_\alpha^* + S_\beta^* - I)\left[\frac{\mathrm{d}}{\mathrm{d}E}\hat{T}_{\alpha\beta}\right]_r, \tag{7.98}$$

$$\tilde{B}_R = \sum_{\alpha\neq\beta} - \left[\hat{T}_{\alpha\beta}\right]_r^* \partial\frac{\mathrm{d}}{\mathrm{d}E}S_\alpha S_\beta - (S_\alpha^* + S_\beta^* - I)\left[\partial\frac{\mathrm{d}}{\mathrm{d}E}\hat{T}_{\alpha\beta}\right]_r,. \tag{7.98'}$$

Therefore, we described all operators in the trace formula (7.94). Let us briefly sketch the scheme of derivation of this formula. First, we isolate from the operator $S_0^* \frac{\mathrm{d}S_0}{\mathrm{d}E}$ in equation (7.90) regular terms having trace. We obtain the formula

$$\Omega_+(E) = \frac{1}{2i}\mathrm{Sp}\left(\left(S^*\frac{\mathrm{d}S}{\mathrm{d}E}\right)_c - A^{(1)}(E)\right) + \frac{1}{2i}\Delta^{(1)}(E). \tag{7.99}$$

At the first step we retain in the regularizing operator $A^{(1)}(E)$ the terms expressible by pair S-matrices and terms containing kernels $Q^{(0)}_{\alpha\beta}(P, P', E)$

which cannot be reduced to the former. In the next step, we compute the quantity $\Delta^{(1)}(E)$ defined as

$$\Delta^{(1)}(E) = \Delta(E) - \mathrm{Sp}\left(S_0^* \frac{\mathrm{d}\, S_0}{\mathrm{d}\, E} - \sum_\alpha S_\alpha^* \frac{\mathrm{d}\, S_\alpha}{\mathrm{d}\, E} - A^{(1)}(E)\right). \tag{7.100}$$

This quantity, as well as $A^{(1)}(E)$, requires off-shell pair T-matrices for its description. Then we introduce the operator $A(E)$ explicitly expressed in terms of pair S-matrices and turn to equation (7.94) in which the operator $\tilde{A}(E)$ is defined as

$$\mathrm{Sp}\,\tilde{A}(E) = \mathrm{Sp}\,(A(E) - A^{(1)}(E)) + \Delta^{(1)}(E). \tag{7.101}$$

Finally, we show that the trace $\mathrm{Sp}\,(A(E) - A^{(1)}(E))$ is expressible up to the term $-\Delta^{(1)}(E)$ in terms of two-particle scattering matrices.

During realization of this program, we will only present examples of the most typical transformations omitting all calculations which reduce to the repetition of already discussed examples.

7.4.5 Calculation of $\Delta^{(1)}(E)$

Consider the operator $S_0^* \frac{\mathrm{d}\, S_0}{\mathrm{d}\, E}$ regularizing the trace in equation (7.90). This operator is a linear combination of products of pair T-matrices T_α and their conjugates up to four-linear terms. Such m-linear products will be called the homogeneous T-polynomials and their linear combination for $m \leq n$, the T-polynomials of degree n.

Let us consider the operator $A^{(1)}(E)$ containing the "true" singular terms of the operator $S_0^* \frac{\mathrm{d}\, S_0}{\mathrm{d}\, E}$. This operator contains all homogeneous T-polynomials of first and second degree and all homogeneous T-polynomials of third and fourth degree except for those having three different subscripts $\alpha \neq \beta \neq \gamma$. Thus, $A^{(1)}(E)$ can be written analogously to equation (7.96), as the sum

$$A(E) = 2\pi i B^{(1)}(E) + C(E),$$

where

$$B^{(1)}(E) = \sum_{\alpha \neq \beta} \left[\hat{Q}^*_{\alpha\beta} \frac{\mathrm{d}}{\mathrm{d}\, E} S_\alpha S_\beta - (S_\alpha^* + S_\beta^* - I) \frac{\mathrm{d}}{\mathrm{d}\, E} \hat{Q}_{\alpha\beta}\right].$$

Here $\hat{Q}_{\alpha\beta}$ denotes the contribution to the S-matrix from the kernel $Q^{(0)}_{\alpha\beta}(P, P', z)$ defined by equation (7.86).

Note that the fourth degree terms in $A^{(1)}(E)$ are taken in the form $\hat{Q}^*_{\alpha\beta}\frac{\mathrm{d}}{\mathrm{d}E}\hat{T}_\alpha\hat{T}_\beta$, where the operator $\hat{T}_\alpha\hat{T}_\beta$ replacing the operator $\hat{Q}_{\alpha\beta}$, is explicitly given in terms of the pair S-matrices and the difference $Q^*_{\alpha\beta}\frac{\mathrm{d}}{\mathrm{d}E}Q_{\alpha\beta} - Q^*_{\alpha\beta}\frac{\mathrm{d}}{\mathrm{d}E}\hat{T}_\alpha\hat{T}_\beta$ having the trace is included in the first term of (7.99).

The kernel of the operator $A^{(1)}(E)$ is singular on the diagonal and therefore this operator has no trace. The other terms have singularities of the following types. Kernels of the second order terms $\hat{T}^*_\alpha\frac{\mathrm{d}}{\mathrm{d}E}\hat{T}_\beta$ diverge as $(\delta(x))^2$. All other terms behave as $(x \pm i0)^{-1}\delta(x)$. In general, these singular terms do not mutually cancel. (An exception is the model mentioned above where the variables separate and all operators mutually commute).

Let us now return to the preparatory formula (7.82) in which the passage to the limit $\epsilon \to 0$ is still to be carried out. We denote by $\Delta_1(E + i\epsilon)$ the terms $\mathrm{Sp}\,[K_0(T)]_c$ generating in the limit $\epsilon \to 0$ the quantity $\Delta^{(1)}(E)$ from equation (7.99). This quantity has the form of the trace of the T-polynomial of fourth degree:

$$\Delta_1 = \sum_{i=2}^{4}\Delta^{(i)} + \sum_{i=2}^{3}\tilde{\Delta}^{(i)}, \tag{7.102}$$

where $\Delta^{(4)}$ is a trace of the homogeneous fourth degree T-polynomials

$$\Delta^{(4)} = (2i\epsilon)^3 \sum_{\alpha\neq\beta} \mathrm{Sp}\, R_0R_0^*Q^*_{\alpha\beta}R_0R_0^*\nabla_0(T_\alpha R_0R_0^*T_\beta), \tag{7.103}$$

$\Delta^{(3)}$ and $\tilde{\Delta}^{(3)}$ are traces of the T-polynomials of third degree of two types

$$\Delta^{(3)} = -(2i\epsilon)^2 \sum_{\alpha\neq\beta} \mathrm{Sp}\, R_0R_0^*(T^*_\alpha R_0R_0^*\nabla_0Q_{\alpha\beta} + \\ + T^*_\beta R_0R_0^*\nabla_0Q_{\alpha\beta}),$$

$$\tilde{\Delta}^{(3)} = -(2i\epsilon)^2 \sum_{\alpha\neq\beta} \mathrm{Sp}\, R_0R_0^*(Q^*_{\alpha\beta}R_0R_0^*\nabla_0T_\alpha + \\ + Q^*_{\alpha\beta}R_0R_0^*\nabla_0T_\beta),$$

and $\Delta^{(2)}$ and $\tilde{\Delta}^{(2)}$ are traces of second degree terms of two types

$$\Delta^{(2)} = -2i\epsilon \sum_{\alpha\neq\beta} \mathrm{Sp}\, R_0R_0^*\nabla_0Q_{\alpha\beta},$$

$$\tilde{\Delta}^{(2)} = -(2i\epsilon)^2 \sum_{\alpha \neq \beta} \mathrm{Sp}\, R_0 R_0^* T_\alpha R_0 R_0^* \nabla_0 T_\beta .$$

Had it been possible to pass formally to the limit $\epsilon \to 0$, the quantity $\Delta_1(E + i\epsilon)$ would turn into trace of the operator $A^{(1)}(E)$. However, we cannot pass to the limit because each term of (7.102) contains terms taking, in appropriate variables, the form

$$\frac{\epsilon^2}{(x^2+\epsilon^2)^2} \quad \text{and} \quad \frac{\epsilon}{(x^2+\epsilon^2)(x \pm i\epsilon)}. \tag{7.104}$$

In the course of formal passage to the limit, these terms generate the singularities $(\delta(x))^2$ and $\delta(x)(x \pm i0)^{-1}$ of the operator $A^{(1)}(E)$ mentioned above. In spite of this, as we already noted, the function $\Delta_1(E + i\epsilon)$ has a finite limit for $\epsilon \downarrow 0$. To calculate this limit, we must get rid of the singular terms (7.104). This will be done as follows.

To each term we apply the identity

$$\frac{2i\epsilon}{x^2+\epsilon^2} = \frac{1}{x - i\epsilon} - \frac{1}{x + i\epsilon}$$

and consider separately terms containing the denominators $\frac{1}{(x-i\epsilon)(x+i\epsilon)}$ and $\frac{1}{(x \pm i\epsilon)^2}$. Expressions of the second type contain for $\epsilon \downarrow 0$ the well defined distribution $\frac{1}{(x \pm i0)^2}$ and obviously have finite limits. All the terms not having limits belong to the first type. It is possible to verify, by using the cyclic permutation of the operators under the trace symbol and by making use of the Hilbert identity that all terms containing the function $\frac{1}{x^2+\epsilon^2}$ which mutually cancel for $\epsilon \neq 0$. It is important that such cancellations occur among homogeneous T-polynomials of the same degree as well as among T-polynomials of different degrees. The lowering of order of T-polynomials is attained with the help of the Hilbert identity. We stress that this mechanism of cancellation does not work in case of the operator $A^{(1)}(E)$ when the limiting process $\epsilon \to 0$ has already been performed. In this case, we cannot integrate the kernels over the diagonal; all of them become infinite. Therefore, one cannot cyclically transpose and rearrange these operators so that the Hilbert identity can be used. For $\epsilon \neq 0$, however, such operations are legitimate.

Carrying out the calculations along this scheme, we obtain a representation of $\Delta^{(1)}(E)$ in each term of whose the limiting process $\epsilon \downarrow 0$ can be performed. This representation reads

$$\Delta_1(E) = \tilde{\Delta}_1^{(3)} + \tilde{\Delta}_2^{(3)} + \tilde{\Delta}_1^{(2)} + \tilde{\Delta}_2^{(2)}. \tag{7.105}$$

Here $\tilde{\Delta}_i^{(3)}$ $(i = 1, 2)$ denote traces of the third degree homogeneous T-polynomials defined as

$$\tilde{\Delta}_1^{(3)} = 2\pi i \sum_{\alpha\neq\beta} \mathrm{Sp}\, T_\alpha^{(-)} \varphi_0 [\nabla_0 (T_\alpha^{(+)} R_+ T_\beta^{(+)}) R_+ -$$

$$- \nabla_0 (T_\alpha^{(+)} R_- T_\beta^{(+)}) R_-],$$

$$\tilde{\Delta}_2^{(3)} = 2\pi i \sum_{\alpha\neq\beta} \mathrm{Sp}\, \varphi_0 T_\beta^{(-)} [R_+ \nabla_0 (T_\alpha^{(+)} R_+ T_\beta^{(+)}) -$$

$$- R_- \nabla_0 (T_\alpha^{(+)} R_- T_\beta^{(+)})],$$

and are traces of the second degree T-polynomials

$$\tilde{\Delta}_1^{(2)} = \sum_{\alpha\neq\beta} \mathrm{Sp}\, [R_+ \nabla_0 (T_\alpha^{(+)} R_+ T_\beta^{(+)}) - R_- \nabla_0 (T_\alpha^{(+)} R_- T_\beta^{(+)})],$$

$$\tilde{\Delta}_2^{(2)} = \mathrm{Sp}\, [R_- T_\beta^{(-)} R_- \nabla_0 T_\alpha^{(+)} - R_+ T_\beta^{(-)} R_+ \nabla_0 T_\alpha^{(+)}].$$

In these equations $T_\alpha^{(\pm)}$ and $R_\pm$ indicate the limits for $\epsilon = 0$ of the operators $T_\alpha(E \pm i\epsilon)$ and $R_0(E \pm i\epsilon)$. φ_0 denotes multiplication by the δ-function $\delta(p^2 - E)$, to wit

$$(\varphi_0 f)(p) = \delta(p^2 - E) f(p).$$

Let us now proceed to justification of the representation (7.105). First, we show how the fourth degree T-polynomials are transformed. Let us differentiate term by term the last factor under the trace symbol in (7.93)

$$\nabla_0 (T_\alpha R_0 R_0^* T_\beta) = \left(D_1 T_\alpha R_0 R_0^* T_\beta + \frac{\partial}{\partial E} (T_\alpha R_0 R_0^*) T_\beta \right) +$$

$$+ \left(D_2 (T_\alpha R_0 R_0^* T_\beta) + T_\alpha R_0 R_0^* \frac{\partial}{\partial E} T_\beta \right). \tag{7.106}$$

Here, in analogy with equation (7.78), $D_k Q$ denotes the operator given by the kernel $Q(P, P', z)$ differentiated with respect the k-th energy argument P^2 or P'^2, for example,

$$D_2 Q(P, P', z) \equiv \frac{\partial}{\partial P'^2} Q(P, P', z).$$

Following (7.106), we split $\Delta^{(4)}$ into two groups of terms. As for the terms generated by the first factor of (7.106), the Hilbert identity can be applied to the operator T_β

$$2i\epsilon T_\beta R_0 R_0^* T_\beta^* = T_\beta^* - T_\beta,$$

if we arrange beforehand operators under the trace symbol appropriately. This group of terms reduces to the trace of a homogeneous T-polynomial of third degree

$$-(2i\epsilon)^2 \sum_{\alpha\neq\beta} \mathrm{Sp}\, R_0^* T_\alpha \left[\left(D_1 + \frac{\partial}{\partial E}\right) T_\alpha R_0 R_0^*\right] (T_\beta^* - T_\beta).$$

In the terms of the second group the Hilbert identity can be applied to the operator T_α by representing it as

$$-(2i\epsilon)^2 \sum_{\alpha\neq\beta} \mathrm{Sp}\, T_\beta^* R_0^* (T_\alpha^* - T_\alpha) R_0 R_0^* \left(D_2 + \frac{\partial}{\partial E}\right) T_\beta.$$

Let us rewrite further derivatives of the operators so as to obtain again expressions of the type $\nabla_0 Q$:

$$\left(D_1 + \frac{\partial}{\partial E}\right) T_\alpha R_0 R_0^* T_\beta =$$

$$\nabla_0(T_\alpha R_0 R_0^* T_\beta) - T_\alpha R_0 R_0^* \left(D_2 + \frac{\partial}{\partial E}\right) T_\beta,$$

$$\left(D_2 + \frac{\partial}{\partial E}\right) T_\alpha R_0 R_0^* T_\beta =$$

$$\nabla_0(T_\alpha R_0 R_0^* T_\beta) - \left(\left(D_2 + \frac{\partial}{\partial E}\right) T_\alpha R_0 R_0^*\right) T_\beta.$$

The Hilbert identity can be applied to the operators T_α and T_β respectively in factors generated by the second terms on the right hand sides of these equations. As a consequence, these terms reduce to traces of the second degree T-polynomials.

A completely analogous procedure should be used to transform the trace $\tilde{\Delta}^{(3)}$.

It is possible to verify that the sum $\Delta^{(4)} + \tilde{\Delta}^{(2)}$ equals the sum of the remaining terms in (7.102) up to permutation of the operator R_0 and $R_0 R_0^*$. Exceptions are the terms $\epsilon\nabla_0(T_\alpha R_0 R_0^* T_\beta)$ identical in both cases. It is clear therefore that by using the elementary identity $2i\epsilon R_0 R_0^* = R_0 - R_0^*$ all terms containing mixed combinations of R_0 and R_0^* cancel. These are the terms which generate divergences in equation (7.104) at $\epsilon \downarrow 0$. The terms which survive contain a pair of the operators R_0 and R_0^* separated by the T-matrices. In the limit $\epsilon \downarrow 0$, these terms can be collected in the representation (7.105).

Let us transform further this representation further. Note that the operator kernels appearing under the trace symbol have, in appropriate variables, a form of the difference $\frac{f(x)}{(x-i0)^n} - \frac{f(x)}{(x+i0)^n}$, $n = 2, 3$. With the help of the identity

$$\frac{1}{(x-i0)^n} - \frac{1}{(x+i0)^n} = (-1)^{n-1} 2\pi i \frac{\mathrm{d}^{n-1}}{\mathrm{d}\,dx^{n-1}} \delta(x)$$

it is therefore possible to localize arguments of these kernels in vicinity of energy shell. We will transform the right hand side of equation (7.105) in this way in the next stage. Before describing results of such transformation, we introduce some new symbols. Consider the kernel $\frac{\partial}{\partial E} S_\alpha(\hat{P}, \hat{P}', E)$ of the operator $\frac{\mathrm{d}}{\mathrm{d}_E} S_\alpha(E)$. The following representation holds

$$\frac{\partial}{\partial E} S_\alpha(\hat{P}, \hat{P}', E) = \delta(p_\alpha - p'_\alpha) \frac{k_\alpha^2}{E} \nabla_E t_\alpha(k_\alpha, k'_\alpha, E - p_\alpha^2 + i0),$$

where the arguments of the kernel t_α lie on the energy shell $K_\alpha^2 = k_\alpha'^2 = E - p_\alpha^2$. ∇_E denotes a differentiation defined as

$$\nabla_E t_\alpha(k_\alpha, k'_\alpha, E - p_\alpha^2 + i0) = \left(\frac{1}{2E} + \frac{k_\alpha^2}{E} \frac{\partial}{\partial k_\alpha^2} + \right.$$

$$\left. + \frac{k_\alpha'^2}{E} \frac{\partial}{\partial k_\alpha'^2} + \left(1 - \frac{p_\alpha^2}{E}\right) \frac{\partial}{\partial E} \right) t_\alpha(k_\alpha, k'_\alpha, E - p_\alpha^2 + i0).$$

The kernels $\nabla_E t_\alpha$ and $\nabla_E t_\alpha$ are related on energy shell $P^2 = P'^2 = E$ by the relation

$$\nabla_E t_\alpha(k_\alpha, k'_\alpha, E - p_\alpha^2 + i0) =$$

$$= \left(-\frac{3}{2E} + \nabla_0 \right) t_\alpha(k_\alpha, k'_\alpha, E - p_\alpha^2 + i0).$$

The symbol ∇_E is convenient because it allows to continue naturally the operator $\frac{\mathrm{d}}{\mathrm{d}_E} S_\alpha(E)$ away from the energy shell. Namely, with this operator we will associate the operator $\nabla_E T_\alpha^{(+)}$ defined on the Hilbert space $L_2(R^6)$ as an integral operator with the kernel

$$\nabla_E T_\alpha^{(+)}(P, P', E) = \delta(p_\alpha - p'_\alpha) \nabla_E t_\alpha(k_\alpha, k'_\alpha, E - p_\alpha^2 + i0).$$

It is possible then to describe any combination of the operators $S_\alpha(E)$ and $\frac{\mathrm{d}}{\mathrm{d}E}S_\alpha(E)$ with the help of the symbols $T_\alpha^{(\pm)}$ and $\nabla_E T_\alpha^{(\pm)}$. For example, kernel of the product $\hat{T}_\beta \frac{\mathrm{d}}{\mathrm{d}E} S_\alpha$ can be written as

$$\hat{T}_\beta \frac{\partial}{\partial E} S_\alpha(P,P',E) = -i\pi E^2 T_\beta^{(+)} \varphi_0 \nabla_E T_\alpha^{(+)}(P,P',E),$$

where the arguments on the right hand side must be taken on energy shell $P^2 = P'^2 = E$. Next we denote $T_\alpha^{(+)} \equiv T_\alpha$, $T_\alpha^{(-)} \equiv T_\alpha^*$. This cannot lead to confusion since the operators $T_\alpha(E \pm i\epsilon)$ for $\epsilon \neq 0$ do not appear in what follows.

The following formula holds in this notation

$$\Delta_1 = \delta\Delta^{(3)} + \delta\Delta^{(2)}, \tag{7.107}$$

where $\delta\Delta^{(3)}$ is a trace of the homogeneous T-polynomial of third degree

$$\delta\Delta^{(3)} =$$
$$= (2\pi i) \sum_{\alpha\neq\beta} \mathrm{Sp}\left(T_\alpha^* \varphi_0 \nabla_E T_\alpha \frac{\mathrm{d}\,\varphi_0}{\mathrm{d}\,E} T_\beta + \varphi_0 T_\beta^* \frac{\mathrm{d}\,\varphi_0}{\mathrm{d}\,E} T_\alpha \nabla_E T_\beta \right) \tag{7.108}$$

and $\delta\Delta^{(2)}$ is a trace of the homogeneous T-polynomial of second degree

$$\delta\Delta^{(2)} = (2\pi i) \sum_{\alpha\neq\beta} \mathrm{Sp}\, T_\beta^* \frac{\mathrm{d}\,\varphi_0}{\mathrm{d}\,E} \nabla_E T_\alpha. \tag{7.109}$$

Let us stress that the quantity $\Delta_1(E)$ cannot be expressed in terms of pair S-matrices even though the arguments of T-matrix kernels and their derivatives are located in vicinity of the energy shell. The point is the occurrence of derivatives of δ-functions $\delta(p^2 - E)$ under the trace symbol. This leads to appearance of derivatives of the pair T-matrices with respect to the third argument $\frac{\partial}{\partial E} t(k, k', E + i0)$ and requires knowledge of the T-matrix off-shell.

To obtain representation (7.107), one has to follow the derivation of equation (7.105). The only difference is that here the parameter ϵ is already set to zero. All the operators have a trace however, so we can arrange appropriately the factors under the trace symbol and employ the Hilbert identity. Each term of (7.95) should be first split into the sum of terms containing derivatives of only one of the kernels t_α or R_0 (R_0^*); then the expression $\nabla_E T_\alpha$ is constructed from derivatives of the kernels t_α. The degree of the resulting

T-polynomials is reduced with the help of the Hilbert identity. As a result, the quantities $\delta\Delta^{(3)}$ and $\delta\Delta^{(2)}$ are formed and finally one finds the remaining trace of the homogeneous T-polynomials of second degree. The sum of the remaining terms is however equal to zero. We are not going to carry out the cumbersome though simple calculations to prove equations (7.107). All necessary procedures have been describes above.

Therefore, we described all terms in the intermediate formula (7.99). Now, in order to proceed to the final representation (7.94), we must compute the operator $\tilde{A}(E)$ defined by equation (7.101). Of course, this identity does not allow to define $\tilde{A}(E)$ uniquely. The expression appearing in (7.47) is selected as to make the form of the operator $\tilde{A}(E)$ resembling that of the operator $A(E)$ as much as possible.

Let us compute the trace $\mathrm{Sp}\,(A(E)-A^{(1)}(E))$. To reveal the uncertainties in the kernel $(A-A^{(1)})(\hat{P},\hat{P}',E)$, we write the trace

$$\mathrm{Sp}\,(A(E)-A^{(1)}(E)) = \int d\hat{P}(A-A^{(1)})(\hat{P},\hat{P}',E)$$

in the form of the sum $\pi i \mathrm{Sp}\, A^{(+)} + \pi i \mathrm{Sp}\, A^{(-)}$ where the sign $\pm$ corresponds to the sign of the operator $T^{(\pm)}_{\alpha\beta} - \hat{Q}_{\alpha\beta}$ entering the difference $A - A^{(1)}$. The following relation holds

$$A^{(\pm)} = \sum_{\alpha\neq\beta} \mathrm{Sp}\,\left(\left[\partial^{(\pm)}_{\alpha\beta} T^{(\pm)}_{\alpha\beta}\right]^*_{\tau} \frac{\mathrm{d}}{\mathrm{d}\,E} S_\alpha S_\beta - \right.$$
$$\left. - (S^*_\alpha + S^*_\beta - I)\left[\partial^{(\pm)}_{\alpha\beta} \frac{\mathrm{d}}{\mathrm{d}\,E} T^{(\pm)}_{\alpha\beta}\right]_{\tau}\right). \tag{7.110}$$

The differentiation symbol $\partial^{(\pm)}_{\alpha\beta}$ is defined as

$$\partial^{(+)}_{\alpha\beta} = -\partial_{\alpha 2} + \partial_{\beta 2} + \partial_{\beta 3}, \quad \partial^{(-)}_{\alpha\beta} = -\partial_{\alpha 1} + \partial_{\alpha 2} - \partial_{\beta 1}.$$

Here the first subscript $\alpha(\beta)$ indicates that the differentiation refers only to the kernel $t_\alpha(t_\beta)$ and the second subscript 1, 2 and 3 shows with respect to which argument of the kernel $t(k,k',z)$ (k^2, k'^2, or E) the differentiation is carried out. Let us sketch the scheme of calculating the trace $\mathrm{Sp}\,(A - A^{(1)})$. It is convenient to consider the terms $\mathrm{Sp}\, A^{(+)}$ and $\mathrm{Sp}\, A^{(-)}$ separately. In each term $\mathrm{Sp}\, A^{(\pm)}$ one must consider traces of homogeneous T-polynomials of subsequently decreasing degrees. In these factors, it is easy to isolate the terms containing the total derivatives ∂ or $\partial\frac{d}{dE}$. Such terms can apparently be expressed in terms of the S-matrix these terms do contribute to (7.97). The other "non S-matrix" terms are of two types. To the terms of the first type the Hilbert identity can be applied and degree of the T-polynomials can lowered. Such terms cancel analogous terms of lower degrees. The Hilbert identity cannot be applied to the terms of the second group; they contain derivatives of the T-matrices with respect to the integration variables. Such terms cancel either between themselves or with analogous terms entering the definition of the quantity $\Delta_1(E)$. As a result, only the terms which can be described by means of the pair S-matrices remain. They can be written in the compact form (7.97). In order to verify equation (7.97), it is convenient to operate in the opposite direction. Namely, one first specifies the right hand side of (7.97) and simplifies the obtained expressions with the help of unitarity condition of pair S-matrices. Then one compares this expression with the result of transformations carried out in accordance with the procedure described above.

Let us demonstrate the realization of this procedure as applied to the first term of (7.110). We will consider only the trace Sp $A^{(+)}$. The calculations of Sp $A^{(-)}$ are completely analogous.

First we transform the trace of the term generated by the symbol $\partial_{\alpha 2}$:

$$F_2 = -\sum_{\alpha\neq\beta} \mathrm{Sp}\ [\partial_{\alpha 2}T_{\alpha\beta}]^*_\tau \frac{d}{dE} S_\alpha S_\beta .$$

We will rewrite it in the form

$$F_2 = -\sum_{\alpha\neq\beta} \mathrm{Sp}\ \left[\hat{T}^*_\beta \partial_{\alpha 2}\hat{T}^*_\alpha\right]_\tau \left(S_\alpha \frac{d}{dE} S_\beta + \frac{d\,S_\alpha}{dE} S_\beta\right). \tag{7.111}$$

Here we made use of definition of the symbol $\partial_{\alpha 2}$ and we differentiated the product $S_\alpha S_\beta$ term by term. In both terms (7.111) it is possible to lower the degree of the T-polynomials. In the first term, which corresponds to the operator $S_\alpha \frac{d}{dE} S_\beta$, we apply the Hilbert identity to the operators T_α. We obtain

$$\left[\hat{T}^*_\beta \partial_{\alpha 2}\hat{T}^*_\alpha\right]_\tau S_\alpha = \left[\hat{T}^*_\beta \partial_{\alpha 2}\hat{T}^*_\alpha S_\alpha\right]_\tau = \left[\hat{T}^*_\beta \partial_{\alpha 1}\hat{T}_\alpha\right]_\tau .$$

In the second term we lower degree of the T-polynomial in the following way

$$S_\beta \left[\hat{T}^*_\beta \partial_{\alpha 2}\hat{T}^*_\alpha\right]_\tau = \left[S_\beta \hat{T}^*_\beta \partial_{\alpha 2}\hat{T}^*_\alpha\right]_\tau = \left[\hat{T}_\beta \partial_{\alpha 1}\hat{T}^*_\alpha\right]_\tau .$$

As a result, we find that $F2$ can be written as a trace of a third degree homogeneous T-polynomial

$$F_2 = -2\pi i \sum_{\alpha\neq\beta} \mathrm{Sp}\left[\partial_{\alpha 2}\hat{T}_2^* \frac{\mathrm{d}\hat{T}_\alpha}{\mathrm{d}E}\hat{T}_\beta + \partial_{\alpha 1}\hat{T}_\alpha \frac{\mathrm{d}\hat{T}_\beta}{\mathrm{d}E}\hat{T}_\beta^*\right]_r . \tag{7.112}$$

Let us consider now the remaining factors of the first term of equation (7.110)

$$F_3 = -\sum_{\alpha\neq\beta} \mathrm{Sp}\left[(\partial_{\beta 2}+\partial_{\beta 3})\hat{T}_\beta^*\hat{T}_\alpha^*\right]_r \left(\frac{\mathrm{d}S_\alpha}{\mathrm{d}E}S_\beta + S_\alpha\frac{\mathrm{d}S_\beta}{\mathrm{d}E}\right).$$

It is again possible to lower degree of the T-polynomial in the second group which corresponds to the operators $S_\alpha \frac{\mathrm{d}s_\beta}{\mathrm{d}E}$. Let us denote this group by δA_3. Then

$$\delta A_3 = 2\pi i \sum_{\alpha\neq\beta} \mathrm{Sp}\left[\hat{T}_\alpha \frac{\mathrm{d}\hat{T}_\beta}{\mathrm{d}E}(\partial_{\beta 2}+\partial_{\beta 3})\hat{T}_\beta^*\right]_r .$$

The terms belonging to the first group cannot be reduced to terms of third degree. We will follow a different procedure: we isolate the traces of the homogeneous T-polynomials of fourth degree which can be expressed in terms of the scattering matrix and reduce the remainder to T-polynomials of the third degree. To this end, the following relation can be used

$$S_\beta(\partial_{\beta 2}+\partial_{\beta 3})\hat{T}_\beta^* = 2\pi i\partial\hat{T}_\beta\hat{T}_\alpha^* + (\partial_{\beta 1}+\partial_{\beta 3})T_\beta . \tag{7.113}$$

Let us prove this relation. We will examine the equivalent formula

$$T_\beta\varphi_0\left(\partial_{\beta 2}+\frac{\mathrm{d}}{\mathrm{d}E}\right)T_\beta^* = 2\pi i\left(\partial_{\beta 1}+\partial_{\beta 2}+\partial_{\beta 3}+\frac{1}{2}h_\beta^{-2}\right)T_\beta\varphi_0 T_\beta^* -$$

$$-\left(\partial_{\beta 1}+\frac{\mathrm{d}}{\mathrm{d}E}\right)T_\beta - \left(\partial_{\beta 2}+\frac{\mathrm{d}}{\mathrm{d}E}\right)T_\beta^*,$$

where $\partial_{\beta 2}T_2^* \equiv (\partial_{\beta 2}T_\beta)^*$ and h_β^{-2} is the operator of multiplication by the function k_β^2.

First we write the operator $T_\beta\varphi_0\frac{\mathrm{d}}{\mathrm{d}E}T_\beta^*$ as a sum

$$T_\beta\varphi_0\frac{\mathrm{d}}{\mathrm{d}E}T_\beta^* = \frac{\mathrm{d}}{\mathrm{d}E}(T_\beta\varphi_0 T_\beta^*) - \frac{\mathrm{d}T_\beta}{\mathrm{d}E}\varphi_0 T_\beta^* - T_\beta\frac{\mathrm{d}\varphi_0}{\mathrm{d}E}T_\beta^* . \tag{7.114}$$

The first term on the right hand side can be simplified by the Hilbert identity

$$2\pi i\frac{\mathrm{d}}{\mathrm{d}E}(T_\beta\varphi_0 T_\beta^*) = \frac{\mathrm{d}}{\mathrm{d}E}(T_\beta - T_\beta^*).$$

Consider further the kernel of the operator $T_\beta \frac{d}{dE} \varphi_0 T_\beta^*$ having the following form

$$\delta(p_\beta - p'_\beta) \int dk''_\beta t_\beta(k_\beta, k''_\beta, E - p_\beta^2 + i0) \times$$
$$\times \overline{t_\beta(k'_\beta, k''_\beta, E - p_\beta^2 + i0)} \frac{d}{dE} \delta(k''^2_\beta + p_\beta^2 - E).$$

Performing the differentiation of the δ-function and integrating by parts with respect to k''^2_β, we obtain

$$T_\beta \frac{d\varphi_0}{dE} T_\beta^* = (\partial_{\beta 2} T_\beta) \varphi_0 T_\beta^* + T_\beta \varphi_0 \partial_{\beta 2} T_\beta^* + \frac{1}{2} T_\beta h_\beta^{-2} \varphi_0 T_\beta^*.$$

Let us insert this equality into (7.114) and subtract on the right hand side the operator

$$\partial_{\beta 1} T_\beta \varphi_0 T_\beta^* = \frac{1}{2\pi i} (\partial_{\beta 2} T_\beta^* - \partial_{\beta 1} T_\beta).$$

By performing obvious cancellation, we obtain (7.113).

Thus, using the described scheme we completed the transformation of the trace of the homogeneous T-polynomials of fourth degree. We represented them in the form of the sum of terms of two types. The first type, i.e., the trace

$$-(2\pi i)^2 \sum_{\alpha \neq \beta} \mathrm{Sp}\, [T_{\alpha\beta}]_r^* \frac{d\hat{T}_\alpha}{dE} \partial \hat{T}_\beta$$

which can be expressed in terms of the S-matrix. All remaining terms belong to the second type: they are traces of T-polynomials of third degree and for their description the off-shell T-matrix is necessary.

All remaining terms in (7.110) can be transformed analogously. As it was already mentioned, all terms containing off-shell T-matrices arising from various terms in (7.110) cancel each other. The remaining terms which can be expressed in terms of pair S-matrices, can be written in the compact form (7.97).

This concludes justification of the trace formula (7.94).

7.4.6 A simple model

To concluding this section, we will consider the trace formula in the model three-particle problem: two particles interacting with an infinitely heavy centre. Scattering in this system has been thoroughly treated in Section 4 of Chapter 3.

The variables in this problem separate and the study of the function $\Omega_+(E)$ reduces to considering two-particle problems. The resolvent $R(z)$ can be expressed in terms of two-particle resolvent by means of a contour integral (2.12). By making use of the trace formula (7.75), we obtain the following relation

$$\Omega_+(E) =$$
$$-\frac{1}{4\pi}\int_0^E dE' \mathrm{Sp}\, s_1^*(E')\frac{\mathrm{d}\, s_1}{\mathrm{d}\, E} \cdot \mathrm{Sp}\, s_2^*(E-E')\frac{\mathrm{d}\, s_2(E-E')}{\mathrm{d}\, E}. \tag{7.115}$$

This relation will be obtained below directly from the general trace formula (7.94) by taking the limit $m_3 \to \infty$. It should be noted that the considered model is clearly the only three-particle example for which trace of the operator $\left(S^* \frac{\mathrm{d}\, s}{\mathrm{d}\, E}\right)_c$ exists.

Since the S-matrix can be factorized in this model $S = S_1 S_2$, the trace Sp $\left(S^* \frac{\mathrm{d}\, s}{\mathrm{d}\, E}\right)_c$ equals zero. We show that the operator $A(E)$ in (7.94) containing "true" singular terms also equals zero.

First we verify that in this model the following formula holds

$$T_{\alpha\beta} + T_{\beta\alpha} = (2\pi)^{-2}(S_\alpha - I)(S_\beta - I), \quad \alpha, \beta = 1, 2. \tag{7.116}$$

As we already noted, in the considered model the variables separate and therefore the operators S_α and S_β commute. Note further that for $m_3 \to \infty$ we have

$$k'_{\alpha\beta} = k'_\alpha, \quad k_{\beta\alpha} = k_\beta, \quad P^2 = k_\alpha^2 + k_\beta^2, \quad P'^2 = k'^2_\alpha + k'^2_\beta.$$

Hence, the operators $T_{\alpha\beta}^{(\pm)}$ can be represented as

$$T_{\alpha\beta}^{(+)}(\hat{P}, \hat{P}, E) = -\frac{2E^2}{(E_\alpha E_\beta)^{1/2}} \frac{\hat{t}_\alpha(\hat{k}_\alpha, \hat{k}'_\alpha, E_\alpha)\hat{t}_\beta(\hat{k}_\beta, \hat{k}'_\beta, E_\beta)}{E'_\alpha - E_\alpha - i0},$$

$$T_{\alpha\beta}^{(-)}(\hat{P}, \hat{P}, E) = -\frac{2E^2}{(E'_\alpha E'_\beta)^{1/2}} \frac{\hat{t}_\alpha(\hat{k}_\alpha, \hat{k}'_\alpha, E'_\alpha)\hat{t}_\beta(\hat{k}_\beta, \hat{k}'_\beta, E'_\beta)}{E'_\alpha - E_\alpha - i0},$$

where by virtue of the fact that $P^2 = P'^2$, the denominator can be written as $E_\beta - E'_\beta - i0$. Using the formula

$$\frac{1}{E'_\alpha - E_\alpha - i0} - \frac{1}{E'_\alpha - E_\alpha + i0} = 2\pi i\delta(E'_\alpha - E_\alpha)$$

we obtain the desired relation

$$T^{(\pm)}_{\alpha\beta} + T^{(\pm)}_{\beta\alpha} = -\hat{T}_\alpha \hat{T}_\beta.$$

Now we write the operator $A(E)$ in the form

$$A(E) =$$

$$= \frac{1}{2}\left[(T_{\alpha\beta} + T_{\beta\alpha})^* \frac{\mathrm{d}}{\mathrm{d}E} S_\alpha S_\beta - (S_\alpha + S_\beta - I)^* \frac{\mathrm{d}}{\mathrm{d}E}(T_{\alpha\beta} + T_{\beta\alpha})\right]$$

and make use of equation (7.116). Differentiating the products $S_\alpha S_\beta$ and $(S_\alpha - I)(S_\beta - I)$ term by term and simplifying the resultant expression with the help of the unitarity condition, we find that $A(E) \equiv 0$.

Next, we consider the operator $\tilde{A}(E)$. First we show that in our model the following relation holds

$$\mathrm{Sp}\left(\frac{1}{E}\frac{\mathrm{d}}{\mathrm{d}E}E - \frac{1}{2}\partial\right) B_R = 0. \tag{7.117}$$

To this end, we note that derivatives of pair T-matrices with respect to E have the form

$$\frac{\mathrm{d}}{\mathrm{d}E}\hat{t}_\alpha(E_\alpha) = \frac{E_\alpha}{E}\frac{\mathrm{d}}{\mathrm{d}E_\alpha}\hat{t}_\alpha(E_\alpha).$$

Therefore,

$$\frac{\mathrm{d}}{\mathrm{d}E}E_\alpha \frac{\mathrm{d}}{\mathrm{d}E_\alpha}\hat{t}_\alpha(E_\alpha) = \frac{E_\alpha}{E}\frac{\mathrm{d}}{\mathrm{d}E_\alpha}\hat{t}_\alpha(E_\alpha) + \frac{E_\alpha^2}{E}\frac{\mathrm{d}^2}{\mathrm{d}E_\alpha^2}\hat{t}_\alpha(E_\alpha),$$

or, by virtue of the definition of the symbol $\partial \frac{\mathrm{d}}{\mathrm{d}E}$, the right hand side can be written as

$$E_\alpha \frac{\mathrm{d}}{\mathrm{d}E}\hat{t}_\alpha.$$

By differentiating with respect to E under the trace symbol and by making use of the relations for kernels $\hat{t}_\alpha$ and $\hat{t}_\beta$, we obtain the desired equation (7.117).

Now let us transform the trace $\mathrm{Sp}\,\tilde{B}_R$. Note that the following formula results from the definition of the symbol ∂:

$$\left[\partial \hat{T}_\alpha \frac{\mathrm{d}}{\mathrm{d}E}\hat{T}_\beta + \frac{\mathrm{d}\hat{T}_\alpha}{\mathrm{d}E}\partial \hat{T}_\beta\right]_r = E^2 F_\alpha(E_\alpha) F_\beta(E_\beta),$$

where

$$F_\alpha(E_\alpha) = \left(\frac{E_\alpha}{E}\frac{\mathrm{d}}{\mathrm{d}\,E_\alpha} + \frac{1}{2E}\right) E_\alpha^{-1/2}\hat{t}_\alpha(E_\alpha). \tag{7.118}$$

By differentiating the operators in (7.98′) term by term and simplifying the resultant equations with the help of the unitarity of operators S_α and S_β, we find

$$\mathrm{Sp}\,\tilde{B}_R = \Omega_M + 2\pi i \mathrm{Sp}\, B_R^{(2)},$$

where

$$B_R^{(2)} = \sum_{\alpha\neq\beta}\left[\hat{T}_\beta^*\partial\frac{\mathrm{d}}{\mathrm{d}\,E}\hat{T}_\alpha\right]_r \tag{7.119}$$

and

$$\Omega_M =$$

$$= -\pi\int_0^E dE_\alpha \mathrm{Sp}\,(s_\alpha^*(E_\alpha)E_\alpha^{1/2}F_\alpha(E_\alpha))\mathrm{Sp}\,(s_\beta^*(E_\beta)E_\beta^{1/2}F_\beta(E_\beta))\;\; E_\beta = E - E_\alpha.$$

It is easy to verify that

$$\frac{E_\alpha}{E}\frac{\mathrm{d}}{\mathrm{d}\,E_\alpha}E_\alpha^{-1/2}\hat{t}_\alpha(E_\alpha) = E_\alpha^{-1/2}\frac{\mathrm{d}}{\mathrm{d}\,E_\alpha}s_\alpha(E_\alpha) - (2E_\alpha)^{-1}\hat{t}_\alpha(E_\alpha).$$

It follows that the quantity Ω_M coincides with the integral (7.115).

Thus, it remains to prove that

$$2\pi i \mathrm{Sp}\, B_R^{(2)} + \frac{1}{4\pi i}\mathrm{Sp}\,\partial[C]_r = 0. \tag{7.120}$$

Let us write the trace $\mathrm{Sp}\, B_R^{(2)}$ in the form

$$\mathrm{Sp}\, B_R^{(2)} = \sum_{\alpha\neq\beta} -\frac{1}{E}\int_0^E dE_\beta \int d\hat{k}_\beta \hat{t}_\beta^*(\hat{k}_\beta, \hat{k}_\beta, E_\beta)\times$$

$$\times \int d\hat{k}_\alpha \frac{\partial}{\partial E_\beta}(E - E_\beta)\frac{\partial}{\partial E}\hat{t}_\alpha(\hat{k}_\alpha, \hat{k}_\alpha, E - E_\beta).$$

Integrating over E_β by parts and taking into account the boundary conditions

$$\hat{t}_\beta|_{E_\beta=0} = \hat{t}_\alpha|_{E_\alpha=0} = 0,$$

we find that

$$\mathrm{Sp}\,\tilde{B}_R^{(2)} = \sum_{\alpha\neq\beta}\left[\partial\hat{T}_\beta^*\frac{\mathrm{d}\,\hat{T}_\alpha}{\mathrm{d}\,E}\right]_r,$$

i.e.,

$$2\pi i \mathrm{Sp}\, \tilde{B}_R^{(2)} = \pi i \sum_{\alpha \neq \beta} \left(\mathrm{Sp} \left[\partial \hat{T}_\beta^* \frac{\mathrm{d}\hat{T}_\alpha}{\mathrm{d}E} \right]_r + \mathrm{Sp} \left[\hat{T}_\beta^* \partial \frac{\mathrm{d}}{\mathrm{d}E} \hat{T}_\alpha \right]_r \right).$$

The right hand side coincides with the quantity $-(4\pi i)^{-1}\mathrm{Sp}\,[C]_r$ which appears in equation (7.97). Now equation (7.120) follows.

Therefore, we see that in the considered model the operator $\left(S^* \frac{\mathrm{d}}{\mathrm{d}E}\right)_c$ and the regularizing operator $A(E)$ both equal zero. The nontrivial contribution to the trace formula comes only from the term $\mathrm{Sp}\, \tilde{B}_R$ in equation (7.97). In this sense we can say that the choice of the regularizing operator $A(E)$ in formula (7.94) is optimal.

CHAPTER 8

Comments on Literature

Since the current literature on the scattering theory is very rich we will refer here only to the immediate sources to our book. From a large number of monographs devoted to the quantum scattering theory we are most closely related to books of the following authors: R. Newton [1], E. Schmidt and H. Ziegelmann [2], who treat the many-body scattering theory from the physical point of view; M. Reed and B. Simon [3] and W. O. Amrein, J. M. Jauch, K.B. Sinkgh [4] which are more mathematically oriented. The mathematical works of one of the authors [5] determined to large extent the choice of the material and of the methods of this book although the present explanation is much less formal.

Chapter 1

The scattering theory was formulated first in terms of wave operators by C. Möller [6] (see also K. O. Friedrichs [7]). These authors considered only the simplest problem of one-channel scattering. At the beginning of the 50's, the series of papers appeared associated with the development of the quantum field theory devoted to the so called "formal scattering theory"; see for example B. A. Lippmann and J. Schwinger [8] and M. Gell-Mann and M. Goldberger [9]. The many-channel formalism of the scattering theory was first developed by H. Eckstein [10]. The recent development of the nonstationary scattering theory is due to V. Enss [11].

At the beginning of the 60's, the scattering theory has gained popularity in mathematical physics. We do not intend to give here an exhaustive review of the literature. We confine ourselves to quoting the names of only few scientists that are spiritually closest to us: A. J. Povzner, T. Kato, M. S. Birman, S. Kuroda, T. Ikebe, V. S. Buslaev. Their works are described in the monographs mentioned above.

The basic ideas of the general scattering theory have been applied to the case of Coulomb interaction by J. Dollard [12].

N-body kinematic concepts like partition, chain of partitions and so on were introduced by O. A. Yakubovski [13]. In the present book, we have introduced such designation of the kinematic variables which is convenient for the description of singularities and asymptotics of the resolvent and of the wave functions.

The definition of the many-body S-matrix used here was first formulated by F. A. Berezin, P. A. Minlos and L. D. Faddeev [14].

Chapter 2

Transition to the stationary formalism is a standard tool of the formal scattering theory (see for example [9, 10]). The structure of poles of the many-body resolvent and the relation of the corresponding residue with the scattering amplitude is considered in [5] on the example of three particles; see also the paper by K. Hepp [15]. A generalization of these formulae to the particular case of Coulomb interaction was attempted by A. M. Veselova [16].

Chapter 3

The compact integral equations were used to justify the scattering problem first by A. J. Povzner [17]; see also a paper by T. Ikebe [18]. Our explanation follows Section 4 of [5]; see also [19].

General information of the compact integral equations can be found for example in the book by S. G. Michlin [20]. The singular integral lemma can be found in the monograph by N. I. Muschelishvili [21]; it is also given in [5].

The use of integral equations to solve the three-body problem was first proposed by G. V. Skornyakov and K. A. Ter-Martirosyan [22]. The interaction was assumed to be of zero range and the derivation of equations heavily relied on this fact. A general form of three-body compact equations was proposed in [23] and their detailed study were presented given in [5].

In a paper by O. A. Yakubovski, the kinematic part of the N-body integral equations is discussed; see also K. Hepp [15].

The nontrivial structure of the three-body discrete spectrum discussed in Section 4 was discovered by V. Efimov [24]. The presentation given in Section 4 follows the lectures [25]. For a detailed study of the homogeneous integral equations see D. R. Yafaev [26].

A modification of the integral equations for Coulomb interactions is considered in the paper by J. Noble [27] and A. M. Veselova [28].

Chapter 4

In the Povzner's work [17] discussed above, the two-body integral equations were considered in the configuration space. The scheme of investigation of many-body equations follows [29]. An alternative configuration space study in terms of Hilbert space is given in [30] and [31].

Asymptotics of the three-body wave functions were discussed by J. Nuttal [32] and in a more detailed way in [33]. The N-body differential equations for the components were formulated by S. P. Merkuriev and S. L. Yakovlev [34]; see also P. Benua and M. L'Huillier [35].

A detailed description of the multidimensional stationary phase method is given in M. V. Fedoryuk's book [36]. Asymptotics of the singular integrals can be found, for example, in the review article by D. Jones [37].

Chapter 5

The Coulomb two-body wave functions were known in the golden age of quantum mechanics – in twenties. The Green function of the Schrödinger operator with Coulomb interaction was calculated much later; see L. Hostler [38], V. F. Bratcev and E. D. Trifonov [39], and J. Schwinger [40]. The well-known Fock method of four dimensional harmonics was used in [39].

The book by H. Bateman and A. Erdelyi [41] serves as a good source of information on hypergeometric functions.

The N-body Coulomb wave function were used by R. K. Peterkop [42].

The eikonal asymptotic formulae are developed in [43] and [44]. The method of the standard equation was first formulated by V. A. Fock [45].

Chapter 6

Asymptotic completeness for the case of three particles was first proved in [5] and our explanation will follow this work. A proof of the asymptotic completeness which would be satisfactory in all respects has not yet been found. Asymptotic completeness in the case of three particles with Coulomb interactions is proved in [29, 44]. A construction of the Coulomb S-matrix in the form of an integral operator for two particles can be found, for example, in [46] and for three particles in [44].

A detailed description of generalized functions used here can be found in [47].

Chapter 7

There exist many papers on separable approximations in many-body problems and we can not give an exhaustive review here. The pioneering works on this field are due to A. Mitra [48], R. Amado [49], A. G. Sitenko, and V. F. Kcharchenko [50]; further development can be found in the detailed paper by S. Lovelace [51], in the review by A. G. Sitenko and V. F. Kcharchenko [52], and in the paper of E. Alt, P. Grassberger and W. Sandhas [53]. A generalization of the latter to the Coulomb case is given in [54], [55]. An alternative method of computing terms of modified integral equations [27] was developed in the papers published by Kcharchenko's group (see for example [56]).

An alternative way of separating angular variables in compact integral equations is discussed by R. Omnes [57].

A modification of the separable interaction in presence of the Coulomb interaction was discussed in numerous papers by H. van Haeringen, R. van Wageningen, and L. Kok; see for example [58, 59]. These papers contain also the most detailed discussion of the properties of the two-body Coulomb T-matrix.

Differential equations for partial components in the case of three particles were given by H. P. Noyes [60]. They were used to calculate bound states for the first time by K. Ginoux and A. Laverne [61] and for the scattering states in [62]. These equations were generalized to the case of Coulomb interactions by Y. A. Kuperin, A. A. Kvitsinski, and S. P. Merkuriev [63]. A considerable amount of work in this field was done by J. Payne, G. Friar, B. Gibson and others (see [64, 65] and references therein).

Singularities of the scattering amplitude for long-range forces are discussed in [66].

General features of the virial expansion can be found, for example, in [67]. The problem of calculating the second virial coefficient was solved by E. Bethe and H. Uhlenbeck [68] in 30's (see also [67]). The problem of calculating higher virial coefficients was considered much later, see, for example, F. A. Berezin [69] and R. Dashen [70]). Here we follow the paper by V. S. Buslaev and S. P. Merkuriev [71].

Bibliography

[1] Newton R.G. *Scattering Theory of Waves and Particles* Springer Verlag, New York, Heidelberg, Berlin, 1982

[2] Schmid E.W. and Ziegelman H. *The Quantum Mechanical Three Body Problem* Pergamon Press, Oxford, 1974

[3] Reed M and Simon B. *Methods of Modern Mathematical Physics III: Scattering Theory* Academic Press, N.Y., 1972

[4] Amrein W.O. and Jauch J.M., and Sinkgh K.B. *Scattering Theory in Quantum Mechanics* Benjamin, N.Y., 1977

[5] Faddeev L.D. *Mathematical Aspects of the Three Body Problem in the Quantum Scattering Theory* (Israel Program for Scientific Translation, Jerusalem, 1965); Trudy Mat. Inst. AN USSR **69**, 1963 (in Russian)

[6] Möler C. Kgl. Danske Videnskab Selskab mat.fis. medd., **23**, 1, 1945

[7] Friedrichs K.O. Comm. Pure Appl. Math. **1**, 4, 361, 1948

[8] Lippmann B.A. and Schwinger J. Phys. Rev. **79**, 469, 1950

[9] Gellmann M. and Goldberger M.L. Phys. Rev. **91**, 398, 1953

[10] Ekstein H. Phys. Rev. **101**, 880, 1956

[11] Enss V. Commun. Math. Phys. **61**, 285, 1978; Ibid 65, 151, 1979; Acta Physica Austr. Suppl. XXIII, 29, 1981

[12] Dollard J. J. Math. Phys. **5**, 729, 1964

[13] Yakubovsky O.A. Yad. Phys. **5**, 1312, 1967 (in Russian)

[14] Berezin F.A. and Minlos P.A., and Faddeev L.D. Trudy IV Vses. Mat. Syezda, Nauka **2**, 532, Moscow, 1964 (in Russian)

[15] HEPP K. Helv. Phys. Acta **42**, 425, 1969

[16] VESELOVA A.M. Teor. Mat. Phys. **13**, 369, 1972 (in Russian)

[17] POVZNER A.YA. Mat. Sbor. **32**, 109, 1953 (in Russian)

[18] IKEBE T. Archive for Rat. Mech. Anal. **5**, 1, 1960

[19] FADDEEV L.D. Trudy MIAN USSR, **73**, 292, 1964 (in Russian)

[20] MIKHLIN S.G. *Lectures on Linear Integral Equations* Fizmatgiz, Moscow, 1959 (in Russian)

[21] MUSKHELISHVILI N.I. *Singular Integral Equations* Fizmatgiz, Moscow, 1962 (in Russian)

[22] SKORNYAKOV G.V. AND TER-MARTIROSJAN K.A. ZhETF **31**, 775, 1956 (in Russian)

[23] FADDEEV L.D. ZhETF **39**, 1459, 1960 (in Russian)

[24] EFIMOV G.V. Yad. Phys. **12**, 1080, 1970 (in Russian)

[25] FADDEEV L.D. *Method of Integral Equations in the Theory of Nuclear Reactions* MIFI Publishing, Moscow, 1971 (in Russian)

[26] YAFAEV D.R. Mat. Sbor., **106**, 622, 1978 (in Russian)

[27] NOBLE J.V. Phys. Rev. **161**, 945, 1967

[28] VESELOVA A.M. Teor. Mat. Phys. **3**, 326, 1970; Teor. Mat. Phys., **35**, 180, 1978 (in Russian)

[29] MERKURIEV S.P. Zapiski Nautsch. Sem. LOMI, **77**, 148, 1978 (in Russian)

[30] GINIBRE J. AND MOULIN M. Ann. Inst. H. Poincar, Sect. A, **21**, 97, 1974

[31] YAFAEV D.R. Teor. Mat. Phys., **37**, 48, 1978 (in Russian)

[32] NUTTALL J. J. Math. Phys., **12**, 1896, 1971

[33] MERKURIEV S.P. Teor. Mat. Phys., **8**, 235, 1971 (in Russian)

[34] MERKURIEV S.P. AND YAKOVLEV S.L. DAN USSR, **262**, 591, 1982; Teor. Mat. Phys., **56**, 60, 1983 (in Russian)

[35] BENOIST-GUETAL P. AND L'HUILLIER M. J. Math. Phys., **23**, 1982

[36] FEDORYUK M.V. *The Saddle-Point Method* Nauka, Moscow, 1977 (in Russian)

[37] JONES D.S. SIAM Rev., **14**, 286, 1972

[38] HOSTLER L. J. Math. Phys., **8**, 642, 1967; 11, 2966, 1970

[39] BRATSEV V.F. AND TRIFONOV E.D. Vestnik LGU, **16**, 36, 1962 (in Russian)

[40] SCHWINGER J. J. Math. Phys., **5**, 1606, 1964

[41] BATEMAN H. AND ERDELYI A. *Higher Transcedental Functions, Vol. 1 and 2* McGraw-Hill, N.Y., 1953

[42] PETERKOP R.K. *Theory of Ionization by Electron Scattering* Zinatne Riga, 1975 (in Russian)

[43] MERKURIEV S.P. Teor. Mat. Phys., **32**, 187, 1977 (in Russian)

[44] MERKURIEV S.P. Ann. Phys., **130**, 395, 1980

[45] FOCK V.A. *Diffraction and Propagation of Electromagnetic Waves* Sov. Radio, Moscow, 1970 (in Russian)

[46] HERBST I. Commun. Math. Phys., **35**, 181, 1974

[47] GELFAND I.M. AND SCHILOV G.E. *Generalized Function* Fizmatgiz, Moscow, 1959 (in Russian)

[48] MITRA A.N. Advances in Nucl. Phys., **3**, 1, 1969

[49] AMADO R.D. Phys. Rev., **132**, 485, 1963

[50] SITENKO A.G. AND KHARCHENKO V.F. Nucl. Phys., **49**, 15, 1963

[51] LOVELACE C. Phys. Rev., **B 135**, 1225, 1964

[52] SITENKO A.G. AND KCHARCHENKO V.F. Usp. Fiz. Nauk **103**, 469, 1971 (in Russian)

[53] ALT E.O. AND GRASSBERGER P. AND SANDHAS W. Phys. Rev., **C 1**, 85, 1970

[54] ALT E.O. AND SANDHAS W. Phys. Rev., **C 21**, 1733, 1980

[55] ALT E.O. AND SANDHAS W. AND ZIEGELMANN H. Phys. Rev. **C 17**, 1981, 1978

[56] KCHARCHENKO V.F. AND SCHADCHIN S.A. Ukr. Fiz. Zh., **23**, 1651, 1978 (in Russian)

[57] OMNES R.L. Phys. Rev., **B 134**, 1358, 1964

[58] VAN HAERINGEN H. AND VAN WAGENINGEN R. J. Math. Phys., **16**, 1441, 1975

[59] KOK L.P. AND VAN HAERINGEN H. Phys. Rev., **C 21**, 512, 1980

[60] NOYES H.P. *In: Three-Body Problem*, North-Holland, Amsterdam, 1970

[61] LAVERNE A. AND GIGNOUX C. Nucl. Phys., **A 203**, 597, 1973

[62] MERKURIEV S.P. AND GIGNOUX C., AND LAVERNE A. Ann. Phys., **99**, 30, 1976

[63] KUPERIN YU. A. AND KVITSINSKI A.A., AND MERKURIEV S.P. Yad. Fiz., **37**, 1440, 1983 (in Russian)

[64] PAYNE G.L. ET AL. Phys. Rev., **C 22**, 823, 1980

[65] FRIAR J.L. AND GIBSON B.F. AND PAYNE G.L. Phys. Rev., **C 28**, 983, 1983

[66] KVITSINSKI A.A. AND KOMAROV I., AND MERKURIEV S.P. Yad. Fiz. **38**, 101, 1983 (in Russian)

[67] HUANG K. *Statistic mechanics* Mir, Moscow, 1966 (Russian translation)

[68] BETHE E. AND UHLENBECK G.E. Physica, **4**, 915, 1937

[69] BEREZIN F.A. DAN USSR, **157**, 1069, 1964 (in Russian)

[70] DASHEN R., S.-KENG MA, AND BERNSTEIN H. Phys. Rev., **187**, 345, 1969

[71] BUSLAEV V.S. AND MERKURIEV S.P. Teor. Mat. Phys., **5**, 372, 1970 (in Russian)

Index

Mathematical Physics and Applied Mathematics

Publications:

1. M. Flato, Z. Maric, A. Milojevic, D. Sternheimer and J.P. Vigier (eds.): *Quantum Mechanics, Determinism, Causality, and Particles.* An International Collection of Contributions in Honor of Louis de Broglie on the Occasion of the Jubilee of His Celebrated Thesis. 1976 ISBN 90-277-0623-9
2. K. Maurin and R. Rączka (eds.): *Mathematical Physics and Physical Mathematics.* Proceedings of an International Symposium held in Warsaw (1974). 1976 ISBN 90-277-0537-2
3. M. Cahen and M. Flato (eds.): *Differential Geometry and Relativity.* A Volume in Honour of André Lichnerowicz on His 60th Birthday. 1976 ISBN 90-277-0745-6
4. M. Flato, C. Fronsdal and K.A. Milton (eds.): *Selected Papers (1937-1976) of Julian Schwinger.* 1979 ISBN Hb: 90-277-0974-2; Pb 90-277-0975-0
5. J.A. Wolf, M. Cahen and M. de Wilde (eds.): *Harmonic Analysis and Representations of Semisimple Lie Groups.* 1980 ISBN 90-277-1042-2
6. E. Tirapegui (ed.): *Field Theory, Quantization and Statistical Physics.* In Memory of Bernard Jouvet. 1981 ISBN 90-277-1128-3
7. V.P. Maslov and M.V. Fedoriuk: *Semi-classical Approximation in Quantum Mechanics.* 1981 ISBN 90-277-1219-0
8. V.N. Popov: *Functional Integrals in Quantum Field Theory and Statistical Physics.* 1983 ISBN 90-247-1471-1
9. F.A. Berezin: *Introduction to Superanalysis.* 1987 ISBN 90-277-1668-4
10. N.N. Bogolubov, A.A. Logunov, A.I. Oksak and I.T. Todorov: *General Principles of Quantum Field Theory.* 1990 ISBN 0-7923-0540-X
11. L.D. Faddeev and S.P. Merkuriev: *Quantum Scattering Theory for Several Particle Systems.* 1993 ISBN 0-7923-2414-5

Kluwer Academic Publishers – Dordrecht / Boston / London

GPSR Compliance
The European Union's (EU) General Product Safety Regulation (GPSR) is a set of rules that requires consumer products to be safe and our obligations to ensure this.

If you have any concerns about our products, you can contact us on

ProductSafety@springernature.com

In case Publisher is established outside the EU, the EU authorized representative is:

Springer Nature Customer Service Center GmbH
Europaplatz 3
69115 Heidelberg, Germany

www.ingramcontent.com/pod-product-compliance
Ingram Content Group UK Ltd.
Pitfield, Milton Keynes, MK11 3LW, UK
UKHW061817190726
13853UKWH00006B/2194
* 9 7 8 9 4 0 1 7 2 8 3 3 1 *